A Student's Manual for *A First Course in General Relativity*

This comprehensive student manual has been designed to accompany the leading textbook by Bernard Schutz, *A First Course in General Relativity*, and uses detailed solutions, cross-referenced to several introductory and more advanced textbooks, to enable self-learners, undergraduates, and postgraduates to master general relativity through problem solving. The perfect accompaniment to Schutz's textbook, this manual guides the reader step-by-step through over 200 exercises, with clear easy-to-follow derivations. It provides detailed solutions to almost half of Schutz's exercises, and includes 125 brand-new supplementary problems that address the subtle points of each chapter. It includes a comprehensive index and collects useful mathematical results, such as transformation matrices and Christoffel symbols for commonly studied spacetimes, in an appendix. Supported by an online table categorizing exercises, a Maple worksheet, and an instructors' manual, this text provides an invaluable resource for all students and instructors using Schutz's textbook.

Robert B. Scott is a Senior Lecturer with CNRS Chaire d'Excellence at the Université de Bretagne Occidentale, France where he specializes in relativity and geophysical fluid dynamics and turbulence.

A Student's Manual for *A First Course in General Relativity*

Robert B. Scott

Université de Bretagne Occidentale

CAMBRIDGE
UNIVERSITY PRESS

CAMBRIDGE
UNIVERSITY PRESS

University Printing House, Cambridge CB2 8BS, United Kingdom

One Liberty Plaza, 20th Floor, New York, NY 10006, USA

477 Williamstown Road, Port Melbourne, VIC 3207, Australia

314-321, 3rd Floor, Plot 3, Splendor Forum, Jasola District Centre, New Delhi - 110025, India

79 Anson Road, #06-04/06, Singapore 079906

Cambridge University Press is part of the University of Cambridge.

It furthers the University's mission by disseminating knowledge in the pursuit of
education, learning and research at the highest international levels of excellence.

www.cambridge.org
Information on this title: www.cambridge.org/9781107638570

© Cambridge University Press 2016

First published 2016
Reprinted 2016
5th printing 2018

A catalogue record for this publication is available from the British Library

Library of Congress Cataloging in Publication data
Scott, Robert B., 1965– author.
A student's manual for A first course in general relativity / Robert B. Scott,
Université de Bretagne Occidentale.
pages   cm
Includes bibliographical references and index.
ISBN 978-1-107-63857-0 (Paperback)
1. General relativity (Physics)–Problems, exercises, etc.   2. Astrophysics–Problems, exercises, etc.
I. Schutz, Bernard F. First course in general relativity.   II. Title.
III. Title: First course in general relativity.
QC173.6.S37 2015
530.11076–dc23   2015020004

ISBN 978-1-107-63857-0 Paperback

Contents

Preface

General relativity is a beautiful theory, our standard theory of gravity, and an essential component of the working knowledge of the theoretical physicist, cosmologist, and astrophysicist. It has the reputation of being difficult but Bernard Schutz, with his groundbreaking textbook, *A First Course in General Relativity* (first edition published in 1984, current edition in 2009), demonstrated that GR is actually quite accessible to the undergraduate physics student. With this solution manual I hope that GR, using Schutz's textbook as a main resource and perhaps one or two complementary texts (see recommendations at the end of this preface), is accessible to all "technically minded self-learners" e.g. the retired engineer with some time to devote to a dormant interest, a philosopher of physics with a serious interest in deep understanding of the subject, the mathematics undergraduate who wants to become comfortable with the language of the physicist, etc.

You can do it too!

I'm speaking with some experience when I say that an engineer can learn GR and in particular starting with Schutz's textbook. My bachelor's and master's degrees are in engineering and I started learning GR on my own when my academic career had gained enough momentum that I could afford a bit of time to study a new area in my free time. I must admit it wasn't always easy. I personally found the explanations of mathematics in the excellent textbook by Misner, Thorne and Wheeler (1973) more confusing then helpful. (In retrospect I'm at a loss to explain why; in no way do I blame the authors.) Soon two children arrived miraculously in our household, free time became an oxymoron, but with the constant reward I found from beavering away at Schutz's exercises I continued to learn GR, albeit slowly and with screaming (not always my own) interruptions. In his autobiography John A. Wheeler explains that he started learning GR in the 1940s when he finally got the chance to teach the subject. Similarly the real breakthrough for me came when I was offered the possibility to teach the subject to third-year undergraduate students at the Université de Bretagné Occidentale in Brest, France. Suddenly my hobby became my day job, fear of humiliation became my motivation, and most significantly I was forced to view the subject from the student's point of view. I also had to learn French. I can honestly state with no exaggeration that, even with Canadian high-school French instruction and years living in Montreal, it was much harder to learn the local language than to learn GR! *Vraiment!*

Is it better to start with a popular level book?

Popular level books are for people who want a superficial overview. There's nothing wrong with that if that's your ultimate goal. But if you want to really understand GR, skip the popular books for now. By the way, the French word for popular level books is *vulgarisation*. If they're good then they'll be at least as difficult, probably much more difficult, to understand then the real thing. Why? Because the author is obliged to explain mathematical concepts in an artificial language, divorced from the logical and precise language in which the ideas were developed. Popularizers are forced to come up with creative analogies that are insightful if you already understand the idea but are always somewhat misleading. If you made it through your college calculus class and basic linear algebra course then you'll find learning the tools of GR a natural extension of these ideas. Once you understand what the metric on a Riemann manifold is (say after working through Chapter 6 of Schutz), then you can easily and completely understand the Robertson–Walker metric that explains the expansion of the Universe. You won't need bread pudding analogies but if you want to create them yourself you'll be free to do so; just be sure to explain to your listener all the caveats.

Using this book

Suppose you'd like to learn some to play chess, or guitar, or ice hockey. You go to the library, or amazon.co.uk, or a bookstore and find an instruction manual. Reasonable start, but you wouldn't expect that after reading the manual you would be a chess master, awesome guitar player, or hockey star. You would have to practice first, a lot, and learn from your experience. Learning physics takes practice too and the initial practice comes from doing exercises. So don't make the mistake of reading this solution manual like a novel or recipe book thinking you've bypassed the practice sessions. Schutz's *A First Course in General Relativity* has a lot of exercises (338 in fact), many more than most of the other textbooks at that level. Do them!

Some advice to GR students and self-learners in particular:

1 Don't give up. It's normal to not understand something the first time you read it. Make note of what you don't understand, try to articulate your question to yourself as clearly as possible, then press on.
2 Read several textbooks. See additional resources for suggestions.
3 Work the exercises yourself, using this solution manual as a guide to get you over the hurdles and to verify your answers.

You'll probably find that sometimes you can find a simpler solution to the exercise then the solution I have offered. Your solution is probably fine. I have followed the idea that the exercises are designed to teach you to use the new mathematical techniques of differential geometry and tensor calculus (the "big machinery") in a simple setting where you might guess the answer or find it easily with simpler tools. For example you certainly don't need the Minkowski space metric tensor to know that the unit basis vector $\vec{e}_x$ is orthogonal to $\vec{e}_y$. But when you learn that in fact the component of the metric $\eta_{xy} = \vec{e}_x \cdot \vec{e}_y$, it's nice to notice that of course $\eta_{xy} = 0$.

You'll also find that I explain the solution steps in more detail than you really need. I certainly don't mean to insult you! My aim was to be complete, to spell it all out. I endeavor to explain the steps with brief comments to the right of most equation lines that anticipate and answer your question: "what did he do to get this line from the previous line?" If you find it too easy, read it quickly!

To distinguish between references to my equations and those in Schutz's book I use the form Schutz Eq. $(n.m)$ for his equations and eqn.$(n.m)$ for equations in this book. If you see something like

$$\bar{t} = (t - vx)\left(1 - v^2\right)^{-1/2} \qquad\qquad \text{used Schutz Eq. (1.12)}$$

$$= (t - vx)\left(1 + \frac{1}{2}v^2 + O(v^4)\right), \qquad\qquad \text{used eqn.(B.2)} \qquad (0.1)$$

this means that the first line follows from Eq. (1.12) in Schutz's textbook, while the RHS of the second line used the equation eqn.(B.2), which in this case is found in Appendix B of the book you're holding. Some of you might not have seen $O(x^2)$ before; look in Table A.2 because it's an equation symbol. For abbreviations and acronyms in the text, like "RHS" in the sentence before the previous one, look in Table A.1.

From time to time I make reference to an accompanying Maple^{TM} worksheet. This is available for free download from the Cambridge University Press website. Please also visit the authors website for this book at http://stockage.univ-brest.fr/~scott/Books/Schutz/index_schutz.html.

Additional resources

There are many good introductory resources for learning GR and throughout this manual you'll find references to them. Eric Poisson (Poisson, 2004, preface) recommends you read Schutz's textbook to get started, then Misner, Thorne, and Wheeler's mammoth tome (Misner et al., 1973) for breadth, and finally Robert Wald's monograph (Wald, 1984) for rigor. It would be hard to improve upon that advice. I suggest if you have time and find you can read Misner, Thorne, and Wheeler (1973) straight off you could even skip Schutz and this solution manual. Otherwise I agree with Eric, start here. But to complement Schutz's book I recommend books at a similar level, for example either Hobson et al. (2006) or Rindler (2006). The first is similar to Schutz's book but at times may be a bit more challenging to the reader. Rindler is a bit weaker on tensor analysis, but great for geometrical and physical insight. Sean Carroll (2004) has a flair for clear explanation and has covered a lot of the material in Wald (1984) in a more concrete fashion.

If you find you are struggling with Schutz's book you are probably missing some basic background. The most important background is a working knowledge of basic differential calculus, for which there are countless good begining university level books. If you have this but your math skills need polishing, you could work through the first six chapters of Felder and Felder (2014) concurrently with Schutz and this solution manual. After completing a good number of Schutz's exercises you'll be ready for advanced books (Misner et al., 1973; Hawking and Ellis, 1973; Wald, 1984; Poisson, 2004) and can even read some of the literature, especially *American Journal of Physics*, *European Journal of Physics*, and *Foundations of Physics* articles.

Thanks

I would especially like to thank Gary Felder who read carefully the first six chapters of this textbook and offered valuable suggestions for improvement. Jean-Philippe Nicolas, Jose Luis Jaramillo, Richard Tweed, and Fred Taylor also had helpful input. I dedicate this book to Dr. Donald Taylor, who was my first instructor in relativity, my first physics supervisor, and the first to encourage me in a career in physics.

If you find any errors, or have suggestions for learning GR, you can first check this book's website: http://stockage.univ-brest.fr/~scott/Books/Schutz/index_schutz.html and, if it is not already there, contact the author via email: robert.scott@univ-brest.fr.

1 Special relativity

The essence of a physical theory expressed in mathematical form is the identification of mathematical concepts with certain physically measurable quantities. This must be our first concern . . .

Bernard Schutz, §7.1

Minkowski pointed out that it is very helpful to regard (t, x, y, z) as simply four coordinates in a four-dimensional space [that] we now call spacetime. This was the beginning of the geometrical point of view, which led directly to general relativity in 1914–16.

Bernard Schutz, §1.1

1.1 Exercises

1.1 Convert the following to [natural] units in which $c = 1$, expressing everything in terms of m and kg:

(a) Worked example: 10 J.

Solution:

$$10\,\mathrm{J} = 10\,\mathrm{N\,m} = 10\,\mathrm{kg\,m^2s^{-2}} = \frac{10\,\mathrm{kg\,m^2s^{-2}}}{(3 \times 10^8\,\mathrm{m\,s^{-1}})^2} = 1.11 \times 10^{-16}\,\mathrm{kg}.$$

(c) Planck's reduced constant, $\hbar = 1.05 \times 10^{-34}$ J s. (Note the definition of $\hbar$ in terms of Planck's constant h: $\hbar \equiv h/2\pi$.)

Solution:

$$\hbar = 1.05 \times 10^{-34}\,\mathrm{J\,s} = \frac{1.05 \times 10^{-34}\,\mathrm{kg\,m^2s^{-1}}}{3 \times 10^8\,\mathrm{m\,s^{-1}}} = 3.52 \times 10^{-43}\,\mathrm{kg\,m}.$$

(e) Momentum of a car.

Solution:

$$p = \frac{30\ \text{ms}^{-1}}{3 \times 10^8\ \text{ms}^{-1}} \times 1000\ \text{kg} = 10^{-4}\ \text{kg}.$$

(g) Water density, 10^3 kg m^{-3}.

Solution:

$$10^3\ \text{kg m}^{-3}.$$

We will learn in Chapter 8 how to express mass in terms of meters, see in particular eqn. (8.8).

1.2   Convert from natural units ($c = 1$) to SI units

(a) Velocity, $v = 10^{-2}$:

Solution:

$$v = 10^{-2} \times c[\text{m s}^{-1}] = 3 \times 10^6\ [\text{m s}^{-1}].$$

(c) Time, 10^{18} [m]:

Solution:

$$\frac{10^{18}\ [\text{m}]}{c[\text{m s}^{-1}]} = 3.3 \times 10^9\ [\text{s}].$$

(e) Acceleration, 10 [m^{-1}]:

Solution:

$$10\ [\text{m}^{-1}] \times c^2[\text{m}^2\ \text{s}^{-2}] = 9 \times 10^{17}\ [\text{m s}^{-2}].$$

1.3   Draw the t and x axes of the spacetime coordinates of an observer $\mathcal{O}$ and then draw:

(c) The $\bar{t}$ and $\bar{x}$ axes of an observer $\bar{\mathcal{O}}$ who moves with velocity $v = 0.5$ in the positive x-direction relative to $\mathcal{O}$ and whose origin ($\bar{t} = \bar{x} = 0$) coincides with that of $\mathcal{O}$.

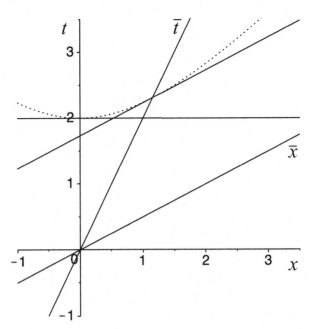

Figure 1.1
The $\bar{x}$ and $\bar{t}$ axes are the solution to Exercise 1.3(c). The dotted line is the invariant hyperbola with $\Delta s^2 = -4$. The solution to 1.3(h) is the horizontal line. The solution to 1.3(i) is the sloping line, parallel to the $\bar{x}$-axis. It is tangent to the invariant hyperbola at the $\bar{t}$-axis. These plots were made using the Maple^{tm} worksheet that accompanies this book.

Solution: Recall from Schutz §1.5 that the $\bar{t}$-axis follows from simple kinematics; it is just the line $t = x/v$, so here $t = 2x$. Recall also from §1.5 (see Schutz Fig. 1.5) that the $\bar{x}$-axis was a straight line with slope equal to the inverse of that of the $\bar{t}$-axis, $x = t/v$. (In SP1.3 you will prove this.) Here $t = x/2$. The solution was plotted in fig. 1.1.

(h) The locus of events, all of which occur at the time $t = 2$ m (simultaneous as seen by $\mathcal{O}$).

Solution: See fig. 1.1.

(i) The locus of events, all of which occur at the time $\bar{t} = 2$ m (simultaneous as seen by $\overline{\mathcal{O}}$).

Solution: The locus of events, all of which occur at the time $\bar{t} = 2$ m, have arbitrary $\bar{x}$, and so the solution is a straight line parallel to the $\bar{x}$-axis. The coordinates in the

$\mathcal{O}$ frame are easily found with the Lorentz transformation. (See SP1.13 for a different approach.) From Schutz Eq. (1.12) we have

$$\bar{t} = 2 = \frac{t - vx}{\sqrt{1 - v^2}} \quad \Rightarrow \quad t = vx + 2\sqrt{1 - v^2} = x/2 + \sqrt{3}.$$

The solution was plotted in fig. 1.1.

1.5   (c) A second observer $\overline{\mathcal{O}}$ moves with speed $v = 0.75$ in the negative x-direction relative to $\mathcal{O}$. Draw the spacetime diagram of $\overline{\mathcal{O}}$ and in it depict the experiment performed by $\mathcal{O}$. Does $\overline{\mathcal{O}}$ conclude that the particle detectors sent out their signals simultaneously? If not, which signal was sent first?

> Hint: See Schutz Fig. 1.5(b) for how the time and space axes look for a reference frame moving in the negative x-direction. Think carefully about what the $\bar{t}$ and $\bar{x}$ mean.

(d) Compute the interval Δs^2 between the events at which the detectors emitted their signals, using both the coordinates of $\mathcal{O}$ and those of $\overline{\mathcal{O}}$.

> Hint: Use the Lorentz transformation for a velocity boost to obtain the coordinates of the events in $\overline{\mathcal{O}}$.

1.6   Show that the interval

$$\Delta \bar{s}^2 = \sum_{\alpha=0}^{3} \sum_{\beta=0}^{3} M_{\alpha\beta} (\Delta x^\alpha)(\Delta x^\beta), \qquad \text{Schutz Eq. (1.2)} \qquad (1.1)$$

contains only $M_{\alpha\beta} + M_{\beta\alpha}$ when $\alpha \neq \beta$, not $M_{\alpha\beta}$ and $M_{\beta\alpha}$ independently. Argue that this allows us to set $M_{\alpha\beta} = M_{\beta\alpha}$ without loss of generality.

Solution: Pick a pair of indices, $\alpha = \alpha*$ and $\beta = \beta*$ say, with $\alpha* \neq \beta*$, and where $\alpha*$ and $\beta*$ are fixed integers in the set $\{0, 1, 2, 3\}$. So $\Delta \bar{s}^2$ contains a term like,

$$M_{\alpha*\beta*} (\Delta x^{\alpha*})(\Delta x^{\beta*}).$$

But $\Delta \bar{s}^2$ also contains a term like,

$$M_{\beta*\alpha*} (\Delta x^{\beta*})(\Delta x^{\alpha*}) = M_{\beta*\alpha*} (\Delta x^{\alpha*})(\Delta x^{\beta*}).$$

The equality follows because of course the product does not depend upon the order of the factors. So we can group these two terms and factor out the $(\Delta x^{\alpha*})(\Delta x^{\beta*})$ leaving,

$$(\Delta x^{\alpha*})(\Delta x^{\beta*})(M_{\alpha*\beta*} + M_{\beta*\alpha*}).$$

Because the off-diagonal terms always appear in pairs as above, we could without changing the interval (and therefore without loss of generality) replace them with their mean value

$$\tilde{M}_{\alpha\beta} \equiv (M_{\alpha\beta} + M_{\beta\alpha})/2.$$

Thus the new tensor $\tilde{M}_{\alpha\beta}$ is by construction symmetric. The RHS of eqn. (1.1) is called a *quadratic form*, and thus the interval of SR can be written as a symmetric quadratic form.

1.8   (a) Derive,

$$\Delta \bar{s}^2 = M_{00}(\Delta r)^2 + 2M_{0i}\Delta x^i \Delta r + M_{ij}\Delta x^i \Delta x^j, \quad \text{Schutz Eq. (1.3)} \quad (1.2)$$

where $\Delta r = \sqrt{(\Delta x)^2 + (\Delta y)^2 + (\Delta z)^2}$, from eqn. (1.1) for general $M_{\alpha\beta}$. [You can assume $\Delta s^2 = 0$ and $\Delta t > 0$.]

Solution: Start with eqn. (1.1), and partially expand the summations

$$\Delta \bar{s}^2 = M_{00}(\Delta t)^2 + \sum_{i=1}^{3} M_{0i}\Delta t \Delta x^i + \sum_{i=1}^{3} M_{i0}\Delta x^i \Delta t + \sum_{i=1}^{3}\sum_{j=1}^{3} M_{ij}\Delta x^i \Delta x^j$$

$$= M_{00}(\Delta t)^2 + 2\sum_{i=1}^{3} M_{0i}\Delta t \Delta x^i + \sum_{i=1}^{3}\sum_{j=1}^{3} M_{ij}\Delta x^i \Delta x^j. \qquad \text{used } M_{i0} = M_{0i}$$

Consider the case $\Delta s^2 = 0$, so from Schutz Eq. (1.1), $\Delta t = \pm \Delta r = \pm\sqrt{(\Delta x)^2 + (\Delta y)^2 + (\Delta z)^2}$. Then, when $\Delta t > 0$,

$$\Delta \bar{s}^2 = M_{00}(\Delta r)^2 + 2\Delta r \sum_{i=1}^{3} M_{0i}\Delta x^i + \sum_{i=1}^{3}\sum_{j=1}^{3} M_{ij}\Delta x^i \Delta x^j,$$

which is eqn. (1.2).

(b) Since $\Delta \bar{s}^2 = 0$ in eqn. (1.2) for any $\{\Delta x^i\}$, replace Δx^i by $-\Delta x^i$ in eqn. (1.2) and subtract the resulting equations from eqn. (1.2) to establish that $M_{0i} = 0$ for $i = 1, 2, 3$.

Solution: Let us first recall why $\Delta \bar{s}^2 = 0$ in eqn. (1.2) for any $\{\Delta x^i\}$. We have set $\Delta s^2 = 0$ (because we were considering the path of a light ray) and it followed, based upon the universality of the speed of light, that we required also $\Delta \bar{s}^2 = 0$. Now why does $\Delta s^2 = 0$ for any Δx^i? Because we have imposed that we are considering the path of a light ray, and regardless of the spatial point x^i on the light ray path we choose, it always has $(\Delta t)^2 = (\Delta r)^2$, so $\Delta s^2 = -(\Delta t)^2 + (\Delta r)^2 = 0$.

Now note that changing Δx^i to $-\Delta x^i$ does not change

$$\Delta r = \sqrt{(\Delta x)^2 + (\Delta y)^2 + (\Delta z)^2}.$$

Thus the only term in eqn. (1.2) to change sign when changing Δx^i to $-\Delta x^i$ is the middle term, the sum over $2 M_{0i} \Delta x^i \Delta r$. The final term does not because changing Δx^i to $-\Delta x^i$ also changes Δx^j to $-\Delta x^j$; the i and j are just dummy indices. So when we subtract $\Delta \bar{s}^2(\Delta t, \Delta x^i) - \Delta \bar{s}^2(\Delta t, -\Delta x^i)$ as instructed, using eqn. (1.2), we find:

$$0 = 0 - 0 = \Delta \bar{s}^2(\Delta t, \Delta x^i) - \Delta \bar{s}^2(\Delta t, -\Delta x^i)$$

$$= M_{00}(\Delta r)^2 + 2\Delta r \sum_{i=1}^{3} M_{0i} \Delta x^i + \sum_{i=1}^{3} \sum_{j=1}^{3} M_{ij} \Delta x^i \Delta x^j$$

$$- \left(M_{00}(\Delta r)^2 + 2\Delta r \sum_{i=1}^{3} M_{0i}(-\Delta x^i) + \sum_{i=1}^{3} \sum_{j=1}^{3} M_{ij}(-\Delta x^i)(-\Delta x^j) \right)$$

$$= 4\Delta r \sum_{i=1}^{3} M_{0i} \Delta x^i. \tag{1.3}$$

This must be true for arbitrary Δx^i so $M_{0i} = 0$.

(c) Derive

$$M_{ij} = -M_{00} \delta_{ij}, \quad (i, j = 1, 2, 3) \qquad \text{Schutz Eq. (1.4b)} \tag{1.4}$$

using eqn. (1.2) with $\Delta \bar{s}^2 = 0$. Hint: $\Delta x, \Delta y,$ and Δz are arbitrary.

Solution: Recall from Exercise 1.8(b) that adding to eqn. (1.2) the following

$$0 = \Delta \bar{s}^2 = M_{00}(\Delta r)^2 - 2\Delta r \sum_{i=1}^{3} M_{0i} \Delta x^i + \sum_{i=1}^{3} \sum_{j=1}^{3} M_{ij} \Delta x^i \Delta x^j$$

gives

$$0 = M_{00}(\Delta r)^2 + \sum_{i=1}^{3} \sum_{j=1}^{3} M_{ij} \Delta x^i \Delta x^j. \tag{1.5}$$

Suppose, $\Delta x = \Delta r, \Delta y = \Delta z = 0$. Substituting into eqn. (1.5) then gives $M_{00} = -M_{11}$. Or, when $\Delta y = \Delta r, \Delta x = \Delta z = 0$, we see that $M_{00} = -M_{22}$. Similarly, $M_{00} = -M_{33}$. To see that the off-diagonal terms are zero, note that it is also possible that $\Delta x = \Delta y = \Delta r / \sqrt{2}$ and $\Delta z = 0$. Substitution into eqn. (1.5) gives that

$$
\begin{aligned}
0 &= (M_{12} + M_{21})(\Delta r)^2/2 + M_{11}(\Delta r)^2/2 + M_{22}(\Delta r)^2/2 + (\Delta r)^2 M_{00} \\
&= (M_{12} + M_{21})(\Delta r)^2/2 - M_{00}(\Delta r)^2/2 - M_{00}(\Delta r)^2/2 + (\Delta r)^2 M_{00} \\
&= (M_{12} + M_{21})(\Delta r)^2/2 = M_{21}(\Delta r)^2.
\end{aligned}
\tag{1.6}
$$

The final step used $M_{\alpha\beta} = M_{\beta\alpha}$, as proved in Exercise 1.6. And since $(\Delta r)^2$ was arbitrary, we have $M_{21} = 0 = M_{12}$. Similarly, $M_{13} = M_{31} = 0 = M_{23} = M_{32}$. In summary,

$$
M_{ij} = -M_{00}\delta_{ij}, \quad (i, j = 1, 2, 3),
$$

which is eqn. (1.4).

1.9   Explain why the line $\mathcal{PQ}$ in Schutz Fig. 1.7 is drawn in the manner described in the text. [Note that in Schutz Fig. 1.7 the $\mathcal{F}$ should be a $\mathcal{Q}$ to be consistent with the text and with the corresponding figure in the first edition (Schutz, 1985, Fig. 1.7).]

Solution: The line $\mathcal{PQ}$ is described in the paragraph after Schutz Eq. (1.5) as perpendicular to the y-axis, parallel to the t–x plane, and parallel to the $\bar{t}$-axis in Schutz Fig. 1.5(a). The line $\mathcal{PQ}$ represents the path of a clock that is stationary in the $\overline{\mathcal{O}}$ frame. Because the $\overline{\mathcal{O}}$ frame moves in the x-direction its path must be orthogonal to the y-axis. And furthermore it must be parallel to the t–x plane, as argued for a clock at the origin of the $\overline{\mathcal{O}}$ frame in Schutz §1.5. In fact the clock is simply displaced a fixed distance from $y = 0$ along the y- or $\bar{y}$-axis and moves parallel to the $\bar{t}$-axis.

1.11   Show that the hyperbolae $-t^2 + x^2 = a^2$ and $-t^2 + x^2 = -b^2$ are asymptotic to the lines $t = \pm x$, regardless of a and b.

Hint: Regardless of how large a and b are, consider the approximate behavior when $|x|$ and $|t|$ are much greater than $|a|$ and $|b|$.

1.12   (a) Use the fact that the tangent to the hyperbola $\mathcal{DB}$ in Schutz Fig. 1.14 is the line of simultaneity for $\overline{\mathcal{O}}$ to show that the time interval $\mathcal{AE}$ is shorter than the time recorded on $\overline{\mathcal{O}}$'s clock as it moved from $\mathcal{A}$ to $\mathcal{B}$.

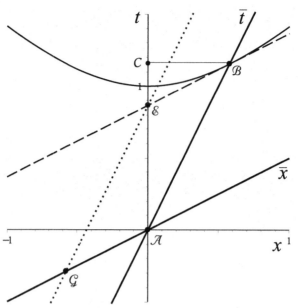

Figure 1.2 Similar to Schutz Fig. 1.14. The dotted line is the path of a second clock at rest in $\overline{\mathcal{O}}$ needed to infer that the moving clock along the t-axis runs slowly.

Solution: This example shows that time dilation is self-consistent. From the perspective of an observer in $\overline{\mathcal{O}}$, the time interval $\mathcal{AE} = \Delta\tau$ corresponds to the proper time of a moving clock, whose world line in Schutz Fig. 1.14 is the t-axis, see fig. 1.2. An observer at rest in $\overline{\mathcal{O}}$ needs two clocks to record the time interval $\Delta\bar{t} = \bar{t}_{\mathcal{E}} - \bar{t}_{\mathcal{A}}$ corresponding to the proper time interval $\Delta\tau$. The clock moving from $\mathcal{A}$ to $\mathcal{B}$ is one of those two clocks, recording $\bar{t}_{\mathcal{A}}$. The other is drawn as a dotted line (fig. 1.2) that passes through $\mathcal{E}$, recording $\bar{t}_{\mathcal{E}}$. The fact that the line of simultaneity in $\overline{\mathcal{O}}$ passes through $\mathcal{B}$ and $\mathcal{E}$ means that $\bar{t}_{\mathcal{E}} = \bar{t}_{\mathcal{B}}$, and hence $\Delta\bar{t} = \bar{t}_{\mathcal{B}} - \bar{t}_{\mathcal{A}}$. Recall the time dilation formula,

$$\Delta\tau = \Delta t \sqrt{1 - v^2}. \qquad \text{Schutz Eq. (1.10)} \qquad (1.7)$$

where Δt was the so-called improper time, an interval measured by two clocks. Here $\Delta\bar{t}$ plays the role of Δt (improper time measured by two clocks):

$$\Delta\tau = \Delta\bar{t} \sqrt{1 - v^2}, \qquad (1.8)$$

implying $\Delta\tau < \Delta\bar{t}$ for $|v| > 0$.

Don't be thrown off by the Δt in eqn. (1.7) not having a bar above it, while it does in eqn. (1.8) above. *It is not the symbol that is important but the role played by the thing it depicts.* The roles of the $\mathcal{O}$ and $\overline{\mathcal{O}}$ frame have been reversed in this exercise, which was the point of discussion around Schutz Fig. 1.14.

1.12 (b) Calculate that[1]

$$(\Delta s^2)_{A\mathcal{E}} = (1 - v^2)(\Delta s^2)_{AB}. \qquad (1.9)$$

Solution: Start with the LHS of eqn. (1.9):

$$
\begin{aligned}
(\Delta s^2)_{A\mathcal{E}} &\equiv -(t_\mathcal{E} - t_A)^2 + (x_\mathcal{E} - x_A)^2 \qquad \text{definition of the interval} \\
&= -t_\mathcal{E}^2 + x_\mathcal{E}^2 \qquad\qquad\qquad\qquad A \text{ is the origin} \\
&= -t_\mathcal{E}^2. \qquad\qquad\qquad\qquad\qquad \mathcal{E} \text{ on } t\text{-axis} \qquad (1.10)
\end{aligned}
$$

From fig. 1.2 herein it is clear that

$$
\begin{aligned}
t_\mathcal{E} &= t_B - x_B v \qquad \text{dashed line parallel to } \bar{x}\text{-axis, slope is } v \\
&= t_B - (t_B v)v \\
&= t_B(1 - v^2). \qquad\qquad\qquad\qquad\qquad\qquad (1.11)
\end{aligned}
$$

Now consider the RHS of eqn. (1.9),

$$(\Delta s^2)_{AB} = -t_B^2 + x_B^2 = -t_B^2 + (vt_B)^2 = -t_B^2(1 - v^2). \qquad (1.12)$$

Combining eqns. (1.10, 1.11, 1.12) one finds,

$$(\Delta s^2)_{A\mathcal{E}} = -t_B^2(1 - v^2)^2 = (1 - v^2)(\Delta s^2)_{AB}. \qquad (1.13)$$

1.12 (c) Use (b) to show that $\overline{\mathcal{O}}$ regards $\mathcal{O}$'s clocks to be running slowly, at just the right rate.

Solution: This corresponds to verfying eqn. (1.8) above; recall $\Delta\tau = t_\mathcal{E}$ and $\Delta\bar{t} = \bar{t}_B$. To find $\bar{t}_B$ use the fact that the interval is invariant between Lorentz frames,

$$
\begin{aligned}
(\Delta s^2)_{AB} &= -t_B^2 + x_B^2 = -\bar{t}_B^2 + \bar{x}_B^2 \\
&= -\bar{t}_B^2. \qquad\qquad\qquad B \text{ on } \bar{t}\text{-axis} \qquad (1.14)
\end{aligned}
$$

Combining eqns. (1.10, 1.13, 1.14)

$$
\begin{aligned}
-t_\mathcal{E}^2 = (\Delta s^2)_{A\mathcal{E}} &= (1 - v^2)(\Delta s^2)_{AB} = -(1 - v^2)\bar{t}_B^2 \\
t_\mathcal{E} &= \bar{t}_B\sqrt{1 - v^2}. \qquad\qquad\qquad \text{took square root} \quad (1.15)
\end{aligned}
$$

1.13 The half-life of the elementary particle called the pi meson (or pion) is 2.5×10^{-8} s when the pion is at rest relative to the observer measuring its decay time. Show, by

---

[1] We had corrected a typo in the original question, replacing $A\mathcal{C}$ with $A\mathcal{E}$. SP1.15 explores the other possible interpretation.

the principle of relativity, that pions moving at speed $v = 0.999$ must have a half-life of 5.6×10^{-7} s, as measured by an observer at rest.

> Hint: Study the solution to Exercise 1.12, and make the analogy with the situation here. Think of the pion as a clock of sorts; its birth is say at time zero and its decay is another tick of the clock. In making the analogy with Exercise 1.12, pay attention to which time intervals are measured by one clock (proper time intervals) and which involve two physically separated clocks.

1.14 Suppose that the velocity v of $\overline{\mathcal{O}}$ relative to $\mathcal{O}$ is small, $v = |\mathbf{v}| \ll 1$. Show that the time dilation, Lorentz contraction, and velocity-addition formulae can be approximated by, respectively:

$$\text{(a)} \quad \Delta t \approx \left(1 + \frac{1}{2}v^2\right) \Delta \bar{t}, \tag{1.16}$$

$$\text{(b)} \quad \Delta x \approx \left(1 - \frac{1}{2}v^2\right) \Delta \bar{x}, \tag{1.17}$$

$$\text{(c)} \quad w' \approx w + v - wv(w + v), \text{ (with } |w| \ll 1 \text{ as well).} \tag{1.18}$$

What are the relative errors in these approximations when $v = w = 0.1$.

(a) Solution: Recall the time dilation formula was given in eqn. (1.7), with here $\Delta \tau = \Delta \bar{t}$. Solving for Δt, and expanding the RHS in a Taylor series in the small parameter v we obtain

$$\Delta t = \Delta \bar{t} \frac{1}{\sqrt{1 - v^2}} = \Delta \bar{t} \, (1 - v^2)^{-1/2}$$

$$= \Delta \bar{t} \left(1 + \frac{1}{2}v^2 + \frac{3}{8}v^4 + \cdots\right) \qquad \text{used eqn. (B.2)}$$

$$\simeq \Delta \bar{t} \left(1 + \frac{1}{2}v^2\right). \tag{1.19}$$

For the Taylor series we have used the binomial series, eqn. (B.2) of Appendix B, a result well worth remembering! The largest term we ignored was $\frac{3}{8}v^4$. You will often see this written as $O(v^4)$, read "of order v to the fourth." This means that we are focusing attention on the important part, i.e. v^4, and ignoring the irrelevant numerical factor $3/8$ that is close to unity. The higher order terms in the series were $O(v^6)$ and these are clearly much smaller since $v \ll 1$. The relative error is then

$$\frac{\frac{3}{8}v^4}{(1 - v^2)^{-1/2}} \approx \frac{3}{8}v^4 = 3.75 \times 10^{-5}.$$

In fact the relative error can be calculated exactly to be 3.76×10^{-5}, see accompanying Maple™ worksheet.

(b) Solution: Recall the Lorentz contraction formula was given in Schutz Eq. (1.11), which we can write as

$$\Delta x = l\sqrt{1-v^2} = \Delta\bar{x}\sqrt{1-v^2}, \qquad \text{cf. Schutz Eq. (1.11)} \qquad (1.20)$$

where $\Delta\bar{x}$ is the so-called proper length of the rod, i.e. as measured in a frame in which the rod is at rest, Δx is the length of the rod measured in a frame in which the rod has speed v. Using again the binomial series we have immediately

$$\Delta x = \Delta\bar{x}\,(1-v^2)^{1/2} \simeq \Delta\bar{x}\left(1 - \frac{1}{2}v^2\right), \qquad \text{used eqn. (B.2)} \qquad (1.21)$$

where we have dropped the terms $O(v^4)$ and higher order terms in the binomial series because $v \ll 1$. The largest error term here is $\frac{1}{8}v^4$, which gives a relative error of about 1.25×10^{-5}. The exact calculation of relative error gives -1.26×10^{-5}, see accompanying Maple™ worksheet.

(c) Solution: Finally the Einstein law of composition of velocities was

$$w' = \frac{w+v}{1+wv}, \qquad \text{Schutz Eq. (1.13)} \qquad (1.22)$$

where w' is the speed of a particle measured in some inertial frame $\mathcal{O}$, v is the speed of an observer $\mathcal{A}$ measured in $\mathcal{O}$, and w is the speed of the particle, in the direction as v, measured by observer $\mathcal{A}$. Using again the binomial series we have immediately

$$w' = (w+v)(1+wv)^{-1} \simeq (w+v)(1-wv), \qquad \text{used eqn. (B.2)} \qquad (1.23)$$

where we have dropped terms $O(w^2v^2)$ and higher order terms in the binomial series because $wv \ll 1$. More precisely, in fact the largest term we dropped was w^2v^2 leading to a relative error of

$$\frac{w^2v^2}{1/(1+wv)} \approx w^2v^2 = 1 \times 10^{-4},$$

which is a very good estimate, agreeing with the exact relative error to one part in 10^{-10}; see accompanying Maple™ worksheet.

1.16 Use the Lorentz transformations,

$$\bar{t} = \gamma t - v\gamma x, \qquad\qquad \text{with } \gamma \equiv \frac{1}{\sqrt{1-v^2}}$$

$$\bar{x} = -v\gamma t + \gamma x,$$

$$\bar{y} = y,$$

$$\bar{z} = z, \qquad\qquad\qquad\qquad \text{Schutz Eq. (1.12)} \qquad (1.24)$$

to derive (a) the time dilation, and (b) the Lorentz contraction formulae. Do this by identifying the pairs of events where the separations (in time or space) are to be compared, and then using the Lorentz transformation to accomplish the algebra that the invariant hyperbolae had been used for in the text.

Solution: It is helpful to have short catchphrases to orient you, e.g. "Moving clocks run slowly." More precisely, in SR time dilation occurs when a clock is moving at constant velocity as observed from an inertial reference frame $\mathcal{O}$. This situation was depicted in Schutz Fig. 1.14, (see fig. 1.2) with the "moving clock" following the $\bar{t}$-axis, passing through events $\mathcal{A}$ and $\mathcal{B}$ during proper time

$$\Delta\tau = \bar{t}_B - \bar{t}_A = \bar{t}_B. \tag{1.25}$$

Note we chose the origins to coincide in the two frames so that the algebra is simplified, i.e. $t_A = \bar{t}_A = 0$. We want to relate $\Delta\tau$ to the time between these same events observed in $\mathcal{O}$, wherein:

$$\Delta t = t_B - t_A = t_B. \tag{1.26}$$

With eqns. (1.25, 1.26) we have the two time intervals we want to relate expressed in terms of the same event. The Lorentz transformations eqn. (1.24) give the $\overline{\mathcal{O}}$ coordinates in terms of the $\mathcal{O}$ coordinates when $\overline{\mathcal{O}}$ is moving at speed v along the x axis. Substituting $v \to -v$ then gives us the transformation back to $\mathcal{O}$:

$$t_B = \frac{\bar{t}_B}{\sqrt{1 - (-v)^2}} - \frac{(-v)\bar{x}_B}{\sqrt{1 - (-v)^2}} = \frac{1}{\sqrt{1 - v^2}}\bar{t}_B.$$

Using eqn. (1.25) and eqn. (1.26) we obtain

$$\Delta\tau = \bar{t}_B = \sqrt{1 - v^2}t_B = \sqrt{1 - v^2}\Delta t,$$

in agreement with the time dilation formula, cf. eqn. (1.7).

"Moving rods contract." More precisely, the Lorentz contraction (or Lorentz–Fitzgerald contraction) occurred when the clock was replaced by a rod. The geometry and algebra are simplified when the rod lies along the $\bar{x}$-axis as depicted in Schutz Fig. (1.13). The proper length of the rod is the length observed in the frame wherein the rod is stationary,

$$l = \bar{x}_C - \bar{x}_A = \bar{x}_C = \bar{x}_B, \tag{1.27}$$

where the final equality holds because the trajectory of the tip of the rod through $\bar{x}_B$ and $\bar{x}_C$ is parallel to the $\bar{t}$-axis. We want to relate the proper length l to that observed in $\mathcal{O}$. The length in $\mathcal{O}$ is the distance between the ends of the rod measured at a given instant in $\mathcal{O}$, e.g. at $t = 0$ the length is

$$\Delta x = x_B - x_A = x_B. \tag{1.28}$$

With (1.27) and (1.28) we have the two lengths we want to relate expressed in terms of the same event. We can use the Lorentz transformations eqn. (1.24) to transform the $\mathcal{O}$ coordinates of event $\mathcal{B}$ to that of $\overline{\mathcal{O}}$

$$\bar{x}_B = \frac{-vt_B}{\sqrt{1 - v^2}} + \frac{x_B}{\sqrt{1 - v^2}} = \frac{x_B}{\sqrt{1 - v^2}}.$$

Using (1.27, 1.28) we find

$$\Delta x = l\sqrt{1 - v^2},$$

in agreement with the Lorentz contraction formula eqn. (Schutz Eq 1.11).

Time dilation and Lorentz contraction were derived using the Lorentz transformation by Hobson et al. (2006, §1.7).

1.18 (a) The Einstein velocity-addition law, eqn. (1.22), has a simpler form if we introduce the concept of the velocity parameter V, defined by the equation $v = \tanh V$. Notice that for $-\infty < V < \infty$, the velocity is confined to the acceptable limits $-1 < v < 1$. Show that if $u = \tanh U$ and $w = \tanh W$, then eqn. (1.22) implies

$$w' = \tanh(W + U).$$

This means that velocity parameters add linearly.

Solution: Simply substitute the definition of velocity parameter into eqn. (1.22):

$$w' = \frac{\tanh(U) + \tanh(W)}{1 + \tanh(U)\tanh(W)} \tag{1.29}$$

$$= \frac{(\tanh(U) + \tanh(W))\cosh(W)\cosh(U)}{\cosh(W)\cosh(U) + \sinh(U)\sinh(W)}. \tag{1.30}$$

The numerator can be written as,

$$N = \sinh(W)\cosh(U) + \cosh(W)\sinh(U),$$

so that

$$w' = \frac{\sinh(W)\cosh(U) + \cosh(W)\sinh(U)}{\cosh(W)\cosh(U) + \sinh(U)\sinh(W)}.$$

The following identities are useful:

$$\cosh(a)\cosh(b) = \left(\frac{\exp(a) + \exp(-a)}{2}\right)\left(\frac{\exp(b) + \exp(-b)}{2}\right)$$

$$= \frac{\exp(a+b) + \exp(-(a+b))}{4} + \frac{\exp(a-b) + \exp(-(a-b))}{4}$$

$$= \frac{\cosh(a+b)}{2} + \frac{\cosh(a-b)}{2}, \tag{1.31}$$

$$\sinh(a)\sinh(b) = \left(\frac{\exp(a) - \exp(-a)}{2}\right)\left(\frac{\exp(b) - \exp(-b)}{2}\right)$$

$$= \frac{\exp(a+b) + \exp(-(a+b))}{4} - \frac{\exp(a-b) + \exp(-(a-b))}{4}$$

$$= \frac{\cosh(a+b)}{2} - \frac{\cosh(a-b)}{2}, \tag{1.32}$$

and

$$\sinh(a)\cosh(b) = \left(\frac{\exp(a) - \exp(-a)}{2}\right)\left(\frac{\exp(b) + \exp(-b)}{2}\right)$$

$$= \frac{\exp(a+b) - \exp(-(a+b))}{4} + \frac{\exp(a-b) - \exp(-(a-b))}{4}$$

$$= \frac{\sinh(a+b)}{2} + \frac{\sinh(a-b)}{2}. \tag{1.33}$$

Using eqns. (1.31) and (1.32) the denominator above simplifies to $D = \cosh(U + W)$. Using (1.33) the numerator simplifies to $N = \sinh(U + W)$. So,

$$w' = \tanh(U + W),$$

which reveals that we can linearly add velocity parameters, then apply tanh to reduce the final parameter to the final velocity.

1.18 (b) Use this to solve the following problem. A star measures a second star to be moving away at speed $v = 0.9c$. The second star measures a third to be receding in the same direction at $0.9c$. Similarly, the third measures a fourth, and so on, up to some large number N of stars. What is the velocity of the Nth star relative to the first? Give an exact answer and an approximation useful for large N.

Solution: The velocity of second star relative to first is $u_2 = 0.9$. The velocity of Nth star relative to $(N-1)$th, $u_N - u_{N-1} = 0.9$. So the velocity of the Nth star relative to the first is,

$$u'_N = \tanh[(N-1)U],$$

where $0.9 = \tanh(U)$, so $U \approx 1.47222$. For large N the argument of the tanh becomes large, so

$$\tanh(z) = \frac{e^z - e^{-z}}{e^z + e^{-z}} = \frac{1 - e^{-2z}}{1 + e^{-2z}} \simeq [1 - e^{-2z}]^2 \simeq 1 - 2e^{-2z}.$$

The first approximation used the binomial series eqn. (B.2) to first order, and the second ignored the square of the small number, i.e. e^{-4z}. So $u'_N \simeq 1 - 2e^{[-2(N-1)U]}$.

1.20 Write the Lorentz transformation equations in matrix form.

Solution: The Lorentz transformation equations were given in eqn. (1.24). These can be written in matrix for as:

$$\bar{\mathbf{x}} = \mathbf{A}x,$$

where

$$\bar{\mathbf{x}} = \begin{pmatrix} \bar{t} \\ \bar{x} \\ \bar{y} \\ \bar{z} \end{pmatrix}, \quad \mathbf{x} = \begin{pmatrix} t \\ x \\ y \\ z \end{pmatrix}$$

and

$$\mathbf{A} = \begin{pmatrix} \gamma & -v\gamma & 0 & 0 \\ -v\gamma & \gamma & 0 & 0 \\ 0 & 0 & 1 & 0 \\ 0 & 0 & 0 & 1 \end{pmatrix} = \begin{pmatrix} \cosh(V) & -\sinh(V) & 0 & 0 \\ -\sinh(V) & \cosh(V) & 0 & 0 \\ 0 & 0 & 1 & 0 \\ 0 & 0 & 0 & 1 \end{pmatrix}, \quad (1.34)$$

where $\gamma = 1/\sqrt{1-v^2}$. The second matrix above used the velocity parameter defined as $v = \tanh(V)$ in Exercise 1.18.

1.2 Supplementary problems

SP 1.1 Construct an experiment to show that one is in an inertial reference frame. Assume that there are no gravitational forces on your test object. (See next problem for complications arising from gravity!)

Solution

When gravity is not present one can use Newton's laws to argue that test particles with no external forces on them should have zero acceleration and therefore move at constant velocity relative to each other in an inertial frame. So any experiment that tests for these conditions suffices. For example, if you let go of an object at rest with respect to you, it should appear to hover unmoving in front of you. A more elaborate experimental test of an inertial frame, that works also in presence of gravity, was presented by Misner et al. (1973, Fig. 1.7).

SP 1.2 Answer the question posed in footnote 4 on page 3 of Schutz: An astronaut is in orbit about Earth, holding a bowl of soup. Does the soup climb up the side of the bowl? You can make some idealizations: assume that the astronaut has a lid on his soup so that the surface of the soup is initially flat, and he removes the lid without accelerating the bowl relative to him; the bowl is small relative to the radius of the orbit and the experiment short enough that the astronaut and all of his soup can be treated as being essentially in the same orbit. (Calculating the effects associated with nonuniform gravitational fields will be an important consideration much later when we get into general relativity.) The footnote alludes to problems gravity poses for SR revealed by considering two astronauts in different orbits. Explain these problems, and in particular, address these questions: (i) Does the astronaut in orbit pass the test for a local inertial reference frame? (ii) Can two inertial reference frames accelerate relative to each other? (iii) Argue that two non-coincident orbits have non-zero relative acceleration.

Philosophers and historians of physics will recognize this footnote as bearing on the famous *bucket experiment* of Isaac Newton (e.g. Maudlin, 2012).

SP 1.3 Referring to Schutz Fig. 1.5, explain why the angle of the $\bar{x}$-axis to the x-axis is $\phi = \arctan(v)$, where $v = |\mathbf{v}|$ is the magnitude of the velocity of $\bar{\mathcal{O}}$ along the x-axis. The result follows from the construction of the $\bar{x}$-axis, but the steps involved are not trivial. Following the spirit of Schutz §1.5, try to find a geometric argument rather than using the Lorentz transformation.

Solution

Refer to Schutz Fig. 1.4. We will find the coordinates of the point $\mathcal{P}$, an arbitrary point on the $\bar{x}$-axis. This can be found from the intersection of the line passing through events $\mathcal{E}$ and $\mathcal{P}$ and the line passing through $\mathcal{R}$ and $\mathcal{P}$ as follows. Let the equation for the $\bar{t}$-axis be $t = mx$. (Any non-vertical line through the origin can be written in this form.) The events $\mathcal{E}$ and $\mathcal{R}$ then have coordinates $(-x_0, -mx_0)$ and (x_0, mx_0), where x_0 is a parameter related to a. The line through $\mathcal{EP}$ has slope unity and contains the point $(-x_0, -mx_0)$, so its equation is $t = x + x_0 - mx_0$. Similarly, the line through $\mathcal{RP}$ has equation $t = -x + x_0 + mx_0$. By setting those equal to each other you can easily find that the point of intersection $\mathcal{P}$ is (mx_0, x_0), which shows that the $\bar{x}$-axis is $t = x/m$.

Using the Lorentz transformation one can obtain the same result more easily.

SP 1.4 A particle that follows the curve $t = x$ in the S coordinate system has the speed of light in the x-direction of that frame. Based upon that fact, what do you anticipate for the equation of that curve under the Lorentz transformations of a velocity boost in the x-direction? Verify your prediction using the Lorentz transformations eqn. (1.24).

SP 1.5 The special theory of relativity has led to a revision in our notion of space and time from that of Euclidean space and absolute time used for Newtonian mechanics to that of Minkowski spacetime. Is Newton's first law of motion consistent with Special Relativity?

Solution

Newton's first law of motion is consistent with Special Relativity. For a free particle moves at uniform speed along a straight line in SR as well as in Newtonian mechanics. Newton's second law can also be made to be consistent with the two hypotheses of

Special Relativity, but that requires a reinterpretation of mass, and a transformation law of forces that would have surprised Newton.

SP 1.6 Prove that the Lorentz transformation for a boost of velocity, which is linear by construction, transforms straight lines to straight lines so that a particle of constant velocity in one frame also has a constant velocity in the other frame. Thus the Lorentz transformation for a boost of velocity respects Newton's first law of motion.

SP 1.7 In Schutz §1.6 when deriving the transformation from the $\mathcal{O}$ frame to the $\overline{\mathcal{O}}$ frame it was assumed that the transformation must be linear.

(a) If we exclude the transformation that reduces all curves to a point, prove that only a linear transformation, possibly followed by a translation, (i.e. an *affine transformation*) is consistent with Newton's first law that a particle subjected to zero net external force must travel in a straight line at constant speed in all inertial reference frames.

Solution

Let $x^\mu(t)$ be the world line of a free particle in inertial frame $\mathcal{O}$, parameterized by the time coordinate t. For a general transformation $F^{\tilde{\alpha}}$ to another inertial frame $\overline{\mathcal{O}}$, the world line of the particle in $\overline{\mathcal{O}}$ will be

$$x^{\tilde{\alpha}}(\tilde{t}) = F^{\tilde{\alpha}}(x^\mu) \circ t(\tilde{t}), \qquad (1.35)$$

where initially we entertain the possibility that $F^{\tilde{\alpha}}$ is a *nonlinear* function of x^μ. To help keep track of dependencies, we explicitly parameterize the world line coordinates in $\overline{\mathcal{O}}$ with the time coordinate $\tilde{t}$. Because the particle is free of external forces Newton's first law requires zero acceleration,

$$\frac{d^2 x^\mu}{dt^2} = 0. \qquad (1.36)$$

To be consistent with Newton's first law this must hold true in all inertial reference frames. Consider the inertial frame $\overline{\mathcal{O}}$ moving at constant velocity relative to $\mathcal{O}$. Applying transformation (1.35) and the chain rule the first derivative is

$$\frac{dx^{\tilde{\alpha}}}{d\tilde{t}} = \frac{\partial F^{\tilde{\alpha}}}{\partial x^\mu} \frac{dx^\mu}{dt} \frac{dt}{d\tilde{t}}, \qquad (1.37)$$

so Newton's first law requires

$$\frac{d^2 x^{\tilde{\alpha}}}{d\tilde{t}^2} = \frac{\partial^2 F^{\tilde{\alpha}}}{\partial x^\mu \partial x^\nu} \frac{dx^\mu}{dt} \frac{dx^\nu}{dt} \left(\frac{dt}{d\tilde{t}}\right)^2 + \frac{\partial F^{\tilde{\alpha}}}{\partial x^\mu} \frac{dx^\mu}{dt} \frac{d^2 t}{d\tilde{t}^2} = 0, \qquad (1.38)$$

where we have used the chain rule and eqn. (1.36). Arbitrary free particles satisfying eqn. (1.36) have constant but otherwise arbitrary velocities dx^μ/dt. Furthermore we exclude the possibility that $dt/d\tilde{t} = 0$, otherwise, from eqn. (1.37), we would have stationary particles in $\overline{\mathcal{O}}$ for arbitrary free particles in $\mathcal{O}$. (This requirement has been overlooked by several authors according to Berzi and Gorini (1969).) So eqn. (1.38) can only be true if both

$$\frac{\partial^2 F^{\bar{\alpha}}}{\partial x^\mu \partial x^\nu} = 0 \qquad \text{and} \qquad \frac{d^2 t}{d\bar{t}^2} = 0. \qquad (1.39)$$

Integrating the first equation we obtain

$$\frac{\partial F^{\bar{\alpha}}}{\partial x^\mu} = \Lambda^{\bar{\alpha}}{}_\mu = \text{constant} \qquad (1.40)$$

$$F^{\bar{\alpha}} = \Lambda^{\bar{\alpha}}{}_\mu x^\mu + \Gamma^{\bar{\alpha}}, \qquad (1.41)$$

where $\Gamma^{\bar{\alpha}}$ is a set of four constants representing a translation. $F^{\bar{\alpha}}$ is an affine transformation, and becomes a linear transformation when $\Gamma^{\bar{\alpha}} = 0$. Note that $dt/d\bar{t} = \Lambda^0{}_{\bar{0}}$, a constant, so the second equation in eqn. (1.39) is implied by the first.

(b) Prove that the assumption of homogeneous spacetime also leads to an affine transformation. This was in fact the argument used by Einstein in his original 1905 article that introduced special relativity (Einstein, 1905), although he apparently did not distinguish between linear and affine.

Solution

Use the general nonlinear transformation introduced above in eqn. (1.35) and consider the slope of the curve of a coordinate in $\overline{\mathcal{O}}$ plotted against a coordinate in $\mathcal{O}$:

$$\frac{\partial x^{\bar{\alpha}}}{\partial x^\mu} = \frac{\partial F^{\bar{\alpha}}}{\partial x^\mu}.$$

For spacetime to be homogeneous this slope must not depend upon position, so we would require the partial derivatives to be constant, as in eqn. (1.40). Integrating this leads to the affine transformation eqn. (1.41).

SP 1.8   Recall in Schutz §1.6 that it was determined that $\phi(\mathbf{v})$ was the ratio of the length of the rod observed in the $\mathcal{O}$ and $\overline{\mathcal{O}}$ frames, where $\mathbf{v}$ was the velocity between the two frames and the rod was by construction orthogonal to $\mathbf{v}$. See Schutz Fig. 1.7 where the rod had ends at $\mathcal{A}$ and $\mathcal{B}$. It was argued further that ϕ could not depend upon the direction of the velocity $\mathbf{v}$ because there were assumed to be no preferred directions in space. Spell out this argument more completely.

SP 1.9   It is useful to have a rulebook for using the spacetime diagrams for special relativity (do not use these in general relativity!). Answer the following questions to test your rulebook. Assume we have two inertial frames of reference, the $\mathcal{O}$ frame and an $\overline{\mathcal{O}}$ moving at constant positive velocity relative to $\mathcal{O}$ along the x-axis, and refer to the diagram in the x–t plane shown in fig. 1.3.

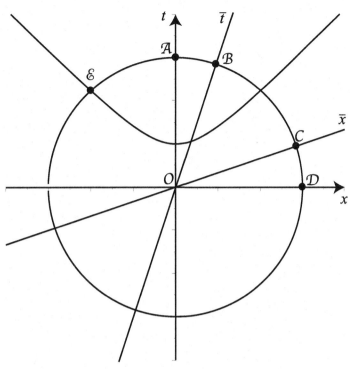

Figure 1.3   The points $\mathcal{A}, \mathcal{B}, \mathcal{C}, \mathcal{D}, \mathcal{E}$ lie on the circle with radius 3 units.

(a) Two events occur at the same point, say $\mathcal{A}$, in a spacetime diagram as depicted in the $\mathcal{O}$ frame. Do they necessarily occur at the same point in the $\overline{\mathcal{O}}$ frame? How many points are needed on the diagram to depict these events in all the frames?

(b) Does a line that appears straight in $\mathcal{O}$'s frame necessarily have to be straight for the reference frame $\overline{\mathcal{O}}$ moving at constant velocity relative to $\mathcal{O}$?

(c) If two points are equidistant to the origin of reference frame $\mathcal{O}$, are they still equidistant from the origin in another reference frame $\overline{\mathcal{O}}$, moving relative to $\mathcal{O}$ but with the same origin?

(d) The events $\mathcal{A}, \mathcal{B}, \mathcal{C}, \mathcal{D}, \mathcal{E}$ in fig. 1.3 all lie on a circle in the $t-x$ plane centered at the origin of frame $\mathcal{O}$. Which has the greatest interval from the origin? What about in the $\overline{\mathcal{O}}$ frame?

(e) Is the x component of $\mathcal{D}$ the same, greater or less than the $\bar{x}$ component of $\mathcal{C}$? Is the t component of $\mathcal{A}$ the same, greater or less than the $\bar{t}$ component of $\mathcal{B}$?

(f) The event $\mathcal{E}$, with $x = -2$ in $\mathcal{O}$, lies on a hyperbola that cuts the t-axis at $t = 1$. Is there a reference frame, say $\overline{\overline{\mathcal{O}}}$ with the same origin as $\mathcal{O}$ but in which $\bar{\bar{x}} = 0$? If so, what is the value of $\bar{\bar{t}}$ of event $\mathcal{E}$?

Solutions

(a) Yes, of course! And we need just one point on the diagram. That is what we mean by an event in 4-dimensional spacetime; it is something that is uniquely determined by four coordinates. While the values of the individual coordinates vary with choice of reference frame, the event itself is fixed. If two events occur at the same point in space and time in one frame they necessarily occur at the same point in space and time in all frames. To prove this mathematically, without loss of generality choose the origin to be the first event. Then the second event has coordinates $(0, 0, 0, 0)$ in $\mathcal{O}$'s frame. Now choose the $\overline{\mathcal{O}}$ frame in the standard configuration with axes and origin aligned with $\mathcal{O}$ at $t' = 0$. The Lorentz transformation immediately gives that the two events have the same coordinates $(0, 0, 0, 0)$ in the $\overline{\mathcal{O}}$ frame.

(b) Yes, a straight line in one frame is straight in all frames. SP1.6 called for a proof of this based on the Lorentz transformation. That straight lines transform to straight in all inertial reference frames relies on the affine structure of the space. Geometries exist without this property, for example the manifolds studied in geometric topology, sometimes called "rubber-sheet geometry." Maudlin (2012) provides an extensive discussion of the hierarchical structures of geometry and their role in the foundations of physics.

(c) Not necessarily. The Lorentz contraction only contracts the component of distance in the direction of the relative motion of the two reference frames.

(d) The event $\mathcal{D}$ has the greatest interval to the origin

$$\Delta s^2 = -(\Delta t)^2 + (\Delta x)^2 = 0 + 3^2 = 9.$$

The interval is a frame-invariant quantity so this conclusion is true in all reference frames.

(e) The x component of $\mathcal{D}$ is greater than the $\bar{x}$ component of $\mathcal{C}$. Recall from (d) above that event $\mathcal{D}$ has $x^2 = 9$. Event $\mathcal{C}$ is simultaneous to the origin in frame $\overline{\mathcal{O}}$ so it has $\bar{x}^2 = \Delta s^2$, but its interval $\Delta s^2 < 9$, as argued in (d) above. By a similar argument the t component of $\mathcal{A}$ is greater than the $\bar{t}$ component of $\mathcal{B}$.

(f) Yes, because the event $\mathcal{E}$ is timelike separated from the origin (that is $(\Delta t)^2 > (\Delta x)^2$) we can find such a reference frame in which $\bar{\bar{x}} = 0$. Then $\bar{\bar{t}} = 1$, since its interval from the origin will remain constant, and this must be $\Delta s^2 = -1$ because the hyperbola passes through the points $x = 0, t = 1$.

SP 1.10   The Lorentz transformations are written above in eqn. (1.24) with dimensionless velocity v, and t has units of length. Reintroduce factors of c to write these equations with t in units of time, say T, and x still in units of length, say L, such that the equations are valid when $c \neq 1[L/T]$ and v has units of $[L/T]$.

Hint

You can multiply or divide by c wherever it is needed. The goal is to get the units to agree on all the terms, where "terms" are quantities that are added or subtracted.

> While it does make sense to multiply factors with different units, it only makes sense to add or subtract terms that have the same units.

SP 1.11   How would you use light signals to synchronize two spatially separate clocks?

Hint

Refer to the construction of the $\bar{x}$-axis in Fig. 1.4 of §1.5 of Schutz. This was how Einstein first defined simultaneity in his 1905 article "On the electrodynamics of moving bodies" (Einstein, 1905).

SP 1.12   Show that the Lorentz transformations of a velocity boost along the x-axis reduce to the Galilean transformations in the limit of small velocity $|\mathbf{v}| \ll 1$.

SP 1.13   Solve Exercise 1.3(i) without using the Lorentz transformations.

Solution

The locus of events, all of which occur at the time $\bar{t} = 2$ m, have arbitrary $\bar{x}$, and so the solution is a straight line parallel to the $\bar{x}$-axis. Thus when plotted on the $\mathcal{O}$ frame the line will have slope $t/x = v = 1/2$, see solution to Exercise 1.3(c) above. But to find the full equation of the line we must also specify the (x, t) coordinates of a point on the line. We can do this most easily by choosing the point where the line crosses the $\bar{t}$-axis, i.e. where $\bar{x} = 0$. Then the interval from the origin is $\Delta s^2 = -\bar{t}^2 + \bar{x}^2 = -\bar{t}^2 = -4$. And of course the squared interval is invariant, so $\Delta s^2 = -t^2 + x^2 = -4$. We solve this equation simultaneously with $t/x = 1/v = 2$ to find $x = \pm 2/\sqrt{3}$. The $\bar{t}$-axis and the hyperbola $\Delta s^2 = -t^2 + x^2 = -4$ cross twice, once with $t > 0$ and once with $t < 0$. And we want to take the positive root, i.e. $(x = 2/\sqrt{3}, t = 2x = 4/\sqrt{3})$ since here clearly we want the $t, \bar{t} > 0$ solution. Using this point and slope the equation for the line is then

$$t = 4/\sqrt{3} + (x - 2/\sqrt{3})/2 = \sqrt{3} + x/2.$$

SP 1.14   In the final paragraph of § 1.7, Schutz states that the tangent to a hyperbola at any event, say $\mathcal{P}$, is a line of simultaneity of the Lorentz frame whose time axis joins event $\mathcal{P}$ to the origin. Prove this. Here Schutz is of course speaking of an invariant hyperbola, centered at the origin, with timelike interval to the origin, for otherwise the $\bar{t}$-axis would not cut the hyperbola.

> **Hint**
>
> Use the fact that the slope of the $\bar{x}$-axis is the inverse of that of the $\bar{t}$-axis, implicit in SP1.3.

SP 1.15   (a) Show that

$$(\Delta s^2)_{AB} = (1 - v^2)\,(\Delta s^2)_{AC}$$

in fig. 1.2, see solution to Exercise 1.12. And (b), use this to show that $\mathcal{O}$ regards $\overline{\mathcal{O}}$'s clocks to be running slowly, at just the "right" rate.

Solution

(a) It is easiest to work in the $\mathcal{O}$ frame.

$$
\begin{aligned}
(\Delta s^2)_{AB} &= -(\Delta t)^2_{AB} + (\Delta x)^2_{AB} \quad \text{definition of interval} \\
&= -t_C^2 + (t_C v)^2 = -t_C^2(1 - v^2). \quad\quad\quad\quad (1.42)
\end{aligned}
$$

Note that we have used

$$
\begin{aligned}
(\Delta t)_{AB} &= t_B - t_A & \text{definition of the time increment} \\
&= t_B & \mathcal{A} \text{ is at the origin so } t_A = 0 \\
&= t_C. & \text{simultaneous events in } \mathcal{O} \quad\quad (1.43)
\end{aligned}
$$

And for

$$
\begin{aligned}
(\Delta s^2)_{AC} &= -(\Delta t)^2_{AC} + (\Delta x)^2_{AC} & \text{definition of interval} \\
&= -t_C^2 = \frac{(\Delta s^2)_{AB}}{(1 - v^2)}. & \text{used eqn. (1.42)} \quad\quad (1.44)
\end{aligned}
$$

(b) Recall from the solution to (a) above that the (proper) time recorded on $\overline{\mathcal{O}}$'s clock as it moved from $\mathcal{A}$ to $\mathcal{B}$ was based on the interval from $\mathcal{A}$ to $\mathcal{B}$,

$$
\begin{aligned}
\bar{t}_B^2 &= -\Delta \bar{s}^2_{AB} & \text{proper time in } \overline{\mathcal{O}} \\
&= -\Delta s^2_{AB} & \text{interval is frame-invariant} \\
&= t_C^2(1 - v^2). & \text{used eqn. (1.42) or eqn. (1.44).} \quad (1.45)
\end{aligned}
$$

So $\mathcal{O}$, who thinks events $\mathcal{C}$ and $\mathcal{B}$ are simultaneous, corresponding to time $t = t_C$, thinks $\overline{\mathcal{O}}$'s clock is slow because it reads only $\bar{t}_B = t_C\sqrt{1 - v^2}$ at this time. This agrees with eqn. (1.7). Note that $\sqrt{1 - v^2} < 1$ for all $|v| > 0$.

2 Vector analysis in special relativity

It is almost always helpful in calculations to use such [frame-invariant] expressions.

Bernard Schutz §2.6

2.1 Exercises

2.2 Identify the free and dummy indices in the following equations and change them into equivalent expressions with different indices. How many different equations does each expression represent?

(a) $A^\alpha B_\alpha = 5$.

Solution: α is the dummy index. There are no free indices and this represents only one equation. Here it could be written equivalently with α replaced by any Greek letter, e.g. $A^\mu B_\mu = 5$ means the same thing, but in general one has to be careful about which letters are being used already in the same term.

(b) $A^{\bar\mu} = \Lambda^{\bar\mu}{}_\nu A^\nu$.

Solution: ν is the dummy index. $\bar\mu$ is the free index, so there are four equations, one for each value of $\bar\mu$. Here it could be written equivalently as $A^{\bar\alpha} = \Lambda^{\bar\alpha}{}_\sigma A^\sigma$. But pay attention when using this expression since now $\bar\alpha$ is playing the role of $\bar\mu$.

(c) $T^{\alpha\mu\lambda} A_\mu C_\lambda{}^\gamma = D^{\gamma\alpha}$.

Solution: Both λ, μ are dummy indices; α, γ free indices; 16 equations, one for each pair of the free indices. One could write this as: $T^{\alpha\nu\sigma} A_\nu C_\sigma{}^\gamma = D^{\gamma\alpha}$, or you could change the free indices too $T^{\beta\nu\sigma} A_\nu C_\sigma{}^\mu = D^{\mu\beta}$, but then β and μ play the roles of α and γ.

(d) $R_{\mu\nu} - \frac{1}{2} g_{\mu\nu} R = G_{\mu\nu}$.

Solution: ν and μ are free indices, and there are 16 equations, one for each pair of the free indices. [We will see later that in this particular case the equations are not all independent because of the symmetry properties of the tensors involved, but that is a separate issue.] Although the indices are repeated, they are not repeated in the same factor. An equivalent expression is: $R_{\alpha\beta} - \frac{1}{2}g_{\alpha\beta}R = G_{\alpha\beta}$.

2.5 A collection of vectors $\{\vec{a}, \vec{b}, \vec{c}, \vec{d}\}$ is said to be linearly independent if no linear combination of them is zero except the trivial one, $0\vec{a} + 0\vec{b} + 0\vec{c} + 0\vec{d} = \vec{0}$.

(a) Show that the basis vectors in

$$\vec{e}_0 \underset{\mathcal{O}}{\rightarrow} (1, 0, 0, 0),$$

$$\vec{e}_1 \underset{\mathcal{O}}{\rightarrow} (0, 1, 0, 0),$$

$$\vec{e}_2 \underset{\mathcal{O}}{\rightarrow} (0, 0, 1, 0),$$

$$\vec{e}_3 \underset{\mathcal{O}}{\rightarrow} (0, 0, 0, 1), \qquad \text{Schutz Eq. (2.9)} \qquad (2.1)$$

are linearly independent.

Solution: Call the basis $\mathcal{O}$ and start with an arbitrary linear combination of these four vectors, $a_0\vec{e}_0 + a_1\vec{e}_1 + a_2\vec{e}_2 + a_3\vec{e}_3 \underset{\mathcal{O}}{\rightarrow} (a_0, a_1, a_2, a_3)$. The only way that vector can equal the vector $(0, 0, 0, 0)$ is if all components are zero, which means all four of the a_μ must be zero, i.e. the trival solution. Thus the basis vectors are linearly independent.

Alternatively, some students will be more comfortable with using linear algebra. One could write this out in matrix notation:

$$\begin{pmatrix} 1 & 0 & 0 & 0 \\ 0 & 1 & 0 & 0 \\ 0 & 0 & 1 & 0 \\ 0 & 0 & 0 & 1 \end{pmatrix} \begin{pmatrix} a_0 \\ a_1 \\ a_2 \\ a_3 \end{pmatrix} = \begin{pmatrix} 0 \\ 0 \\ 0 \\ 0 \end{pmatrix}.$$

It is a result of elementary linear algebra that this system has nontrivial solutions only if the determinant of the matrix is zero. But the determinant is $+1$. So there are no nontrivial solutions and thus the basis vectors must be linearly independent.

Either of the above solutions are fine. But to get some practice with our more general notation, start with a general linear combination a_μ,

$$0 = a_\mu(\vec{e}_\alpha)^\mu = a_\mu\delta_\alpha{}^\mu. \qquad \text{used eqn. (2.3)} \qquad (2.2)$$

Consider the first component, $\alpha = 0$. The equation above is $0 = a_0 \times 1$, so a_0 must be zero; this is the only solution. Similarly for the other components. Since this trivial solution is the only solution, the basis vectors must be linearly independent.

(b) Is the following set linearly independent, $\{\vec{a}, \vec{b}, \vec{c}, 5\vec{a} + 3\vec{b} - 2\vec{c}\}$?

Solution: The given set is not linearly independent, since the linear combination $(-5, -3, 2, 1)$ gives the zero vector.

2.7 (a) Verify
$$(\vec{e}_\alpha)^\beta = \delta_\alpha{}^\beta \qquad\qquad \text{Schutz Eq. (2.10)} \qquad\qquad (2.3)$$
for all α, β.

Solution: Eqn. (2.3) introduces some new notation. As Schutz advised, the superscript and subscript index notation will become clear when we get into differential geometry. The RHS is the Kronecker delta, which after reading Chapter 3 we will interpret as a second-rank tensor. For now note that there are two free indices, one up and one down, and it can be written as the identity matrix. So of course the LHS must also be a second-rank tensor (SP3.14 asks for a proof of this) and must have two free indices. How can that be? Here $\vec{e}_\alpha$ is a set of four vectors. And enclosing this set within parentheses and writing a superscript β implies that we are pulling off the β component for each vector α; this gives us two free indices on the LHS as well.[1]

If we write the basis vectors as row vectors as in eqn. (2.1), then the set form a matrix, and the matrix element is unity when row and column numbers are equal, and zero otherwise, i.e. the identity matrix:

$$\begin{pmatrix} 1 & 0 & 0 & 0 \\ 0 & 1 & 0 & 0 \\ 0 & 0 & 1 & 0 \\ 0 & 0 & 0 & 1 \end{pmatrix}.$$

The RHS of eqn. (2.3) can of course be written as the identity matrix too, which demonstrates the equality.

(b) Prove
$$\vec{A} = A^\alpha \vec{e}_\alpha \qquad\qquad \text{Schutz Eq. (2.11)} \qquad\qquad (2.4)$$
from eqn. (2.1) above.

Solution: In Schutz §2.2 we were told that if
$$\vec{A} \underset{\mathcal{O}}{\rightarrow} (A^0, A^1, A^2, A^3),$$

---

[1] While not explicitly explained in the text, you will find this notation used in other texts, (see Example 1.4.8 in Hassani, 1999). This was also used in Schutz Eq. (2.21) and will be used later in the book, e.g. Schutz Eq. (5.52).

then $\vec{A} = A^\alpha \vec{e}_\alpha$. To demonstrate this we simply substitute the definition of the basis given in eqn. (2.1) and perform the multiplication. For the first component, $\alpha = 0$, A^0 multiplies the basis vector $\vec{e}_0$:

$$A^0 \vec{e}_0 = A^0(1, 0, 0, 0) = (A^0, 0, 0, 0).$$

And this is the only contribution to the first component because all the other basis vectors are zero in the first component. Similarly for the other components:

$$A^1 \vec{e}_1 = A^1(0, 1, 0, 0) = (0, A^1, 0, 0)$$
$$A^2 \vec{e}_2 = A^2(0, 0, 1, 0) = (0, 0, A^2, 0)$$
$$A^3 \vec{e}_3 = A^3(0, 0, 0, 1) = (0, 0, 0, A^3).$$

Summing these we obtain, (A^0, A^1, A^2, A^3).

2.9   Prove, by writing out all the terms, that

$$\sum_{\bar{\alpha}=0}^{3}\left(\sum_{\beta=0}^{3}\Lambda^{\bar{\alpha}}{}_{\beta}A^{\beta}\vec{e}_{\bar{\alpha}}\right) = \sum_{\beta=0}^{3}\left(\sum_{\bar{\alpha}=0}^{3}\Lambda^{\bar{\alpha}}{}_{\beta}A^{\beta}\vec{e}_{\bar{\alpha}}\right). \qquad (2.5)$$

Since the order of summation does not matter, we are justified in using the Einstein summation convention to write simply $\Lambda^{\bar{\alpha}}{}_{\beta}A^{\beta}\vec{e}_{\bar{\alpha}}$ which does not specify the order of summation.

Solution: There are 16 terms to write out. Starting with the LHS of eqn. (2.5) and expanding the inner sum first we have:

$$
\begin{array}{cccc}
\Lambda^0{}_0 A^0 \vec{e}_0 & + \Lambda^0{}_1 A^1 \vec{e}_0 & + \Lambda^0{}_2 A^2 \vec{e}_0 & + \Lambda^0{}_3 A^3 \vec{e}_0 \\
\Lambda^1{}_0 A^0 \vec{e}_1 & + \Lambda^1{}_1 A^1 \vec{e}_1 & + \Lambda^1{}_2 A^2 \vec{e}_1 & + \Lambda^1{}_3 A^3 \vec{e}_1 \\
\Lambda^2{}_0 A^0 \vec{e}_2 & + \Lambda^2{}_1 A^1 \vec{e}_2 & + \Lambda^2{}_2 A^2 \vec{e}_2 & + \Lambda^2{}_3 A^3 \vec{e}_2 \\
\Lambda^3{}_0 A^0 \vec{e}_3 & + \Lambda^3{}_1 A^1 \vec{e}_3 & + \Lambda^3{}_2 A^2 \vec{e}_3 & + \Lambda^3{}_3 A^3 \vec{e}_3
\end{array} \qquad (2.6)
$$

The RHS of eqn. (2.5) gives the same sum of terms but with the order changed so of course the sums are equal:

$$
\begin{array}{cccc}
\Lambda^0{}_0 A^0 \vec{e}_0 & + \Lambda^1{}_0 A^0 \vec{e}_1 & + \Lambda^2{}_0 A^0 \vec{e}_2 & + \Lambda^3{}_0 A^0 \vec{e}_3 \\
\Lambda^0{}_1 A^1 \vec{e}_0 & + \Lambda^1{}_1 A^1 \vec{e}_1 & + \Lambda^2{}_1 A^1 \vec{e}_2 & + \Lambda^3{}_1 A^1 \vec{e}_3 \\
\Lambda^0{}_2 A^2 \vec{e}_0 & + \Lambda^1{}_2 A^2 \vec{e}_1 & + \Lambda^2{}_2 A^2 \vec{e}_2 & + \Lambda^3{}_2 A^2 \vec{e}_3 \\
\Lambda^0{}_3 A^3 \vec{e}_0 & + \Lambda^1{}_3 A^3 \vec{e}_1 & + \Lambda^2{}_3 A^3 \vec{e}_2 & + \Lambda^3{}_3 A^3 \vec{e}_3
\end{array} \qquad (2.7)
$$

To help line up terms note that the way I chose to write it, the rows and columns are interchanged.

Thinking more abstractly, you can convince yourself that for each term in the sum on the LHS of eqn. (2.5) there must be a corresponding term on the RHS, and vice versa. For on both sides these terms look like, $\Lambda^{\bar{\alpha}}{}_{\beta} A^{\beta}\vec{e}_{\bar{\alpha}}$. And the double sums ensure that we have each possible pair of indices. And of course the order does not matter for a finite sum, so the sums are equal.

2.10 Prove

$$\vec{e}_\alpha = \Lambda^{\bar{\beta}}{}_\alpha \vec{e}_{\bar{\beta}} \qquad\qquad \text{Schutz Eq. (2.13)} \qquad\qquad (2.8)$$

from

$$A^\alpha \left(\Lambda^{\bar{\beta}}{}_\alpha \vec{e}_{\bar{\beta}} - \vec{e}_\alpha \right) = 0 \qquad\qquad (2.9)$$

by making specific choices for the components of the arbitrary vector $\vec{A}$. [This is a very important exercise. Note the RHS of eqn. (2.9) is a vector, $\vec{0}$.]

Solution: There are four equations in eqn. (2.8), one for each value of α. The trick here is to choose an A^α with only one non-zero entry to isolate one of these equations. For example, choosing $(1, 0, 0, 0)$, or $(10, 0, 0, 0)$, shows straight away that

$$\Lambda^{\bar{\beta}}{}_0 \vec{e}_{\bar{\beta}} = \vec{e}_0,$$

which is eqn. (2.8) with $\alpha = 0$. Similarly choosing A^α as $(0, 1, 0, 0)$, or $(0, 2, 0, 0)$, shows straight away

$$\Lambda^{\bar{\beta}}{}_1 \vec{e}_{\bar{\beta}} = \vec{e}_1.$$

Repeating this argument gives the result for the other two basis vectors.

2.12 Given $\vec{A} \underset{\mathcal{O}}{\rightarrow} (0, -2, 3, 5)$, find:

(b) the components of $\vec{A}$ in $\overline{\overline{\mathcal{O}}}$, which moves at speed 0.6 relative to $\overline{\mathcal{O}}$ in the positive x-direction. [In part (a) we saw that $\overline{\mathcal{O}}$ moves at speed 0.8 relative to $\mathcal{O}$ also in the positive x-direction.]

Hint: Remember not to add the velocities linearly, but to use the Einstein law of composition of velocities eqn. (1.22), or use the velocity parameters introduced in Exercise 1.18.

(c) the magnitude of $\vec{A}$ from its components in $\mathcal{O}$;

Hint: Note that the definition of the magnitude of the vector is analogous to the interval introduced in Schutz Chapter 1, see eqn. (2.28).

(d) the magnitude of $\vec{A}$ from its components in $\mathcal{O}$;

> Hint: If you got a different answer than in (c), something went wrong!

2.14　The following matrix gives a Lorentz transformation from $\mathcal{O}$ to $\overline{\mathcal{O}}$:

$$\Lambda(\mathbf{v}) = \begin{pmatrix} 1.25 & 0 & 0 & 0.75 \\ 0 & 1 & 0 & 0 \\ 0 & 0 & 1 & 0 \\ 0.75 & 0 & 0 & 1.25 \end{pmatrix}.$$

(a) What is the velocity (speed and direction) of $\overline{\mathcal{O}}$ relative to $\mathcal{O}$?

Solution: $v = -3/5$ in the z-direction (or $3/5$ in the negative z-direction). Because the off-diagonal term is in the z column, this gives the direction. The magnitude is found easily from $-v\gamma = 0.75$, where $\gamma = 1.25$ from the diagonal term. One can confirm that $\gamma = 1/\sqrt{1-v^2}$, once v is found.

(b) What is the inverse matrix to the given one?

Solution: One could in principle use the techniques of linear algebra to find the inverse matrix. But since it is a Lorentz transformation, the inverse can be obtained from the Lorentz transformation from $\overline{\mathcal{O}}$ back to $\mathcal{O}$. That is, one simply changes the sign of the velocity from that in (a), giving:

$$\Lambda(-\mathbf{v}) = \begin{pmatrix} 1.25 & 0 & 0 & -0.75 \\ 0 & 1 & 0 & 0 \\ 0 & 0 & 1 & 0 \\ -0.75 & 0 & 0 & 1.25 \end{pmatrix}.$$

Indeed matrix multiplication confirms this is the inverse.

(c) Find the components in $\mathcal{O}$ of a vector $\vec{A} \underset{\overline{\mathcal{O}}}{\rightarrow} (1,2,0,0)$.

Solution:

$$\begin{pmatrix} 1.25 & 0 & 0 & -0.75 \\ 0 & 1 & 0 & 0 \\ 0 & 0 & 1 & 0 \\ -0.75 & 0 & 0 & 1.25 \end{pmatrix} \begin{pmatrix} 1 \\ 2 \\ 0 \\ 0 \end{pmatrix} = \begin{pmatrix} 1.25 \\ 2 \\ 0 \\ -0.75 \end{pmatrix}.$$

2.16 Derive the Einstein velocity-addition formula by performing a Lorentz transformation with velocity **v** on the four-velocity of a particle whose speed in the original frame was W.

Solution: The particle moves with speed W, say along the $\bar{x}$-axis, in a reference frame $\overline{\mathcal{O}}$ that is moving along the x-axis of $\mathcal{O}$ with speed v. The particle's four-velocity in reference frame $\overline{\mathcal{O}}$, $\vec{U} \underset{\overline{\mathcal{O}}}{\rightarrow} (\gamma(W), \gamma(W)\,W, 0, 0)$ (if that is not obvious then see the example in Schutz §2.4). Be careful here, we want the Lorentz transformation from $\overline{\mathcal{O}}$ to $\mathcal{O}$, which is:

$$[\Lambda(-v)] = \begin{pmatrix} \gamma(v) & v\,\gamma(v) & 0 & 0 \\ v\,\gamma(v) & \gamma(v) & 0 & 0 \\ 0 & 0 & 1 & 0 \\ 0 & 0 & 0 & 1 \end{pmatrix}.$$

Performing the matrix multiplication, $\Lambda(-v)^{\alpha}{}_{\bar{\beta}}\, U^{\bar{\beta}} = U^{\alpha}$, gives the four-velocity components in $\mathcal{O}$: $\vec{U} \underset{\mathcal{O}}{\rightarrow} (\gamma(W)\gamma(v) + vW\gamma(W)\gamma(v), v\gamma(v)\gamma(W) + W\gamma(v)\gamma(W), 0, 0)$. It turns out to be very easy to convert this to the three-velocity. Again we have a four-velocity with the only non-zero spatial component in the x-direction, and from the relation above from the example in Schutz §2.4 it is clear that the only non-zero component of the corresponding three-velocity is $U^1/U^0 = v^x$. So here we have,

$$v^x = \frac{U^1}{U^0} = \frac{\gamma(v)\,\gamma(W)(v + W)}{\gamma(v)\,\gamma(W)(1 + v\,W)}$$
$$= \frac{(v + W)}{(1 + v\,W)}, \tag{2.10}$$

which agrees with eqn. (1.22).

This algorithm to convert a four-velocity to three-velocity turns out to apply more generally, as is clear from the definition of four-velocity:

$$\vec{U} \equiv \frac{\mathrm{d}\vec{x}}{\mathrm{d}\tau}. \qquad \text{Schutz Eq. (2.31)} \tag{2.11}$$

For then we see in particular that

$$U^0 = \frac{\mathrm{d}t}{\mathrm{d}\tau}, \quad \text{and} \quad U^i = \frac{\mathrm{d}x^i}{\mathrm{d}\tau}, \quad \text{so} \quad \frac{U^i}{U^0} = \frac{\mathrm{d}x^i}{\mathrm{d}\tau} \times \frac{\mathrm{d}\tau}{\mathrm{d}t} = \frac{\mathrm{d}x^i}{\mathrm{d}t} \equiv v^i. \tag{2.12}$$

This was derived a different way in Exercise 2.15(c).

2.18 (a) Show that the sum of any two orthogonal spacelike vectors is spacelike.

Solution: Spacelike vectors have positive magnitude so here $\vec{A} \cdot \vec{A} > 0$ and $\vec{B} \cdot \vec{B} > 0$. By definition, orthogonal vectors have $\vec{A} \cdot \vec{B} = 0$, so

$$(\vec{A} + \vec{B}) \cdot (\vec{A} + \vec{B}) = \vec{A} \cdot \vec{A} + \vec{B} \cdot \vec{B} + 2\vec{A} \cdot \vec{B}$$
$$= \vec{A} \cdot \vec{A} + \vec{B} \cdot \vec{B} > 0. \qquad \text{used } \vec{A} \text{ and } \vec{B} \text{ both spacelike}$$

$$(2.13)$$

So $(\vec{A} + \vec{B})$ is also spacelike.

(b) Show that a timelike vector and a null vector cannot be orthogonal.

Solution: Let $\vec{A}$ and $\vec{N}$ be arbitrary timelike and null vectors respectively. Keep the algebra simple and rotate to a coordinate frame such that the spatial part of the null vector $\vec{N}$ is all in one component,

$$\vec{N} \underset{\mathcal{O}}{\rightarrow} (N^0, N^1, 0, 0) = (N^0, N^0, 0, 0),$$

where $N^0 = N^1$ because it is a null vector. The vector $\vec{A}$ has unknown coordinates in this frame, but

$$\vec{A} \cdot \vec{N} = -A^0 N^0 + A^1 N^0 = N^0 (A^1 - A^0).$$

We cannot have $N^0 = 0$, for otherwise $\vec{N}$ would be a zero vector (recall null vectors are not zero vectors). Furthermore $(A^1 - A^0) \neq 0$ because a timelike vector has a dominant time component. More formally, if $(A^1 - A^0) = 0$ then

$$\vec{A} \cdot \vec{A} = (A^2)^2 + (A^3)^2 \geq 0,$$

which contradicts the stipulation that $\vec{A}$ is timelike. We conclude that $\vec{A} \cdot \vec{N} \neq 0$, which proves they are not orthogonal.

2.20 The world line of a particle is described by the equations

$$x = at + b\sin(\omega t), \qquad\qquad y = b\cos(\omega t),$$
$$z = 0, \qquad\qquad |b\,\omega| < 1, \qquad (2.14)$$

in some inertial frame. Describe the motion and compute the components of the particles four-velocity and four-acceleration.

Solution: The particle moves in a circle in the $x-y$ plane of radius b, in a clockwise sense when viewed in the direction of decreasing z. The circle translates along the x-axis at speed a. The three-velocity is computed directly by differentiating the given equations, $\mathbf{v} \underset{\mathcal{O}}{\rightarrow} (\dot{x}, \dot{y}, 0)$, where

$$\dot{x} = a + \omega b\cos(\omega t), \qquad\qquad \dot{y} = -\omega b\sin(\omega t). \qquad (2.15)$$

Note the magnitude of the three-velocity has a maximum when the rotational motion is in the direction of the translational motion, so we require $|a + b\omega| < 1$ for a realistic particle. So the statement in the question that $|b\omega| < 1$ is not quite strict enough. The four-velocity is obtained from the three-velocity using the formula derived in Exercise 2.15(b), see also eqn. (2.12):

$$\vec{U} \underset{\mathcal{O}}{\rightarrow} (\gamma(v), \dot{x}\gamma(v), \dot{y}\gamma(v), 0) = \gamma(v)(1, a + b\omega\cos(\omega t), -\omega b\sin(\omega t), 0), \quad (2.16)$$

where $v = |\mathbf{v}| = \sqrt{(a + \omega b\cos(\omega t))^2 + (\omega b\sin(\omega t))^2} = \sqrt{a^2 + 2a\omega b\cos(\omega t) + \omega^2 b^2}$.

To obtain the particle four-acceleration we require the four-velocity as a function of proper time, τ, not the time in the inertial frame, t. But remember that the proper time is the time measured by a clock stationary in the MCRF. Call this MCRF $\overline{\mathcal{O}}$, and then $\bar{t} = \tau = x^{\bar{0}}$. And $t = \Lambda(-v)^0{}_{\bar{\alpha}} x^{\bar{\alpha}}$. For simplicity we choose the MCRF with origin at the particle location, so $x^{\bar{\alpha}} \underset{\overline{\mathcal{O}}}{\rightarrow} (\tau, 0, 0, 0)$, and $t = \gamma(-v)\tau = \gamma(v)\tau$. Then we obtain the four-acceleration from the given equations in t and the chain rule,

$$\vec{a} \equiv \frac{\mathrm{d}\vec{U}}{\mathrm{d}\tau} = \frac{\mathrm{d}\vec{U}}{\mathrm{d}t}\frac{\mathrm{d}t}{\mathrm{d}\tau} = \gamma(v)\frac{\mathrm{d}\vec{U}}{\mathrm{d}t}.$$

Do not forget that $\gamma(v)$ is a function of t too! In supplementary problem SP2.5 in section 2.2 we derive a general expression for the four-acceleration,

$$\vec{a} = \gamma^3[\dot{x}\ddot{x} + \dot{y}\ddot{y} + \dot{z}\ddot{z}]\vec{U} + \gamma^2(0, \ddot{x}, \ddot{y}, \ddot{z}). \qquad \text{used eqn. (2.52)} \qquad (2.17)$$

Substituting the values for our particular problem we find:

$$\vec{a} = \gamma^3[-\omega^2 ab\sin(\omega t)]\vec{U} + \gamma^2\left(0, -\omega^2 b\sin(\omega t), -\omega^2 b\cos(\omega t), 0\right). \qquad (2.18)$$

2.22 (a) Find the energy, rest mass, and three-velocity $\mathbf{v}$ of a particle whose four-momentum has the components $(4, 1, 1, 0)$ kg.

Solution: Call the reference frame in which the components of the four-momentum were given $\mathcal{O}$.

Energy in $\mathcal{O}$: In general $\vec{p} \underset{\mathcal{O}}{\rightarrow} (E, p^1, p^2, p^3)$, so $E = 4$ kg.

Three-velocity in $\mathcal{O}$: In general $m\vec{U} = \vec{p}$, where m is the rest mass and $\vec{U}$ is the four-velocity. And the three-velocity is related to the four-velocity as inferred in Exercise 2.15(b); for general three-velocity $\mathbf{v} = (u, v, w)$, the corresponding four-velocity is

$$\vec{A} \underset{\mathcal{O}}{\rightarrow} (\gamma(|\mathbf{v}|), u\,\gamma(|\mathbf{v}|), v\,\gamma(|\mathbf{v}|), w\,\gamma(|\mathbf{v}|)), \qquad (2.19)$$

where $|\mathbf{v}| = \sqrt{u^2 + v^2 + w^2}$. So $\vec{p} \underset{\mathcal{O}}{\rightarrow} (m\gamma, mu\gamma, mv\gamma, mw\gamma)$, where $\mathbf{v} \underset{\mathcal{O}}{\rightarrow} (u, v, w)$ are the components of the three-velocity. Note that $E = m\gamma$, and simply dividing $\vec{p}$

through by E reveals that $\mathbf{v} \underset{\mathcal{O}}{\rightarrow} (1/4, 1/4, 0)$. The corresponding Lorentz factor, is

$$\gamma(|\mathbf{v}|) = \frac{1}{\sqrt{1 - \mathbf{v} \cdot \mathbf{v}}} = \frac{4}{\sqrt{14}}.$$

Rest mass: Since $E = m\gamma = 4$, we find $m = \sqrt{14} \approx 3.74\,\text{kg}$.

2.22 (b) The collision of two particles of four-momenta

$$\vec{p}_1 \underset{\mathcal{O}}{\rightarrow} (3, -1, 0, 0)\,\text{kg}, \qquad \vec{p}_2 \underset{\mathcal{O}}{\rightarrow} (2, 1, 1, 0)\,\text{kg}$$

results in the destruction of the two particles and the production of three new ones, two of which have four-momenta

$$\vec{p}_3 \underset{\mathcal{O}}{\rightarrow} (1, 1, 0, 0)\,\text{kg}, \qquad \vec{p}_4 \underset{\mathcal{O}}{\rightarrow} (1, -1/2, 0, 0)\,\text{kg}.$$

Find the four-momentum, energy, rest mass, and three-velocity of the third particle produced. Find the CM frame's three-velocity.

Solution: We must apply the law of conservation of four-momentum, see Schutz Eq. (2.22). Here

$$\vec{p}_I = \vec{p}_1 + \vec{p}_2 \underset{\mathcal{O}}{\rightarrow} (5, 0, 1, 0)\,\text{kg}.$$

By conservation of four-momentum,

$$\vec{p}_F = \vec{p}_I = \vec{p}_3 + \vec{p}_4 + \vec{p}_5,$$

so

$$\vec{p}_5 = \vec{p}_I - \vec{p}_3 - \vec{p}_4 \underset{\mathcal{O}}{\rightarrow} (3, -1/2, 1, 0)\,\text{kg}.$$

Now, like in Exercise 2.22(a), we know the four-momentum. From an analysis just like in (a), we find the fifth particle has in this same reference frame: $E_5 \underset{\mathcal{O}}{\rightarrow} 3\,\text{kg}$, and $\mathbf{v}_5 \underset{\mathcal{O}}{\rightarrow} (-1/6, 1/3, 0)$. Finally, the rest mass is $m = \sqrt{31}/2 \approx 2.78\,\text{kg}$.

The three-velocity of the CM frame is simply the three-velocity associated with the four-momentum $\vec{p}_I$. As in Exercise 2.22(a) we find its components by dividing the spatial components of the four-momentum by the energy:

$$\mathbf{v} \underset{\mathcal{O}}{\rightarrow} (0, 1/5, 0).$$

It is easy to verify that the corresponding Lorentz transformation to the CM frame transforms $\vec{p}_I (= \vec{p}_F)$ to have only a time component, $m(\vec{e}_{\bar{0}})^{\bar{\alpha}}$, via

$$\Lambda^{\bar{\alpha}}{}_{\beta}\, p_F^{\beta} = p_F^{\bar{\alpha}} \underset{\text{CM}}{\rightarrow} (m, 0, 0, 0). \tag{2.20}$$

For example, look at the $\bar{y}$ component:

$$p_F^{\bar{2}} = \Lambda^{\bar{2}}_0 \, p_F^0 + \Lambda^{\bar{2}}_2 \, p_F^2 \qquad\qquad \text{just the } \bar{y} \text{ component}$$
$$= -v\gamma 5 + \gamma = 0. \tag{2.21}$$

2.24   Prove that conservation of four-momentum forbids a reaction in which an electron and positron annihilate and produce a single photon (γ-ray). Prove that the production of two photons is not forbidden.

Solution: Particles come and go, but four-momentum is conserved. Consider the hypothetical situation where a positron and electron annihilate producing a single particle, a γ-ray. Then conservation of four-momentum would require,

$$\vec{p}_{e^+} + \vec{p}_{e^-} = \vec{p}_\gamma, \tag{2.22}$$

which gives two equations. The time component of eqn. (2.22) looks like conservation of energy,

$$p_{e^+}^0 + p_{e^-}^0 = p_\gamma^0 = \hbar\omega,$$

where $\hbar$ is the reduced Planck constant and ω is the angular frequency of the γ-ray. The spatial part of eqn. (2.22) looks like traditional conservation of momentum,

$$p_{e^+}^i + p_{e^-}^i = p_\gamma^i.$$

We can always work in the CM frame of the electron and positron, so that the spatial part of $\vec{p}_{e^+} + \vec{p}_{e^-}$ is zero. But the γ-ray's four-momentum is always a null vector and thus must have a spatial part in any inertial reference frame. For example, rotate the CM coordinates such that the x-axis is aligned with the direction of propagation of the γ-ray, and then

$$\vec{p}_\gamma \underset{\text{CM}}{\rightarrow} (\hbar\,\omega, \hbar\,\omega, 0, 0).$$

The non-zero $p_\gamma^1 = \hbar\,\omega$ violates conservation of momentum in the x-direction.

It is possible to produce two γ-rays while conserving four-momentum. Suppose they travel in opposite directions with equal and opposite momentum in some frame of reference $\mathcal{O}$. Then the final total four-momentum has zero spatial part. To satisfy spatial momentum conservation we only require that the positron and electron have equal and opposite momentum in $\mathcal{O}$, so $\mathbf{v}_{e^+} = -\,\mathbf{v}_{e^-}$ with arbitrary $\mathbf{v}_{e^+}$, which can obviously be satisfied. Conservation of energy dictates the frequency of the pair of γ-rays through $2\gamma(|\mathbf{v}_{e^+}|)m_e = 2\hbar\omega$. In fact, if $\gamma \approx 1$, then $\hbar\omega \approx 511$ keV. Large numbers of collinear γ-ray pairs, with random orientations, allow one to locate the source, a fact exploited in Positron Emission Tomography.

2.25 (a) Let frame $\overline{\mathcal{O}}$ move with speed u in the x-direction relative to $\mathcal{O}$. Let a photon have frequency v in $\mathcal{O}$ and move at an angle θ with respect to $\mathcal{O}$'s x-axis. Show that its frequency in $\overline{\mathcal{O}}$ is

$$\frac{\overline{v}}{v} = \gamma(u)(1 - u\cos(\theta)) = \frac{1}{\sqrt{1-u^2}}(1 - u\cos(\theta)). \quad \text{Schutz Eq. (2.42)}$$

$$(2.23)$$

Solution: In the reference frame $\mathcal{O}$ the photon has four-momentum

$$\vec{p} \underset{\mathcal{O}}{\rightarrow} (hv, hv\cos(\theta), hv\sin(\theta), 0). \tag{2.24}$$

This follows from the fact that the momentum of a photon must be a null vector. Transforming to the frame $\overline{\mathcal{O}}$ moving at speed u along the x-axis, we apply the Lorentz transformation

$$\Lambda(u) = \begin{pmatrix} \gamma & -u\gamma & 0 & 0 \\ -u\gamma & \gamma & 0 & 0 \\ 0 & 0 & 1 & 0 \\ 0 & 0 & 0 & 1 \end{pmatrix}, \qquad \gamma = \frac{1}{\sqrt{1-u^2}},$$

to p^α in eqn. (2.24) to obtain $p^{\bar{\alpha}}$:

$$\vec{p} \underset{\overline{\mathcal{O}}}{\rightarrow} \begin{pmatrix} \gamma hv - u\gamma hv\cos(\theta) \\ -u\gamma hv + \gamma hv\cos(\theta) \\ hv\sin(\theta) \\ 0 \end{pmatrix} = \begin{pmatrix} h\bar{v} \\ p^{\bar{1}} \\ p^{\bar{2}} \\ p^{\bar{3}} \end{pmatrix}.$$

The Doppler shift is obtained from the time component, i.e. the first component $p^{\bar{0}} = h\bar{v} = \gamma hv - u\gamma hv\cos(\theta)$, and can be expressed as,

$$\frac{\overline{v}}{v} = \gamma(1 - u\cos(\theta)) = \frac{1}{\sqrt{1-u^2}}(1 - u\cos(\theta)).$$

2.25 (b) At what angle θ does the photon have to move so that there is no Doppler shift between $\mathcal{O}$ and $\overline{\mathcal{O}}$?

Solution: No Doppler shift occurs when

$$\frac{\overline{v}}{v} = 1 = \frac{1}{\sqrt{1-u^2}}(1 - u\cos(\theta)). \qquad \text{used eqn. (2.23)} \tag{2.25}$$

Solving for $\cos\theta$ gives:

$$\cos\theta = \frac{1 - \sqrt{1-u^2}}{u}.$$

$$\theta = \arccos\left(\frac{1 - \sqrt{1 - u^2}}{u}\right). \tag{2.26}$$

See also the Maple™ worksheet.

2.25 (c) Use

$$-\vec{p} \cdot \vec{U}_{\text{obs}} = \overline{E} \qquad \text{Schutz Eq. (2.35)} \tag{2.27}$$

and $E = h\nu$, Schutz Eq. (2.38), to calculate eqn. (2.23) in part (a) above.

Solution: Eqn. (2.27) is the frame-invarient expression for energy $\overline{E}$ relative to an observer moving with velocity $\vec{U}_{\text{obs}}$ relative to some frame $\mathcal{O}$. This calculation ends up being exactly the same as above, but allows one to focus on the relevant parts, i.e. just the time component. The observer's four-velocity is

$$\vec{U}_{\text{obs}} \underset{\mathcal{O}}{\rightarrow} (\gamma, \gamma u, 0, 0), \qquad \text{with } \gamma = \frac{1}{\sqrt{1 - u^2}},$$

and recall the photon's four-momentum is

$$\vec{p} \underset{\mathcal{O}}{\rightarrow} (h\nu, h\nu \cos(\theta), h\nu \sin(\theta), 0),$$

so we can immediately find

$$\overline{E} = -\vec{p} \cdot \vec{U}_{\text{obs}} = \gamma(u)h\nu - u\gamma(u)h\nu \cos(\theta).$$

This was the time component of the $\vec{p}$ with respect to $\overline{\mathcal{O}}$ found in part (a) above.

2.27 Two identical bodies of mass 10 kg are at rest at the same temperature. One of them is heated by the addition of 100 J of heat. Both are then subjected to the same force. Which accelerates faster, and by how much?

Solution: Each object has rest mass, $m(T_0) = 10$[kg]. Increase the temperature from T_0 to T by heat flux $\delta Q = 100$ J. In the MCRF of the object, $U^0 = 1$ and $mU^0 = p^0 = E$. So

$$m(T) = m(T_0)[\text{kg}] + \delta Q[\text{J}]/c^2[\text{m}^2/\text{s}^2] = 10 + 1.1 \times 10^{-15}[\text{kg}].$$

Increasing the temperature increases the rest mass! With the same applied force of course the cooler (hence lighter) body accelerates faster by about 1 part in 10^{16}.

This problem is interesting to look at from a thermodynamics point of view. The heat flux increases the temperature and enthalpy of the object, which is reflected on a microscopic scale by an increase in the motion, relative to the center of mass of the object, of the elements (atoms or molecules or sea of electrons depending on the material) composing the object. This motion increases the effective mass of the elements. Say an element has rest mass m_i, then when it has thermal speed v_i it has "relativistic mass" $m_{i,\text{rel}} = m_i \gamma(v_i)$. An interesting discussion of this can be found online by searching http://en.wikipedia.org for "mass-energy equivalence."

2.29   Prove, using the component expressions

$$\vec{A}^2 = -(A^0)^2 + (A^1)^2 + (A^2)^2 + (A^3)^2 \qquad \text{Schutz Eq. (2.24)} \qquad (2.28)$$

and

$$\vec{A} \cdot \vec{B} = -A^0 B^0 + A^1 B^1 + A^2 B^2 + A^3 B^3 \qquad \text{Schutz Eq. (2.26)} \qquad (2.29)$$

that

$$\frac{\mathrm{d}}{\mathrm{d}\tau}(\vec{U} \cdot \vec{U}) = 2\vec{U} \cdot \frac{d\vec{U}}{d\tau}.$$

Solution:

$$\frac{\mathrm{d}}{\mathrm{d}\tau}(\vec{U} \cdot \vec{U}) = \frac{\mathrm{d}}{\mathrm{d}\tau}\left(-(U^0)^2 + (U^1)^2 + (U^2)^2 + (U^3)^2\right)$$

$$= -2U^0 \frac{\mathrm{d}U^0}{\mathrm{d}\tau} + 2U^i \frac{\mathrm{d}U^i}{\mathrm{d}\tau}$$

$$= 2\vec{U} \cdot \frac{d\vec{U}}{\mathrm{d}\tau}. \qquad (2.30)$$

2.31   A photon of frequency ν is reflected without change of frequency from a mirror, with an angle of incidence θ. Calculate the momentum transferred to the mirror. What momentum would be transferred if the photon were absorbed rather than reflected?

Solution: This appears to be a straightforward application of conservation of four-momentum, but is fun because it gets us thinking about all four components. Let the mirror lie in the $y-z$ plane, with photon traveling initially in the $x-y$ plane, with angle θ to the x-axis. Then the initial four-momentum of the photon is written

$$\vec{P}_i = (h\nu, \cos(\theta)h\nu, \sin(\theta)h\nu, 0),$$

where h is Planck's constant. First construct the four-momentum of the reflected photon $\vec{P}_r$. Since the photon frequency does not change, we know the time component,

$$P_r^0 = P_i^0 = h\nu.$$

Assume that the momentum transferred is only in the x-direction (say it is a very smooth mirror). So then we can also construct the reflected y and z components:

$$P_r^2 = P_i^2 = \sin(\theta)h\nu, \quad P_r^3 = P_i^3 = 0.$$

Recall that the four-momentum of a photon is orthogonal to itself,

$$\vec{p} \cdot \vec{p} = 0. \qquad\qquad \text{Schutz Eq. (2.37)} \qquad\qquad (2.31)$$

This alone gives us two possibilities for $P_r^1 = \pm\cos(\theta)h\nu$. For the *reflected* photon, we choose the minus sign. In summary,

$$\vec{P}_r = (h\nu, -h\nu\cos(\theta), h\nu\sin(\theta), 0).$$

By conservation of four-momentum, we see that the momentum transferred to the mirror must be $\Delta P_m^1 = 2h\nu\cos(\theta)$ in the x-direction. How did the mirror acquire x-direction momentum without gaining energy? See SP2.4 in section 2.2.

If the photon is absorbed, then the momentum transferred to the mirror has three components,

$$\Delta\vec{P}_m = (\Delta E_m, \Delta P_m^1, \Delta P_m^2, 0) = (h\nu, h\nu\cos(\theta), h\nu\sin(\theta), 0).$$

How did the mirror acquire the extra energy $\Delta E_m = h\nu$? See SP2.4 in section 2.2.

2.32 Let a particle of charge e and rest mass m, initially at rest in the laboratory, scatter a photon of initial frequency ν_i. This is called Compton scattering. Suppose the scattered photon comes off at an angle θ from the incident direction. Use conservation of four-momentum to deduce that the photon's final frequency ν_f is given by

$$\frac{1}{\nu_f} = \frac{1}{\nu_i} + h\left(\frac{1 - \cos(\theta)}{m}\right). \qquad \text{Schutz Eq. (2.43)} \qquad (2.32)$$

Hint: One uses of course conservation of four-momentum giving three equations in three unknowns, ϕ, v, ν_f, where ϕ is the angle of the path of the scattered electron from the initial path of the photon, v is the speed of the electron, and ν_f is the frequency of the photon after scattering. We can treat the angle θ as known because the detector of the scattered radiation is placed at a known angle relative to the initial path of the photon. Eliminating ϕ is straightforward, leaving two equations in ν_f and v. Eliminating v is straightforward but leaves a fourth-order equation for ν_f. It is somewhat easier, although still tedious, to instead eliminate ν_f and solve a quadratic for v^2. Then substituting this v^2 into the energy equation gives a messy expression that finally simplifies to eqn. (2.32). But the algebra is far simpler if one works with the expression for the kinetic energy of the particle K, see eqn. (2.54) in SP2.8. One uses conservation of the spatial components of the four-momentum to solve for the magnitude of the three-momentum, and the energy equation to give K. Substituting these into eqn. (2.54) immediately gives eqn. (2.32), (Eisberg and Resnick, 1985, §2.4). Alternatively one can set up the system of three equations and three unknowns in Maple™, which gives a simple solution that easily simplifies to eqn. (2.32).

2.33   Space is filled with cosmic rays ([most of which are] high-energy protons) and the cosmic microwave background radiation. These can Compton scatter off one another. Suppose a photon of energy $h\nu = 2 \times 10^{-4}$ eV scatters off a proton of energy $10^9 m_P = 10^{18}$ eV, energies measured in the Sun's rest frame. Use eqn. (2.32) in the proton's initial rest frame to calculate the maximum final energy the photon can have in the solar rest frame after the scattering. What energy range is this (X-ray, visible, etc.)?

Solution: This very nice problem at first appears very challenging, but the extreme differences in energy between the two particles simplifies things tremendously. First we note that in the rest frame of the particle, Compton scattering only reduces the frequency and more so for less massive particles. So how can Compton scattering increase the energy of the photon?! The increase in energy is revealed via the Doppler shift.

The key simplification in this problem is that the Compton scattering in the frame of the particle, say $\overline{\mathcal{O}}$, has very little effect on frequency because the rest mass energy of the proton is much greater than the photon energy, even when Doppler shifted to the proton frame. Ignoring the factor $(1 - u\cos\theta)$ factor in the general formula for Doppler shift frequency, see eqn. (2.23) above, we have

$$h\bar{\nu}_i \approx \gamma(u)h\nu_i = 10^9 \cdot 2 \times 10^{-4}\text{eV} \sim 10^5\text{eV} \ll m_p \approx 10^9\text{eV}.$$

We inferred $\gamma(u)$ from the proton energy given in the Sun's frame. This means that the angle of the Compton scattering has very little effect on the final frequency *in the particle's initial rest frame* since

$$\frac{1}{h\bar{\nu}_f} = \frac{1}{h\bar{\nu}_i} + \frac{1 - \cos\theta}{m_p} \approx \frac{1}{h\bar{\nu}_i}.$$

So when maximizing the photon energy over angle we need only consider the effect of angle on the Doppler shift when transforming between the proton and Sun's frames.

Now the problem is easy. The Doppler shift in frequency is given in general by eqn. (2.23) above. Obviously to maximize the frequency in the cosmic ray frame, $\bar{\nu}_i$, we want the photon and cosmic ray traveling in a line in opposite directions, i.e. $\theta = \pi$ radians, for which eqn. (2.23) gives

$$h\bar{\nu}_i = h\nu_i\gamma(u)(1 + u) \approx h\nu_i\gamma(u)2 = h\nu_i 2 \times 10^9 = 4 \times 10^5\text{eV}.$$

We have used the fact that the proton's speed is very close to the speed of light relative to the Sun, $u \approx 1$. The Doppler shift has made a tremendous increase in frequency! The Compton scattering will make very little difference; so to maximize the scattered frequency in the Sun's frame, choose the Compton scattering angle to maximize the Doppler shift again. That is, choose the scattering angle to be π. Eqn. (2.32) gives

$$\frac{1}{h\bar{\nu}_f} = \frac{1}{h\bar{\nu}_i} + \frac{2}{m_p} = 0.25 \times 10^{-5} + 2 \times 10^{-9} \approx 0.25 \times 10^{-5} [\text{eV}]^{-1}.$$

Compton scattering caused negligible decrease in energy in the proton's frame. The proton, like the mirror in Exercise 2.31, is massive enough to cause little change in frequency of the photon in the proton's frame; see also supplementary problem SP2.4. Now Lorentz transform back to the Sun's frame. The photon again gains tremendously from the Doppler shift (that's why we chose the scattering angle to be complete reflection).

$$h\nu_f \approx h\bar{\nu}_f \, 2 \times 10^9 \approx 8 \times 10^{14} \text{eV}.$$

This is a very hard γ-ray. A pair of 511 keV photons arising from annihilation of an electron and positron are considered to be γ-rays. And this is more than a billion times more energetic than that.

2.35 Show that the vectors $\vec{e}_{\bar{\beta}}$ obtained from $\{\vec{e}_\alpha\}$ by

$$\vec{e}_{\bar{\mu}} = \Lambda^\nu_{\ \bar{\mu}}(-\mathbf{v})\vec{e}_\nu \qquad \text{Schutz Eq. (2.15)} \qquad (2.33)$$

satisfy

$$\vec{e}_{\bar{\alpha}} \cdot \vec{e}_{\bar{\beta}} = \eta_{\bar{\alpha}\bar{\beta}}$$

for all $\bar{\alpha}, \bar{\beta}$.

Solution: First make the substitution so we know what we are up against:

$$\vec{e}_{\bar{\alpha}} \cdot \vec{e}_{\bar{\beta}} = \left(\Lambda^\nu_{\ \bar{\alpha}}(-\mathbf{v})\vec{e}_\nu\right) \cdot \left(\Lambda^\mu_{\ \bar{\beta}}(-\mathbf{v})\vec{e}_\mu\right)$$
$$= \Lambda^\nu_{\ \bar{\alpha}} \Lambda^\mu_{\ \bar{\beta}} \, \vec{e}_\nu \cdot \vec{e}_\mu$$
$$= \Lambda^\nu_{\ \bar{\alpha}} \Lambda^\mu_{\ \bar{\beta}} \, \eta_{\nu\mu}. \qquad (2.34)$$

These Lorentz transformations are really quite general, but it is easiest to consider the case of a boost along one of the axes, say the x-axis, and that focuses our attention on what distinguishes SR from Newtonian kinematics. Put more formally, we note that eqn. (2.34) is a vector equation (LHS is a simpe dot product and the RHS is a dot product of linear combinations of vectors) and thus cannot depend upon the orientation of the axes. So without loss of generality we rotate all the axes so that $-\mathbf{v}$ is oriented along the x-axis. Then

$$(\Lambda(-\mathbf{v})^\mu_{\ \bar{\beta}}) = \begin{array}{c} \\ \mu=0 \\ \mu=1 \\ \mu=2 \\ \mu=3 \end{array} \begin{pmatrix} \gamma(-v) & v\gamma(-v) & 0 & 0 \\ v\gamma(-v) & \gamma(-v) & 0 & 0 \\ 0 & 0 & 1 & 0 \\ 0 & 0 & 0 & 1 \end{pmatrix},$$

with column headers $\bar{\beta}=0 \quad \bar{\beta}=1 \quad \bar{\beta}=2 \quad \bar{\beta}=3$

with $\gamma(-v) = 1/\sqrt{1-(-v)^2}$. Note we are interpreting the left index as the row number and right index as the column number. For convenience, we also write out the Minkowski metric:

$$(\eta_{\alpha\beta}) = \begin{array}{c} \\ \alpha = 0 \\ \alpha = 1 \\ \alpha = 2 \\ \alpha = 3 \end{array} \begin{array}{cccc} \beta = 0 & \beta = 1 & \beta = 2 & \beta = 3 \\ \left(\begin{array}{cccc} -1 & 0 & 0 & 0 \\ 0 & 1 & 0 & 0 \\ 0 & 0 & 1 & 0 \\ 0 & 0 & 0 & 1 \end{array}\right) \end{array}.$$

Now we're just practicing linear algebra using the index notation. Stepping through the various terms in eqn. (2.34) we find

$$
\begin{aligned}
\vec{e}_{\bar{0}} \cdot \vec{e}_{\bar{0}} &= \Lambda^{\nu}{}_{\bar{0}} \Lambda^{\mu}{}_{\bar{0}} \eta_{\nu\mu} && \bar{\alpha} = \bar{\beta} = 0 \\
&= \Lambda^{\nu}{}_{\bar{0}} [\Lambda^{0}{}_{\bar{0}} \eta_{\nu 0} + \Lambda^{1}{}_{\bar{0}} \eta_{\nu 1}] && \text{sum over } \mu, \text{ zero terms ignored} \\
&= \Lambda^{0}{}_{\bar{0}} \Lambda^{0}{}_{\bar{0}} \eta_{00} + \Lambda^{1}{}_{\bar{0}} \Lambda^{1}{}_{\bar{0}} \eta_{11} && \text{sum over } \nu, \text{ simple because diagonal} \\
&= \gamma\gamma(-1) + (v\gamma)(v\gamma) \cdot 1 && \text{sub values} \\
&= -(1-v^2)\gamma^2 = -1. && (2.35)
\end{aligned}
$$

$$
\begin{aligned}
\vec{e}_{\bar{i}} \cdot \vec{e}_{\bar{0}} &= \Lambda^{\nu}{}_{\bar{i}} \Lambda^{\mu}{}_{\bar{0}} \eta_{\nu\mu} && \bar{\alpha} = i, \bar{\beta} = 0 \\
&= \Lambda^{\nu}{}_{\bar{i}} [\Lambda^{0}{}_{\bar{0}} \eta_{\nu 0} + \Lambda^{1}{}_{\bar{0}} \eta_{\nu 1}] && \text{sum over } \mu, \text{ zero terms ignored} \\
&= \Lambda^{0}{}_{\bar{i}} \Lambda^{0}{}_{\bar{0}} \eta_{00} + \Lambda^{1}{}_{\bar{i}} \Lambda^{1}{}_{\bar{0}} \eta_{11} && \text{sum over } \nu, \text{ simple because diagonal} \\
\vec{e}_{\bar{1}} \cdot \vec{e}_{\bar{0}} &= (v\gamma)(\gamma)(-1) + (\gamma)(v\gamma) \cdot 1 = 0 && \text{sub values} \\
\vec{e}_{\bar{2}} \cdot \vec{e}_{\bar{0}} &= \vec{e}_{\bar{3}} \cdot \vec{e}_{\bar{0}} = 0. && \text{sub values} \\
&&& (2.36)
\end{aligned}
$$

Because all the matrices (after Chapter 3 we'll think of them as tensors) on the RHS of eqn. (2.34) are symmetric, the LHS must be symmetric and so $\vec{e}_{\bar{0}} \cdot \vec{e}_{\bar{i}} = 0$ as well. Finally,

$$
\begin{aligned}
\vec{e}_{\bar{i}} \cdot \vec{e}_{\bar{j}} &= \Lambda^{\nu}{}_{\bar{i}} \Lambda^{\mu}{}_{\bar{j}} \eta_{\nu\mu} && \bar{\alpha} = i, \bar{\beta} = 0 \\
&= \Lambda^{\nu}{}_{\bar{i}} [\Lambda^{0}{}_{\bar{j}} \eta_{\nu 0} + \Lambda^{1}{}_{\bar{j}} \eta_{\nu 1} + \Lambda^{2}{}_{\bar{j}} \eta_{\nu 2} + \Lambda^{3}{}_{\bar{j}} \eta_{\nu 3}] && \text{sum over } \mu \\
&= \Lambda^{0}{}_{\bar{i}} \Lambda^{0}{}_{\bar{j}} \eta_{00} + \Lambda^{1}{}_{\bar{i}} \Lambda^{1}{}_{\bar{j}} \eta_{11} + \Lambda^{2}{}_{\bar{i}} \Lambda^{2}{}_{\bar{j}} \eta_{22} + \Lambda^{3}{}_{\bar{i}} \Lambda^{3}{}_{\bar{j}} \eta_{33} && \text{sum over } \nu \\
\vec{e}_{\bar{1}} \cdot \vec{e}_{\bar{1}} &= (v\gamma)(v\gamma)(-1) + (\gamma)(\gamma) \cdot 1 = 1 && \text{sub values} \\
\vec{e}_{\bar{2}} \cdot \vec{e}_{\bar{2}} &= \vec{e}_{\bar{3}} \cdot \vec{e}_{\bar{3}} = 1 && \text{sub values} \\
\vec{e}_{\bar{i}} \cdot \vec{e}_{\bar{j}} &= 0 \quad \text{when } i \neq j. && (2.37)
\end{aligned}
$$

In fact eqn. (2.34) is sometimes taken as the definition of a Lorentz transformation.

2.2 Supplementary problems

SP 2.1 Find a general relation between three-velocity and four-velocity, as requested in Exercise 2.15, but now using eqn. (2.11).

Solution

First note that $dt/d\tau = \gamma$, which follows from

$$\lim_{\Delta t \to 0} \Delta\tau = \Delta t\, \sqrt{1 - v^2} \qquad\qquad \text{infinitesimal limit of eqn. (1.7)}$$

$$d\tau = dt/\gamma(v). \tag{2.38}$$

Now we can find the four-velocity from the chain rule,

$$U^\alpha = \frac{dx^\alpha}{d\tau} = \frac{dt}{d\tau}\frac{dx^\alpha}{dt} = \gamma(|\mathbf{v}|)\left(1, \frac{dx}{dt}, \frac{dy}{dt}, \frac{dz}{dt}\right), \tag{2.39}$$

where of course $\mathbf{v} = (\frac{dx}{dt}\mathbf{e}_x + \frac{dy}{dt}\mathbf{e}_y + \frac{dz}{dt}\mathbf{e}_z)$. (Here $\mathbf{e}_i$ are the unit three-vectors along the three spatial axes.)

SP 2.2 Does the matrix of the Lorentz transformations always have to be symmetric?

SP 2.3 Suppose the four-velocity of a rocket ship is $\vec{U} \underset{\mathcal{O}}{\to} (2, 1, \sqrt{2}, 0)$ in some reference frame $\mathcal{O}$.

(a) Show that the given $\vec{U}$ is a legitimate four-velocity. Show that $\vec{V} \underset{\mathcal{O}}{\to} (2, 1, 1, 0)$ is not possible.

(b) Find the three-velocity in $\mathcal{O}$. Hint: see Exercise 2.15 or SP2.1 above. (You will need this for (c).)

(c) Find the matrix that rotates the spatial axes such that the three-velocity has only one non-zero component, in say the x-direction, and verify that this works as expected. What's the corresponding matrix that rotates the axes such that the four-velocity has only one non-zero spatial component?

(d) Find the inverse rotation matrices for above. Hint: Think physically and check mathematically, i.e. $\mathbf{R}_4^{-1}\mathbf{R}_4 = \mathbf{I}$.

(e) Find the Lorentz transformations from $\mathcal{O}$ to the MCRF of the rocket ship. Confirm that it has the correct effect applied to $\vec{U}$ itself. Hint: The problem here is that we have so far only seen the Lorentz transformations when the three-velocity has only one non-zero component. Use your rotation matrix from above and its inverse.

(a) Solution

Normalization of the four-velocity requires that

$$\vec{U} \cdot \vec{U} = -1. \qquad\qquad \text{Schutz Eq. (2.28)} \tag{2.40}$$

Here we find

$$\vec{U} \cdot \vec{U} = -2^2 + 1^2 + (\sqrt{2})^2 = -1,$$

which is consistent with eqn. (2.40). On the other hand,

$$\vec{V} \cdot \vec{V} = -2^2 + 1^2 + 1^2 = -2,$$

which is inconsistent with eqn. (2.40).

(b) Solution

In Exercise 2.15 one derives the four-velocity for a general particle three-velocity, see also SP2.1 where we found eqn. (2.39). Using this we find $\gamma = U^0 = 2$ and

$$\mathbf{v} \underset{\mathcal{O}}{\rightarrow} (1/2, \sqrt{2}/2, 0).$$

(c) Solution

Rotating the axes counterclockwise about the z-axis through angle $\theta = \arccos(1/\sqrt{3})$ aligns the x-axis with the three-velocity. This is accomplished with the matrix $^{(3)}\mathbf{R}$,

$$^{(3)}\mathbf{R} = \begin{pmatrix} \cos(\theta) & \sin(\theta) & 0 \\ -\sin(\theta) & \cos(\theta) & 0 \\ 0 & 0 & 1 \end{pmatrix}.$$

Let's verify this worked. Call the rotated frame $\mathcal{O}'$. Then $\mathbf{v}$ in $\mathcal{O}'$ has coordinates:

$$\mathbf{v} \underset{\mathcal{O}'}{\rightarrow} {}^{(3)}R^{i'}{}_j \, v^j = \begin{pmatrix} \cos(\theta) & \sin(\theta) & 0 \\ -\sin(\theta) & \cos(\theta) & 0 \\ 0 & 0 & 1 \end{pmatrix} \begin{pmatrix} 1/2 \\ \sqrt{2}/2 \\ 0 \end{pmatrix} = \begin{pmatrix} \frac{\sqrt{3}}{2} \\ 0 \\ 0 \end{pmatrix}. \tag{2.41}$$

The computations in eqn. (2.41) are performed in the Maple™ worksheet. Clearly $v^{j'}$ is aligned with the x'-axis of $\mathcal{O}'$, and its norm has not changed $\sqrt{\mathbf{v} \cdot \mathbf{v}} = v = \sqrt{3}/2$, all as required.

For the four-velocity we use

$$^{(4)}\mathbf{R} = \begin{pmatrix} 1 & 0 & 0 & 0 \\ 0 & \cos(\theta) & \sin(\theta) & 0 \\ 0 & -\sin(\theta) & \cos(\theta) & 0 \\ 0 & 0 & 0 & 1 \end{pmatrix}.$$

(d) Solution

To find the inverse of the rotation matrix just change the sign of the angle!

$$^{(4)}\mathbf{R}^{-1} = \begin{pmatrix} 1 & 0 & 0 & 0 \\ 0 & \cos(\theta) & -\sin(\theta) & 0 \\ 0 & \sin(\theta) & \cos(\theta) & 0 \\ 0 & 0 & 0 & 1 \end{pmatrix}.$$

It's easy to verify that this satisfies $^{(4)}\mathbf{R}^{-1} \, {}^{(4)}\mathbf{R} = \mathbf{I}$.

(e) Solution

The Lorentz transformation to the reference frame of the rocket ship can be built from the above tools. Consider transforming a general four-vector, $\vec{V}$. First we rotate axes so that the three-velocity has only an x component; call the rotated frame $\mathcal{O}'$:

$$V^{\alpha'} = {}^{(4)}R^{\alpha'}{}_{\beta}V^{\beta}. \tag{2.42}$$

Now the Lorentz transformation from $\mathcal{O}'$ to the MCRF of the rocket ship is easy because, by construction $\mathbf{v} \underset{\mathcal{O}'}{\rightarrow} (v, 0, 0)$, as we verified in (c) above. Call $\overline{\mathcal{O}'}$ the reference frame that moves with the rocket, but with axes aligned with $\mathcal{O}'$. Then coordinates of $\vec{V}$ in $\overline{\mathcal{O}'}$ are

$$V^{\bar{\alpha}'} = \Lambda^{\bar{\alpha}'}{}_{\alpha'}(v)V^{\alpha'} = \Lambda^{\bar{\alpha}'}{}_{\alpha'}(v) \; {}^{(4)}R^{\alpha'}{}_{\beta} \; V^{\beta}, \tag{2.43}$$

where $\Lambda^{\bar{\alpha}'}{}_{\alpha'}(v)$ is a velocity boost in the standard configuration:

$$\Lambda^{\bar{\alpha}'}{}_{\alpha'}(v) = \begin{pmatrix} \gamma(v) & -v\gamma(v) & 0 & 0 \\ -v\gamma(v) & \gamma(v) & 0 & 0 \\ 0 & 0 & 1 & 0 \\ 0 & 0 & 0 & 1 \end{pmatrix}. \tag{2.44}$$

Finally, to clean up our tools at end of the job, we rotate the axes back to their original orientation with the x, y-axes, calling the final coordinate system $\overline{\mathcal{O}}$,

$$\begin{aligned} V^{\bar{\alpha}} &= \left({}^{(4)}R^{-1})^{\bar{\alpha}}{}_{\bar{\alpha}'}\Lambda^{\bar{\alpha}'}{}_{\alpha'}(v) \; {}^{(4)}R^{\alpha'}{}_{\beta} \right) V^{\beta} \\ &= \Lambda^{\bar{\alpha}}{}_{\beta} \; V^{\beta}, \end{aligned} \tag{2.45}$$

where we have defined a single Lorentz transformation,

$$\Lambda^{\bar{\alpha}}{}_{\beta} \equiv \left({}^{(4)}R^{-1})^{\bar{\alpha}}{}_{\bar{\alpha}'}\Lambda^{\bar{\alpha}'}{}_{\alpha'}(v) \; {}^{(4)}R^{\alpha'}{}_{\beta} \right), \tag{2.46}$$

that transforms the coordinates of a vector in $\mathcal{O}$ to $\overline{\mathcal{O}}$. Applying $\Lambda^{\bar{\alpha}}{}_{\beta}$ to $\vec{U}$, the velocity of the rocket ship, we should end up with $(1, 0, 0, 0)$, since in the MCRF of the rocket ship its own three-velocity is nil. It is straightforward, albeit a bit tedious, to show that indeed this is the case:

$$\Lambda^{\bar{\alpha}}{}_{\alpha}U^{\alpha} = \begin{pmatrix} \gamma(v) & -v\gamma(v)\cos\theta & -v\gamma(v)\sin\theta & 0 \\ -v\gamma(v)\cos\theta & \gamma(v)\cos^2(\theta) + \sin^2(\theta) & (\gamma(v) - 1)\cos(\theta)\sin(\theta) & 0 \\ -v\gamma(v)\sin\theta & (\gamma(v) - 1)\cos(\theta)\sin(\theta) & \gamma(v)\sin^2(\theta) + \cos^2(\theta) & 0 \\ 0 & 0 & 0 & 1 \end{pmatrix}$$

$$\times \begin{pmatrix} 2 \\ 1 \\ \sqrt{2} \\ 0 \end{pmatrix}$$

$$= \begin{pmatrix} 1 \\ 0 \\ 0 \\ 0 \end{pmatrix}. \tag{2.47}$$

SP 2.4 (a) How did the mirror in Exercise 2.31 acquire x-direction momentum without acquiring energy when the photon was reflected? (b) How did it acquire the energy when the photon was absorbed?

SP 2.5 Show that the four-acceleration is orthogonal to the four-velocity. Start with the expression for the four-velocity in terms of the three-velocity with the components of the three-velocity written as a function of time in an inertial frame $\mathcal{O}$:

$$\vec{U} \underset{\mathcal{O}}{\to} \gamma(v)(1, \dot{x}, \dot{y}, \dot{z}),$$

where

$$v = |\mathbf{v}| = \sqrt{\dot{x}^2 + \dot{y}^2 + \dot{z}^2}.$$

Use the chain rule to derive a general expression for the four-acceleration involving derivatives with respect to time.

Solution

The four-acceleration is defined as

$$\vec{a} \equiv \frac{d\vec{U}}{d\tau} \qquad\qquad \text{Schutz Eq. (2.32)} \qquad (2.48)$$

$$= \frac{d}{d\tau}[\gamma(v)(1, \dot{x}, \dot{y}, \dot{z})] \qquad\qquad \text{used eqn. (2.39)}$$

$$= \frac{dt}{d\tau}\frac{d}{dt}[\gamma(v)(1, \dot{x}, \dot{y}, \dot{z})]. \qquad (2.49)$$

Recall from eqn. (2.38) that the differential of coordinate time with respect to proper time gives us,

$$\frac{dt}{d\tau} = \gamma(v). \tag{2.50}$$

For eqn. (2.49) we also require

$$\frac{d}{dt}\gamma(v) = \frac{d}{dt}\left[\frac{1}{\sqrt{1 - \dot{x}^2 - \dot{y}^2 - \dot{z}^2}}\right] = \gamma^3[\dot{x}\ddot{x} + \dot{y}\ddot{y} + \dot{z}\ddot{z}]. \tag{2.51}$$

Substituting eqn. (2.50) and eqn. (2.51) into eqn. (2.49) we find,

$$\vec{a} = \gamma\left(\gamma^3[\dot{x}\ddot{x} + \dot{y}\ddot{y} + \dot{z}\ddot{z}](1, \dot{x}, \dot{y}, \dot{z}) + \gamma(0, \ddot{x}, \ddot{y}, \ddot{z})\right)$$

$$= \gamma \left(\gamma^2 [\dot{x}\ddot{x} + \dot{y}\ddot{y} + \dot{z}\ddot{z}]\vec{U} + \gamma (0, \ddot{x}, \ddot{y}, \ddot{z}) \right)$$

$$= \gamma^3 [\dot{x}\ddot{x} + \dot{y}\ddot{y} + \dot{z}\ddot{z}]\vec{U} + \gamma^2 (0, \ddot{x}, \ddot{y}, \ddot{z}). \tag{2.52}$$

Now take the dot product with $\vec{U}$:

$$\vec{U} \cdot \vec{a} = \vec{U} \cdot \left(\gamma^3 [\dot{x}\ddot{x} + \dot{y}\ddot{y} + \dot{z}\ddot{z}]\vec{U} + \gamma^2 (0, \ddot{x}, \ddot{y}, \ddot{z}) \right)$$

$$= -\gamma^3 [\dot{x}\ddot{x} + \dot{y}\ddot{y} + \dot{z}\ddot{z}] + \gamma^3 (1, \dot{x}, \dot{y}, \dot{z}) \cdot (0, \ddot{x}, \ddot{y}, \ddot{z}) \tag{2.53}$$

$$= 0.$$

SP 2.6   In Exercise 2.25 we found that no Doppler shift occurs when

$$\frac{\bar{\nu}}{\nu} = 1 = \frac{1}{\sqrt{1 - v^2}} (1 - v \cos(\theta)).$$

This implied

$$v = \frac{2 \cos \theta}{1 + \cos^2 \theta}$$

or

$$\theta = \arccos \left(\frac{1 - \sqrt{1 - v^2}}{v} \right).$$

What is the solution in the Newtonian limit $|v| \ll 1$? What is the solution in the relativistic limit $|v| \to 1$? Plot a diagram to determine the maximum angle of no Doppler shift? Show that at $v = 1/2, \theta \approx 74.5°$.

SP 2.7   Suppose we have two particles moving with four-velocity $\vec{V}$ and $\vec{U}$ respectively. Is $\vec{W} = \vec{V} - \vec{U}$ a four-velocity, and if so, of what?

Solution

No, $\vec{W}$ is not necessarily a four-velocity. To be a valid four-velocity we would require the normalization condition, $\vec{W} \cdot \vec{W} = -1$, see eqn. (2.40). But in general this normalization condition will not be true for an arbitrary four-vector, and thus in general is not true for the difference between two arbitrary four-velocities. For a simple example, consider the special case where $\vec{V} = \vec{U}$. Then $\vec{W} = 0 \; (= \vec{0})$, the second equality added to remind you that in this case 0 is a four-vector. The magnitude of $\vec{W}$ would then be nil, so clearly not a four-vector.

SP 2.8   Sometimes we write the total relativistic energy of a particle as the sum of its kinetic energy K and its rest mass energy:

$$E = K + m.$$

Use this to show that, in general,

$$K^2 + 2Km = \sum_{i=1}^{3}(p^i)^2. \tag{2.54}$$

SP 2.9 A timelike or null vector is said to be *future directed* if it points into the future light cone. Show that the sum of a future directed timelike vector t^α and a future directed null vector l^α is timelike and future directed.

Solution

Let $v^\alpha = t^\alpha + l^\alpha$. We wish to show that $v^\alpha v_\alpha < 0$ to verify that it is timelike. We find

$$
\begin{aligned}
v^\alpha v_\alpha &= (t^\alpha + l^\alpha)(t_\alpha + l_\alpha) \\
&= t^\alpha t_\alpha + l^\alpha l_\alpha + t^\alpha l_\alpha + t_\alpha l^\alpha \\
&= t^\alpha t_\alpha + 2t^\alpha l_\alpha.
\end{aligned} \tag{2.55}
$$

We need to show that $t^\alpha l_\alpha < 0$. Without loss of generality, orient the x-axis in the direction of the null vector, so that $\vec{l} \underset{\mathcal{O}}{\rightarrow} (l^0, l^1, 0, 0)$ with $l^0 = l^1$ because it is null. Then

$$t^\alpha l_\alpha = t^0 l_0 + t^1 l_0 = l_0(t^0 + t^1). \tag{2.56}$$

Because t^α and l^α are future directed, $t^0 > 0$ and $l_0 = \eta_{0\beta}l^\beta = -l^0 < 0$. Furthermore, because t^α is timelike, we know the temporal part dominates the spatial part, $(t^0)^2 > \sum_i(t^i)^2$, so certainly $(t^0 + t^1) > 0$. So $t^\alpha l_\alpha < 0$ as well, and $v^\alpha v_\alpha$ consists of the sum of two negative terms, guaranteeing that v^α is timelike. Finally the fact that t^α and l^α are both future directed, $t^0 > 0$ and $l^0 > 0$, immediately implies that their sum is also future directed.

Tensor analysis in special relativity

Notice that the definition of a tensor does not mention components of the vectors. A tensor must be a rule [that] gives the same real number independently of the reference frame in which the vectors' components are calculated.

Bernard Schutz, §3.2

3.1 Exercises

3.2 Prove that the set of all one-forms is a vector space.

Solution: While most instructors will be satisfied with a much more succinct answer, we give a detailed response to help clarify the concepts. One-forms were introduced in Schutz Eqs. (3.6a,b):

$$\left.\begin{array}{l} \tilde{s} = \tilde{p} + \tilde{q}, \\ \tilde{r} = \alpha\tilde{p}, \end{array}\right\} \quad \text{Schutz Eq. (3.6a)} \tag{3.1}$$

where $\tilde{s}$ and $\tilde{r}$ above must obey the following for all arguments $\vec{A}$:

$$\left.\begin{array}{l} \tilde{s}(\vec{A}) = \tilde{p}(\vec{A}) + \tilde{q}(\vec{A}), \\ \tilde{r}(\vec{A}) = \alpha\tilde{p}(\vec{A}). \end{array}\right\} \quad \text{Schutz Eq. (3.6b)} \tag{3.2}$$

To prove that the set of all one-forms is a vector space, we must show that this set meets axioms (1) and (2) given in Schutz Appendix A, p. 374. Axiom (1) states that "[The set] V is an abelian group with operation $+$ $(A + B = B + A \in V)$ and identity 0 $(A + 0 = A)$." For a brief account of group theory see (e.g. Schutz, 1980). Very briefly, a group is a set of objects that can be combined with a binary operation, here "+" but it can be very general, to form new elements that are also in the set. There are four axioms of group theory that must be met: (i) The closure property, $A + B \in V$, is the one we just mentioned that the result of the binary operation (here "+") between two elements results in an element of the set, (ii) Associativity property, $A + (B + C) = (A + B) + C$, (iii) There must be an identity element, i.e. an element that does not alter any elements of the set, $A + 0 = A$ for all $A \in V$, (iv) For every element $A \in V$ there must be an inverse element B that combines to produce the identity element, $A + B = 0$ for any $A \in V$. So applying this to a vector space, Axiom (1) tells us that the binary operation is vector addition and the identity element is the zero vector 0. Some books will emphasize that 0 in this context is a vector by writing $\vec{0}$. We are asked to apply this to one-forms.

The sum of two one-forms must also be a one-form (so closure property (i) met) which is satisfied by eqn. (3.1), $\tilde{s} = \tilde{p} + \tilde{q}$. We require that the order of summation does not matter (this is the "Abelian" property), which is satisfied by eqn. (3.2) because a one-form acting on a vector evaluates to a real and the sum of two reals does not depend upon the order. Similarly the property (ii) of associativity is also met. We also require a zero (there must be an identity element (iii)). The zero one-form gives zero for any vector. So say $\tilde{q}$ is the zero one-form. Then assuming eqn. (3.1) and by eqn. (3.2)

$$\tilde{s}(\vec{A}) = \tilde{p}(\vec{A}) + \tilde{q}(\vec{A}) = \tilde{p}(\vec{A}) + 0 = \tilde{p}(\vec{A}), \tag{3.3}$$

so $\tilde{p} + 0 = \tilde{p}$ and we have a zero. Finally setting $\alpha = -1$ in eqns. (3.1) and (3.2) we see that for each $\tilde{p}$ we can always construct the inverse element $\tilde{r} = -1 \cdot \tilde{p}$ that sums to give the zero element, $\tilde{p} + \tilde{r} = \tilde{p} - 1 \cdot \tilde{p} = 0$. So the set of one-forms with addition as defined in eqns. (3.1, 3.2) satisfies Axiom (1).

Axiom (2) of Appendix A requires that multiplication of an element of the vector space by a real number gives another element of the vector space, with four requirements:

$$\begin{array}{llll} \text{(i)} & a(A + B) = aA + aB, & \text{(ii)} & (a + b)A = aA + bA, \\ \text{(iii)} & (ab)A = a(bA), & \text{(iv)} & 1(A) = A, \end{array} \tag{3.4}$$

with A, B elements of the vector space, a, b reals. Although it is not made explicit in eqns. (3.1) and (3.2), it was clear from the context that $\alpha \in \mathbb{R}$. By eqns. (3.1 and 3.2) it is immediately clear that multiplication of a one-form by a real scalar meets all the requirements of Axiom 2. For instance Axiom 2(i) requires

$$\alpha(\tilde{p} + \tilde{q}) = \alpha\tilde{p} + \alpha\tilde{q}. \tag{3.5}$$

On the LHS of eqn. (3.5) we have a one-form, say $\tilde{s} = \alpha(\tilde{p} + \tilde{q})$, where $\tilde{s}$ is the one-form such that

$$\begin{aligned} \tilde{s}(\vec{V}) &= \alpha \cdot (\tilde{p} + \tilde{q})(\vec{V}) = \alpha \cdot \left(\tilde{p}(\vec{V}) + \tilde{q}(\vec{V}) \right) && \text{used eqn. (3.2)} \\ &= \alpha\tilde{p}(\vec{V}) + \alpha\tilde{q}(\vec{V}). && \text{because } \tilde{p}(\vec{V}) \text{ and } \tilde{q}(\vec{V}) \text{ are reals} \end{aligned} \tag{3.6}$$

On the RHS of eqn. (3.5) we have a one-form, say $\tilde{s}' = \alpha\tilde{p} + \alpha\tilde{q}$, such that

$$\begin{aligned} \tilde{s}'(\vec{V}) &= (\alpha\tilde{p} + \alpha\tilde{q})(\vec{V}) = \alpha\tilde{p}(\vec{V}) + \alpha\tilde{q}(\vec{V}) && \text{used eqn. (3.2)} \\ &= \tilde{s}(\vec{V}). && \text{used eqn. (3.6)} \end{aligned} \tag{3.7}$$

So because $\tilde{s}$ and $\tilde{s}'$ are one-forms that evaluate to the same real number when they operate on the same vector $\vec{V}$, we conclude $\tilde{s} = \tilde{s}'$ and eqn. (3.5) holds, i.e. multiplication by a scalar α is distributive over addition of one-forms. The other three properties in Axiom 2 follow similarly.

3.4   Given the following vectors in $\mathcal{O}$:

$$\vec{A} \underset{\mathcal{O}}{\rightarrow} (2,1,1,0), \quad \vec{B} \underset{\mathcal{O}}{\rightarrow} (1,2,0,0), \quad \vec{C} \underset{\mathcal{O}}{\rightarrow} (0,0,1,1), \quad \vec{D} \underset{\mathcal{O}}{\rightarrow} (-3,2,0,0),$$

(b) find components of $\tilde{p}$ if

$$\tilde{p}(\vec{A}) = 1, \quad \tilde{p}(\vec{B}) = -1, \quad \tilde{p}(\vec{C}) = -1, \quad \tilde{p}(\vec{D}) = 0.$$

Solution: Using the expression

$$\tilde{p}(\vec{A}) = p_\alpha A^\alpha \qquad\qquad \text{Schutz Eq. (3.8)} \qquad\qquad (3.8)$$

we can write a linear system in the four unknown components:

$$\begin{pmatrix} \vec{A} \\ \vec{B} \\ \vec{C} \\ \vec{D} \end{pmatrix} \begin{pmatrix} p_0 \\ p_1 \\ p_2 \\ p_3 \end{pmatrix} = \begin{pmatrix} \tilde{p}(\vec{A}) \\ \tilde{p}(\vec{B}) \\ \tilde{p}(\vec{C}) \\ \tilde{p}(\vec{D}) \end{pmatrix}, \quad \begin{pmatrix} 2 & 1 & 1 & 0 \\ 1 & 2 & 0 & 0 \\ 0 & 0 & 1 & 1 \\ -3 & 2 & 0 & 0 \end{pmatrix} \begin{pmatrix} p_0 \\ p_1 \\ p_2 \\ p_3 \end{pmatrix} = \begin{pmatrix} 1 \\ -1 \\ -1 \\ 0 \end{pmatrix}. \qquad (3.9)$$

Note the matrix is written with the rows given by the vectors (so that one-forms appeared as columns) but this choice was arbitrary. This was solved in the accompanying Maple™ worksheet, giving $\tilde{p} \underset{\mathcal{O}}{\rightarrow} (-2, -3, 15, -23)/8$.

(d) determine whether the one-forms $\tilde{p}, \tilde{q}, \tilde{r}, \tilde{s}$ are linearly independent if

$$\begin{aligned} \tilde{q}(\vec{A}) = 0, \quad \tilde{q}(\vec{B}) = 0, \qquad & \tilde{q}(\vec{C}) = 1, \quad \tilde{q}(\vec{D}) = -1, \\ \tilde{r}(\vec{A}) = 2, \quad \tilde{r}(\vec{B}) = 0, \qquad & \tilde{r}(\vec{C}) = 0, \quad \tilde{r}(\vec{D}) = 0, \\ \tilde{s}(\vec{A}) = -1, \quad \tilde{s}(\vec{B}) = -1, \qquad & \tilde{s}(\vec{C}) = 0, \quad \tilde{s}(\vec{D}) = 0. \end{aligned} \qquad (3.10)$$

Solution: Given the values of the four one-forms, $\tilde{p}, \tilde{q}, \tilde{r}, \tilde{s}$ applied to the four known vectors $\vec{A}, \vec{B}, \vec{C}, \vec{D}$ we can, in principle, find all components of all four one-forms, repeating the procedure we did in Exercise 3.4(b). And then one could write a matrix M where the columns of M are taken from the one-form components. If the determinant of M is zero the one-forms are linearly dependent. But that is a lot of work.

There is a simpler way to test for linear dependence. If the one-forms are linearly dependent, then there are nontrivial real numbers a, b, c, d such that

$$a\tilde{p} + b\tilde{q} + c\tilde{r} + d\tilde{s} = \tilde{t} = 0$$

$$(\tilde{p} \quad \tilde{q} \quad \tilde{r} \quad \tilde{s}) \begin{pmatrix} a \\ b \\ c \\ d \end{pmatrix} = \tilde{t} = \begin{pmatrix} 0 \\ 0 \\ 0 \\ 0 \end{pmatrix}. \qquad (3.11)$$

But then

$$
\begin{pmatrix} \vec{A} \\ \vec{B} \\ \vec{C} \\ \vec{D} \end{pmatrix} \tilde{t} = \begin{pmatrix} 0 \\ 0 \\ 0 \\ 0 \end{pmatrix}. \tag{3.12}
$$

By eqn. (3.2) we have

$$
\begin{pmatrix} \vec{A} \\ \vec{B} \\ \vec{C} \\ \vec{D} \end{pmatrix} \tilde{t} = \begin{pmatrix} \vec{A} \\ \vec{B} \\ \vec{C} \\ \vec{D} \end{pmatrix} \begin{pmatrix} \tilde{p} & \tilde{q} & \tilde{r} & \tilde{s} \end{pmatrix} \begin{pmatrix} a \\ b \\ c \\ d \end{pmatrix} = \begin{pmatrix} \tilde{p}(\vec{A}) & \tilde{q}(\vec{A}) & \tilde{r}(\vec{A}) & \tilde{s}(\vec{A}) \\ \tilde{p}(\vec{B}) & \tilde{q}(\vec{B}) & \tilde{r}(\vec{B}) & \tilde{s}(\vec{B}) \\ \tilde{p}(\vec{C}) & \tilde{q}(\vec{C}) & \tilde{r}(\vec{C}) & \tilde{s}(\vec{C}) \\ \tilde{p}(\vec{D}) & \tilde{q}(\vec{D}) & \tilde{r}(\vec{D}) & \tilde{s}(\vec{D}) \end{pmatrix} \begin{pmatrix} a \\ b \\ c \\ d \end{pmatrix}
$$

$$
= \begin{pmatrix} 1 & 0 & 2 & -1 \\ -1 & 0 & 0 & -1 \\ -1 & 1 & 0 & 0 \\ 0 & -1 & 0 & 0 \end{pmatrix} \begin{pmatrix} a \\ b \\ c \\ d \end{pmatrix} = \begin{pmatrix} 0 \\ 0 \\ 0 \\ 0 \end{pmatrix}. \tag{3.13}
$$

The latter can only be true if the determinant is zero, but

$$
\begin{vmatrix} 1 & 0 & 2 & -1 \\ -1 & 0 & 0 & -1 \\ -1 & 1 & 0 & 0 \\ 0 & -1 & 0 & 0 \end{vmatrix} = -2, \tag{3.14}
$$

so the one-forms must be linearly independent.

3.5   Justify steps from Schutz Eq. (3.10a) to Eq. (3.10d), where

$$
\begin{aligned}
A^{\tilde{\alpha}} p_{\tilde{\alpha}} &= (\Lambda^{\tilde{\alpha}}{}_{\beta} A^{\beta})(\Lambda^{\mu}{}_{\tilde{\alpha}} p_{\mu}), &&\text{Schutz Eq. (3.10a)} \\
&= (\Lambda^{\mu}{}_{\tilde{\alpha}} \Lambda^{\tilde{\alpha}}{}_{\beta}) (A^{\beta} p_{\mu}), &&\text{Schutz Eq. (3.10b)} \\
&= \delta^{\mu}{}_{\beta} A^{\beta} p_{\mu}, &&\text{Schutz Eq. (3.10c)} \\
&= A^{\beta} p_{\beta}. &&\text{Schutz Eq. (3.10d)} \tag{3.15}
\end{aligned}
$$

Solution: We start with the respective Lorentz transformations of the vector and one-form, and then use the inverse property of the two Lorentz transformations. Summing over the dummy index results in the desired expression. In particular,

$$
\begin{aligned}
A^{\tilde{\alpha}} p_{\tilde{\alpha}} &= (\Lambda^{\tilde{\alpha}}{}_{\beta} A^{\beta})(\Lambda^{\mu}{}_{\tilde{\alpha}} p_{\mu}), &&\text{by Schutz Eqs. (2.7) and (3.9) respectively} \\
&= (\Lambda^{\mu}{}_{\tilde{\alpha}} \Lambda^{\tilde{\alpha}}{}_{\beta}) (A^{\beta} p_{\mu}), &&\text{just rearranged the terms} \\
&= \delta^{\mu}{}_{\beta} (A^{\beta} p_{\mu}), &&\text{by eqn. (3.23)} \\
&= A^{\beta} p_{\beta}. &&\text{sum over } \mu, \text{ use properties of the Kronecker delta}
\end{aligned}
$$

3.6 Consider the basis $\{\vec{e}_\alpha\}$ of a frame $\mathcal{O}$ and a basis $\{\tilde{\lambda}^0, \tilde{\lambda}^1, \tilde{\lambda}^2, \tilde{\lambda}^3\}$ for the space of one-forms, with

$$\tilde{\lambda}^0 \underset{\mathcal{O}}{\to} (1, 1, 0, 0), \qquad\qquad \tilde{\lambda}^1 \underset{\mathcal{O}}{\to} (1, -1, 0, 0),$$

$$\tilde{\lambda}^2 \underset{\mathcal{O}}{\to} (0, 0, 1, -1), \qquad\qquad \tilde{\lambda}^3 \underset{\mathcal{O}}{\to} (0, 0, 1, 1). \qquad (3.16)$$

Note that $\{\tilde{\lambda}^\beta\}$ is *not* the basis dual to $\{\vec{e}_\alpha\}$.

(a) Show that $\tilde{p} \neq \tilde{p}(\vec{e}_\alpha)\tilde{\lambda}^\alpha$ for arbitrary $\tilde{p}$.

Solution: Applying a one-form to a basis vector results in the corresponding component of the one-form,

$$p_\alpha = \tilde{p}(\vec{e}_\alpha), \qquad\qquad \text{Schutz Eq. (3.7)} \qquad (3.17)$$

but this component "belongs to" the basis one-form dual to the basis vector $\vec{e}_\alpha$. So using these components with a different basis such as $\tilde{\lambda}^\alpha$ leads to a different one-form. That is, $\tilde{p} = p_\alpha \tilde{\omega}^\alpha$, when and only when $\tilde{\omega}^\alpha$ is dual to $\vec{e}_\alpha$ so that eqn. (3.91) applies. Indeed some texts use the same symbol for both the basis vectors and their corresponding dual basis one-forms, (e.g. Hobson et al., 2006, Eq. (3.2)), which emphasizes this correspondence but de-emphasizes the distinction between one-forms and vectors.

More formally, consider an arbitrary one-form $\tilde{p}$ and vector $\vec{A}$.

$$\tilde{p}(\vec{e}_\alpha)\tilde{\lambda}^\alpha(\vec{A}) = p_\alpha \tilde{\lambda}^\alpha(\vec{A}) = p_\alpha \tilde{\lambda}^\alpha(A^\beta \vec{e}_\beta) = p_\alpha A^\beta \tilde{\lambda}^\alpha(\vec{e}_\beta)$$

$$= p_\alpha A^\alpha \quad \text{iff } \tilde{\lambda}^\alpha(\vec{e}_\beta) = \delta^\alpha{}_\beta. \qquad (3.18)$$

But it is clear that $\tilde{\lambda}^\alpha(\vec{e}_\beta) \neq \delta^\alpha{}_\beta$ by inspection of the given basis. For example, $\delta^0{}_1 = 0$ but

$$\tilde{\lambda}^0(\vec{e}_1) = \begin{pmatrix} 1 & 1 & 0 & 0 \end{pmatrix} \begin{pmatrix} 0 \\ 1 \\ 0 \\ 0 \end{pmatrix} = 1 \cdot 0 + 1 \cdot 1 + 0 \cdot 0 + 0 \cdot 0 = 1. \qquad \text{used eqn. (3.8)}$$

3.7 Prove that the basis one-forms transform under a change of basis as follows:

$$\tilde{\omega}^{\bar{\alpha}} = \Lambda^{\bar{\alpha}}{}_\beta \tilde{\omega}^\beta. \qquad\qquad \text{Schutz Eq. (3.13)} \qquad (3.19)$$

Solution: We were told just before Schutz Eq. (3.13) that its derivation is analogous to the corresponding relation for basis vectors, see derivation of Schutz Eq. (2.13) in §2.2. Imagine that $\tilde{p}$ is an arbitrary one-form. Let $\tilde{\omega}^\alpha$ and $\tilde{\omega}^{\bar{\alpha}}$ be the sets of basis one-forms in frames $\mathcal{O}$ and $\overline{\mathcal{O}}$ respectively. $\tilde{p}$ can be expressed in terms of either basis set and we want this to be the same geometrical object,

$$\tilde{p} = p_\alpha \tilde{\omega}^\alpha = p_{\bar{\alpha}} \tilde{\omega}^{\bar{\alpha}}.$$

Note the analogy with Schutz Eq. (2.12), which specifies that vectors are frame independent. It is important to realize that in general $p_0 \neq p_{\bar{0}}$. Part of the subtlety arises here because the notation, while very standard, is actually misleading. In fact a few textbooks (e.g. Hobson et al., 2006; Weinberg, 1972) use a different symbol for the components of $\tilde{p}$ in the $\overline{\mathcal{O}}$ frame, namely $\bar{p}_\alpha$. Here we will continue with the standard notation used by Schutz and most others

$$p_\alpha \tilde{\omega}^\alpha = p_{\bar{\alpha}} \tilde{\omega}^{\bar{\alpha}} \qquad \text{same geometrical object}$$

$$= \Lambda^\beta_{\ \bar{\alpha}} \, p_\beta \tilde{\omega}^{\bar{\alpha}}. \qquad \text{used eqn. (3.87)} \qquad (3.20)$$

At this point we may relabel dummy indices and replace β with α,

$$p_\alpha \tilde{\omega}^\alpha = \Lambda^\alpha_{\ \bar{\alpha}} \, p_\alpha \tilde{\omega}^{\bar{\alpha}}. \qquad \text{relabeled dummy index} \qquad (3.21)$$

Because the equality above must hold for arbitrary $\tilde{p}$ it is clear that we require

$$\tilde{\omega}^\alpha = \Lambda^\alpha_{\ \bar{\alpha}} \, \tilde{\omega}^{\bar{\alpha}}.$$

If you are uncomfortable with this last step, see SP3.5 below.

3.10 (a) Given a frame $\mathcal{O}$ whose coordinates are $\{x^\alpha\}$, show that

$$\partial x^\alpha / \partial x^\beta = \delta^\alpha_{\ \beta}.$$

Solution: Let's use an asterix to denote an arbitrary but fixed index, like $\alpha*$. When $\alpha* \neq \beta*$, then $x^{\alpha*}$ and $x^{\beta*}$ are *independent variables* and of course $\partial x^{\alpha*}/\partial x^{\beta*} = 0$. But when $\alpha* = \beta*$ then $x^{\alpha*} = x^{\beta*}$ are the very *same variable* and of course $\partial x^{\alpha*}/\partial x^{\alpha*} = 1$. This completes the proof, which is valid in any coordinate system.

To be more concrete, consider pseudo-Cartesian coordinates wherein $x^0 = t, x^1 = x, x^2 = y, x^3 = z$. When $\alpha \neq \beta$ we have terms like, say,

$$\frac{\partial x^0}{\partial x^1} = \frac{\partial t}{\partial x} = 0,$$

because t and x are independent variables. But when $\alpha = \beta$ we have terms like, say,

$$\frac{\partial x^3}{\partial x^3} = \frac{\partial z}{\partial z} = 1.$$

3.10 (b) For any two frames, we have:

$$\frac{\partial x^\beta}{\partial x^{\bar{\alpha}}} = \Lambda^\beta_{\ \bar{\alpha}}. \qquad \text{Schutz Eq. (3.18)} \qquad (3.22)$$

Show that (a) and the chain rule imply

$$\Lambda^{\beta}{}_{\bar{\alpha}} \, \Lambda^{\bar{\alpha}}{}_{\mu} = \delta^{\beta}{}_{\mu}. \tag{3.23}$$

This is the inverse property again.

Solution: We start with the result from Exercise 3.10(a), and apply a Lorentz transformation as follows:

$$\frac{\partial}{\partial x^{\mu}} x^{\beta} = \delta^{\beta}{}_{\mu} \qquad\qquad \text{from Exercise 3.10(a)}$$

$$\frac{\partial}{\partial x^{\mu}} \left(\Lambda^{\beta}{}_{\bar{\alpha}} x^{\bar{\alpha}} \right) = \delta^{\beta}{}_{\mu} \qquad\qquad \text{sub coordinate transform}$$

$$\Lambda^{\beta}{}_{\bar{\alpha}} \frac{\partial x^{\bar{\alpha}}}{\partial x^{\mu}} = \delta^{\beta}{}_{\mu} \qquad\qquad \text{transform is a constant}$$

$$\Lambda^{\beta}{}_{\bar{\alpha}} \Lambda^{\bar{\alpha}}{}_{\mu} = \delta^{\beta}{}_{\mu}. \qquad\qquad \text{from Schutz Eq. (3.18)}$$

Eqn. (3.22) reveals that the Lorentz transformations are a type of coordinate transformation with the special property that they apply globally. Later in the text, when one studies GR, the idea of so-called general covariance will be of fundamental importance and general coordinate transformations $x^{\alpha}(x^{\bar{\alpha}})$ will be considered wherein $\partial x^{\alpha}/\partial x^{\bar{\alpha}}$ is not constant throughout spacetime; however the inverse property eqn. (3.23) will still apply locally.

3.12 Let S be the two-dimensional plane $x = 0$ in three-dimensional Euclidean space. Let $\tilde{n} \neq 0$ be a normal one-form to S.

(a) Show that if $\vec{V}$ is a vector that is not tangent to S, then $\tilde{n}(\vec{V}) \neq 0$.

Solution: Applying the general rule for the contraction of $\vec{V}$ and $\tilde{n}$ to this three-dimensional Euclidean space, we have

$$\tilde{n}(\vec{V}) = n_x V^x + n_y V^y + n_z V^z. \qquad\qquad \text{used eqn. (3.8)} \tag{3.24}$$

A normal one-form must produce zero when contracted with any vector tangent to the surface. It follows that for the $x = 0$ surface in 3D space,

$$\tilde{n} \underset{\mathcal{O}}{\rightarrow} (n_x, 0, 0), \tag{3.25}$$

where any $n_x \neq 0$ will give us a non-zero, normal one-form. So the contraction eqn. (3.24) reduces to

$$\tilde{n}(\vec{V}) = n_x V^x.$$

Because $\vec{V}$ is not tangent to surface S it must have a component in the $\vec{e}_x$-direction, $V^x \neq 0$. Thus the contraction is non-zero.

(c) Show that any normal to S is a multiple of $\tilde{n}$.

Solution: Any normal to the surface S must have $n_y = 0$ and $n_z = 0$, as we argued in Exercise 3.12(a) in deriving eqn. (3.25). To be non-zero it requires $n_x \neq 0$. So any

$$\tilde{n} \underset{\mathcal{O}}{\rightarrow} a(n_x, 0, 0),\qquad(3.26)$$

where $a \neq 0$ and $a \in \mathbb{R}$ will serve also as a non-zero, normal one-form.

3.13 Prove, by geometric or algebraic arguments, that the gradient of f, denoted by the tensor $\tilde{d}f$, is normal to surfaces of constant f.

Solution: Consider an arbitrary point $p = (t, x, y, z)$ where $f(t, x, y, z) = f_p$. Now imagine taking an infinitesimal step $(\Delta t, \Delta x, \Delta y, \Delta z)$ such that the change in value of f,

$$\Delta f = \frac{\partial f}{\partial t}\Delta t + \frac{\partial f}{\partial x}\Delta x + \frac{\partial f}{\partial y}\Delta y + \frac{\partial f}{\partial z}\Delta z = 0.\qquad(3.27)$$

This ensures we do not leave the surface of constant f. So a tangent vector to the surface of constant f is obtained from an arbitrary multiple of such a step:

$$\vec{A} \underset{\mathcal{O}}{\rightarrow} a(\Delta t, \Delta x, \Delta y, \Delta z),\qquad(3.28)$$

where $a \in \mathbb{R}$ and $a \neq 0$. The gradient one-form applied to such a tangent vector is

$$\tilde{d}f(\vec{A}) = a\left(\frac{\partial f}{\partial t}\Delta t + \frac{\partial f}{\partial x}\Delta x + \frac{\partial f}{\partial y}\Delta y + \frac{\partial f}{\partial z}\Delta z\right) \quad \text{used Schutz Eq. (3.15)}$$
$$= 0. \qquad \text{used eqn. (3.27)}\quad(3.29)$$

3.15 Supply the reasoning leading from

$$\mathbf{f} = f_{\alpha\beta}\tilde{\omega}^{\alpha\beta} \qquad \text{Schutz Eqs. (3.23)}$$

to

$$\tilde{\omega}^{\alpha\beta}(\vec{e}_\mu, \vec{e}_\nu) = \delta^\alpha{}_\mu \delta^\beta{}_\nu. \qquad \text{Schutz Eq. (3.24)}$$

Solution: This exercise provides a good opportunity to deepen one's understand of tensor algebra. We proceed step-by-step:

$$\mathbf{f} = f_{\alpha\beta}\tilde{\omega}^{\alpha\beta}, \qquad \text{what we mean by a basis}$$
$$f_{\mu\nu} = \mathbf{f}(\vec{e}_\mu, \vec{e}_\nu), \qquad \text{what we mean by components}$$
$$f_{\mu\nu} = f_{\alpha\beta}\,\tilde{\omega}^{\alpha\beta}(\vec{e}_\mu, \vec{e}_\nu), \qquad \text{sub first line into second line}\quad(3.30)$$
$$\tilde{\omega}^{\alpha\beta}(\vec{e}_\mu, \vec{e}_\nu) = \delta^\alpha{}_\mu \delta^\beta{}_\nu. \qquad \text{solving above}\quad(3.31)$$

The last step deserves a few words of explanation. While $f_{\mu\nu} = f_{\alpha\beta}\delta^\alpha{}_\mu\,\delta^\beta{}_\nu$ is obviously a solution, how do we know this is *the* solution? Here is one way to convince yourself:

$$f_{\mu\nu} = f_{\alpha\beta}\,A^{\alpha\beta}{}_{\mu\nu}, \qquad\qquad \text{re-write eqn. (3.30) with } A^{\alpha\beta}{}_{\mu\nu} \text{ as unknown}$$

$$f_{\sigma\gamma}\,\delta^\sigma{}_\mu\,\delta^\gamma{}_\nu = f_{\alpha\beta}\,A^{\alpha\beta}{}_{\mu\nu}, \qquad\qquad \text{re-write LHS}$$

$$f_{\alpha\beta}\,\delta^\alpha{}_\mu\,\delta^\beta{}_\nu = f_{\alpha\beta}\,A^{\alpha\beta}{}_{\mu\nu}, \qquad\qquad \text{relabel dummy indices, see SP3.5}$$

$$0 = f_{\alpha\beta}\left(\delta^\alpha{}_\mu\,\delta^\beta{}_\nu - A^{\alpha\beta}{}_{\mu\nu}\right). \qquad\qquad \text{rearrange}$$

But $f_{\alpha\beta}$ is arbitrary so the only way this last line above can always be true is for $\delta^\alpha{}_\mu\,\delta^\beta{}_\nu = A^{\alpha\beta}{}_{\mu\nu}$.

3.17 (a) Suppose that $\mathbf{h}$ is a $\binom{0}{2}$ tensor with the property that, for *any* two vectors $\vec{A}$ and $\vec{B}$ (where $\vec{B} \neq 0$),

$$\mathbf{h}(\ ,\vec{A}) = \alpha\,\mathbf{h}(\ ,\vec{B}),$$

where α is a number that may depend on $\vec{A}$ and $\vec{B}$. Show that there exist one-forms $\tilde{p}$ and $\tilde{q}$ such that

$$\mathbf{h} = \tilde{p} \otimes \tilde{q}.$$

Solution: In general,

$$\mathbf{h}(\vec{C},\vec{A}) = h_{\gamma\beta}\,C^\gamma A^\beta.$$

Treat $\vec{C}$ as an arbitrary vector. We were given that for arbitrary vectors $\vec{A}$ and $\vec{B}$, but with $\vec{B} \neq 0$

$$\mathbf{h}(\ ,\vec{A}) = \alpha\mathbf{h}(\ ,\vec{B}). \qquad\qquad (3.32)$$

Suppose $C^\gamma \underset{\mathcal{O}}{\to} (1,0,0,0)$. Then

$$\mathbf{h}(\vec{C},\vec{A}) = \alpha\mathbf{h}(\vec{C},\vec{B})$$

$$h_{0\beta}A^\beta = \alpha h_{0\mu}B^\mu \qquad\qquad (3.33)$$

$$\tilde{q}(\vec{A}) = \alpha\tilde{q}(\vec{B}). \qquad\qquad (3.34)$$

The LHS and RHS of eqn. (3.33) have the form of a one-form contracted with vectors $\vec{A}$ and $\vec{B}$ respectively, so we wrote that explicitly in eqn. (3.34), defining $q_\beta \equiv h_{0\beta}$. So far there is no restriction on $\tilde{q}$; we simply choose

$$\alpha = \frac{\tilde{q}(\vec{A})}{\tilde{q}(\vec{B})}, \qquad\qquad (3.35)$$

and we note the stipulation that $\vec{B} \neq 0$.

Now suppose $C^\gamma \underset{\mathcal{O}}{\rightarrow} (1, 1, 0, 0)$. Then

$$\mathbf{h}(\vec{C}, \vec{A}) = \alpha \mathbf{h}(\vec{C}, \vec{B})$$

$$h_{0\beta} A^\beta + h_{1\beta} A^\beta = \alpha(h_{0\mu} B^\mu + h_{1\mu} B^\mu) \tag{3.36}$$

$$h_{1\beta} A^\beta = \alpha(h_{1\mu} B^\mu). \qquad \text{subtracted eqn. (3.33)} \tag{3.37}$$

Now α is no longer a free variable, being set by eqn. (3.33) or equivalently eqn. (3.35). Both eqns. (3.33) and (3.37) can be satisfied if

$$a\, h_{0\beta} = h_{1\beta} \tag{3.38}$$

for an arbitrary a since then eqn. (3.37) is simply eqn. (3.33) multiplied on both sides by a. And this must be the unique solution as is obvious from considering the case where the vectors have only a single component, e.g. $\vec{A} \underset{\mathcal{O}}{\rightarrow} (0, 1, 0, 0)$ and $\vec{B} \underset{\mathcal{O}}{\rightarrow} (1, 0, 0, 0)$. For then eqns. (3.33) and (3.37) reduce to simple algebraic equations, e.g.

$$\left. \begin{array}{l} h_{01} = \alpha h_{00} \\ h_{11} = \alpha h_{10} \end{array} \right\} \Rightarrow \frac{h_{01}}{h_{11}} = \frac{h_{00}}{h_{10}} = a^{-1}. \tag{3.39}$$

Repeating the argument above for different $\vec{C}$, we find a different constant for each first index of $\mathbf{h}$. In particular, with $C^\gamma \underset{\mathcal{O}}{\rightarrow} (1, 0, 1, 0)$ we conclude that $b h_{0\beta} = h_{2\beta}$, for some constant b. With $C^\gamma \underset{\mathcal{O}}{\rightarrow} (1, 0, 0, 1)$ we find $c h_{0\beta} = h_{3\beta}$. That is, if the tensor $\mathbf{h}$ is written as a matrix, the rows are arbitrary scalar constants $p_\mu = (1, a, b, c)$ of the first row $(h_{0\nu}) \equiv q_\nu$. And so

$$h_{\mu\beta} = p_\mu q_\beta \tag{3.40}$$

and so the tensor $\mathbf{h}$ has the form of an outer product,

$$\mathbf{h} = \tilde{p} \otimes \tilde{q}. \tag{3.41}$$

3.17 (b) Suppose $\mathbf{T}$ is a $\binom{1}{1}$ tensor, $\tilde{\omega}$ a one-form, $\vec{v}$ a vector, and $\mathbf{T}(\tilde{\omega}; \vec{v})$ the value of $\mathbf{T}$ on $\tilde{\omega}$ and $\vec{v}$. Prove that $\mathbf{T}(\ ; \vec{v})$ is a vector and $\mathbf{T}(\tilde{\omega};\ )$ is a one-form, i.e. that a $\binom{1}{1}$ tensor provides a map of vectors to vectors and one-forms to one-forms.

Solution: You can think of a $\binom{1}{1}$ tensor as a machine with two input slots, a slot for one-forms and another for vectors. Inserting a one-form and a vector into their respective slots produces a real number. So once you've filled the one-form slot, this machine takes one more input, a vector, to produce a real. But a machine that takes a vector as input to produce a real is, by definition, a one-form. Now let's go back to the $\binom{1}{1}$ tensor, the machine with two empty slots. We've just seen that the act of inserting the one-form converts the machine to one-form; so we can say a $\binom{1}{1}$ tensor is a map from one-forms to one-forms. By a similar argument, a $\binom{1}{1}$ tensor is also a map from vectors to vectors.

It's also satisfying to see how this plays out using the mathematical symbols. Write **T** acting on the vector alone:

$$\mathbf{T}(\quad;\vec{v}) = \mathbf{T}(\quad;v^{\beta}\vec{e}_{\beta}) \qquad\qquad \text{used eqn. (2.4)}$$
$$= T^{\nu}{}_{\mu}v^{\beta}\,\vec{e}_{\nu}\otimes\tilde{\omega}^{\mu}(\quad;\vec{e}_{\beta}) \qquad \text{used Schutz Eq. (3.61)}$$
$$= T^{\nu}{}_{\mu}v^{\beta}\,\tilde{\omega}^{\mu}(\vec{e}_{\beta})\,\vec{e}_{\nu} \qquad \text{applied basis to argument}$$
$$= T^{\nu}{}_{\mu}v^{\beta}\delta^{\mu}{}_{\beta}\,\vec{e}_{\nu} \qquad\qquad \text{used eqn. (3.91)}$$
$$= \left(T^{\nu}{}_{\mu}v^{\mu}\right)\vec{e}_{\nu}. \qquad\qquad \text{summed over }\beta \qquad (3.42)$$

And now we are done because the RHS of eqn. (3.42) is the product of four reals $\left(T^{\nu}{}_{\mu}v^{\mu}\right)$, i.e. one for each value of ν, times the four basis vectors, $\vec{e}_{\nu}$. This is a vector, cf. eqn. (2.4).

And similarly we can write **T** acting on the one-form alone:

$$\mathbf{T}(\tilde{p};\quad) = \mathbf{T}(p_{\alpha}\tilde{\omega}^{\alpha};\quad) \qquad\qquad \text{used Schutz Eq. (3.11)}$$
$$= T^{\nu}{}_{\mu}\,p_{\alpha}\,\vec{e}_{\nu}\otimes\tilde{\omega}^{\mu}(\tilde{\omega}^{\alpha};\quad) \qquad \text{used Schutz Eq. (3.61)}$$
$$= T^{\nu}{}_{\mu}\,p_{\alpha}\,\tilde{\omega}^{\mu}\,\delta^{\alpha}{}_{\nu} \qquad\qquad \text{used eqn. (3.91)}$$
$$= \left(T^{\nu}{}_{\mu}\,p_{\nu}\right)\tilde{\omega}^{\mu}. \qquad\qquad \text{summed over }\alpha \qquad (3.43)$$

Again we are done here because the RHS is clearly a one-form, i.e. it is the product of four reals $\left(T^{\nu}{}_{\mu}p_{\nu}\right)$, one for each value of μ, and the set of four one-form bases $\tilde{\omega}^{\mu}$.

So a $\binom{1}{1}$ tensor is a map from vectors to vectors in the sense that when you contract it with a vector the result is a vector. Likewise a $\binom{1}{1}$ tensor is a map from one-forms to one-forms because contracting it with a one-form results in a one-form.

3.19 (b) Derive the equation for the dot product of two one-forms:

$$\tilde{p}\cdot\tilde{q} = -p_0 q_0 + p_1 q_1 + p_2 q_2 + p_3 q_3. \qquad \text{Schutz Eq. (3.53)} \qquad (3.44)$$

Solution: To derive the formula for the inner product of one-forms in terms of components, eqn. (3.44), we start with the definition:

$$\tilde{p}\cdot\tilde{q} = \frac{1}{2}[(\tilde{p}+\tilde{q})^2 - \tilde{p}^2 - \tilde{q}^2]. \qquad \text{Schutz Eq. (3.52)} \qquad (3.45)$$

This involves only addition and squares of one-forms. The square of a one-form has been defined in component form in eqn. (3.83). The addition of one-forms was defined abstractly in eqns. (3.1, 3.2), and in SP3.3 we show that this implies that we just add the components; if $\tilde{s} = \tilde{p}+\tilde{q}$ then $s_{\alpha} = p_{\alpha}+q_{\alpha}$. So the square of the sum will be:

$$(\tilde{p}+\tilde{q})^2 = \tilde{s}^2 = \eta^{\alpha\beta}s_{\alpha}s_{\beta}, \qquad\qquad \text{by eqn. (3.83)}$$
$$= \eta^{\alpha\beta}(p_{\alpha}+q_{\alpha})(p_{\beta}+q_{\beta}), \qquad \text{component-wise addition}$$
$$= \eta^{\alpha\beta}(p_{\alpha}p_{\beta} + q_{\alpha}q_{\beta} + p_{\alpha}q_{\beta} + p_{\beta}q_{\alpha}). \qquad \text{components are just reals}$$
$$(3.46)$$

Substituting eqn. (3.46) into eqn. (3.45) and using $\{\eta^{\alpha\beta}\}$ from Schutz Eq. (3.44):

$$\tilde{p} \cdot \tilde{q} = \frac{1}{2}[(\tilde{p} + \tilde{q})^2 - \tilde{p}^2 - \tilde{q}^2]$$

$$= \frac{1}{2}[\eta^{\alpha\beta}(p_\alpha p_\beta + q_\alpha q_\beta + p_\alpha q_\beta + p_\beta q_\alpha) - \eta^{\alpha\beta} p_\alpha p_\beta - \eta^{\alpha\beta} q_\alpha q_\beta],$$

$$= \frac{1}{2}[\eta^{\alpha\beta}(p_\alpha q_\beta + p_\beta q_\alpha)] \qquad \text{cancelled terms}$$

$$= \eta^{\alpha\beta} p_\alpha q_\beta \qquad\qquad\qquad \eta^{\alpha\beta} \text{ symmetric}$$

$$= -p_0 q_0 + p_1 q_1 + p_2 q_2 + p_3 q_3, \qquad\qquad (3.47)$$

which is eqn. (3.44).

3.20 In Euclidean three-space in Cartesian coordinates, we do not normally distinguish between vectors and one-forms, because their components transform identically. Prove this in two steps. Suppose we are in Euclidean three-space in Cartesian coordinates.

(a) Show that

$$A^{\bar{\alpha}} = \Lambda^{\bar{\alpha}}{}_\beta A^\beta$$

and

$$P_{\bar{\beta}} = \Lambda^\alpha{}_{\bar{\beta}} P_\alpha \qquad\qquad (3.48)$$

are the same transformation if the matrix $(\Lambda^{\bar{\alpha}}{}_\beta)$ is equal to the transpose of its inverse. Such a matrix is said to be *orthogonal*.

Hint: Solution available in (Schutz, 1985, Appendix B).

(b) The metric of such a space has components $\{\delta_{ij}, i, j = 1, \ldots, 3\}$. Prove that a transformation from one Cartesian coordinate system to another must obey

$$\delta_{\bar{i}\bar{j}} = \Lambda^k{}_{\bar{i}} \Lambda^l{}_{\bar{j}} \delta_{kl} \qquad\qquad (3.49)$$

and that this implies $(\Lambda^k{}_{\bar{i}})$ is an orthogonal matrix. See Exercise 3.32 for the analog of this in SR.

Solution: All we are given is that the metric tensor for Cartesian three-space is $\{\delta_{ij}, i, j = 1, 2, 3\}$. The metric tensor is used in forming the inner product of vectors, which we know must be frame invariant. So write the inner product between two three-space vectors in two different frames as

$$\delta_{\bar{i}\bar{j}} A^{\bar{i}} B^{\bar{j}} = \vec{A} \cdot \vec{B} = \delta_{kl} A^k B^l$$

$$= \delta_{kl} (A^{\bar{i}} \Lambda^k_{\ \bar{i}})(B^{\bar{j}} \Lambda^l_{\ \bar{j}}), \tag{3.50}$$

and so upon "cancelling" the $A^{\bar{i}} B^{\bar{j}}$ on either side (allowed because these are arbitrary vectors) and rearranging we see that

$$\delta_{\bar{i}\bar{j}} = \Lambda^k_{\ \bar{i}} \Lambda^l_{\ \bar{j}} \delta_{kl}, \tag{3.51}$$

as required. If you were uncomfortable with cancelling the $A^{\bar{i}} B^{\bar{j}}$ on either side, see SP3.5.

Now to show that this implies an orthogonal transformation matrix, sum over l:

$$\delta_{\bar{i}\bar{j}} = \Lambda^k_{\ \bar{i}} \Lambda^l_{\ \bar{j}} \delta_{kl}, \qquad \text{from eqn. (3.51)}$$

$$= \Lambda^k_{\ \bar{i}} \Lambda^k_{\ \bar{j}}. \qquad \text{after summing over } l \tag{3.52}$$

The RHS of eqn. (3.52) is the product of a matrix by its transpose, and for this to equal the identity matrix (i.e. the LHS), we require the matrix to be orthogonal.

And now you see clearly why we never learned about one-forms when working in Cartesians coordinates in Euclidean space, and why we called the gradient of a scalar field a vector.

3.23 (a) Prove that the set of all $\binom{M}{N}$ tensors for fixed M and N forms a vector space. (You must define addition of such tensors and their multiplication by numbers.)

Solution: This is like Exercise 3.2, but now we need to define what we mean by the addition of two $\binom{M}{N}$ tensors and the multiplication of an $\binom{M}{N}$ tensor by a scalar. So we are guided by eqns. (3.5) and (3.6) above. That is, we note that $\binom{M}{N}$ tensors produce real numbers that can be added like real numbers, so the generalization of eqns. (3.1) and (3.2) is trivial. The tensor **S** where

$$\mathbf{S} = \mathbf{P} + \mathbf{Q} \tag{3.53}$$

is defined to be that which gives the sum of the two values obtained by applying the input to **P** and **Q**. That is,

$$\mathbf{S}(\tilde{a}^1, \tilde{a}^2, \ldots, \tilde{a}^M; \vec{b}_1, \vec{b}_2, \ldots, \vec{b}_N) =$$
$$\mathbf{P}(\tilde{a}^1, \tilde{a}^2, \ldots, \tilde{a}^M; \vec{b}_1, \vec{b}_2, \ldots, \vec{b}_N) + \mathbf{Q}(\tilde{a}^1, \tilde{a}^2, \ldots, \tilde{a}^M; \vec{b}_1, \vec{b}_2, \ldots, \vec{b}_N), \tag{3.54}$$

where the notation $(\tilde{a}^1, \tilde{a}^2, \ldots, \tilde{a}^M; \vec{b}_1, \vec{b}_2, \ldots, \vec{b}_N)$ was used to represent the M one-form inputs and N vector inputs. The choice of one-forms first, we will see later, gives the basis in the order Schutz gave in Exercise 3.23(b). We have followed the convention that superscript integers are used as indices of different one-forms. That is, $\tilde{a}^1$ and $\tilde{a}^2$ are two different one-forms, not components of the same one-form.

Similarly, subscripts are used to denote different vectors. Some authors will use a different notation to avoid this ambiguity; Carroll (2004) uses parentheses around the indices, like $\tilde{a}^{(1)}$, to distinguish them from components.

In analogy with eqn. (3.2) we can define multiplication of an $\binom{M}{N}$ tensor by a scalar α

$$\mathbf{R} = \alpha \mathbf{P} \tag{3.55}$$

to be the tensor that, for a given input, gives just α times the real number produced by supplying the input to $\mathbf{P}$:

$$\mathbf{R}(\tilde{a}^1, \tilde{a}^2, \ldots, \tilde{a}^M; \vec{b}_1, \vec{b}_2, \ldots, \vec{b}_N) = \alpha \mathbf{P}(\tilde{a}^1, \tilde{a}^2, \ldots, \tilde{a}^M; \vec{b}_1, \vec{b}_2, \ldots, \vec{b}_N). \tag{3.56}$$

The set of $\binom{M}{N}$ tensors for fixed M and N forms a vector space by the same argument as given for Exercise 3.2. What do we mean by the zero $\binom{M}{N}$ tensor? This is the tensor that gives zero for any input,

$$\mathbf{0}(\tilde{a}^1, \tilde{a}^2, \ldots, \tilde{a}^M; \vec{b}_1, \vec{b}_2, \ldots, \vec{b}_N) = 0.$$

The set of $\binom{M}{N}$ tensors, with addition defined by eqns. (3.53) and (3.54) and scalar multiplication defined by eqns. (3.55) and (3.56) then meets axiom (1) in Appendix A: $\binom{M}{N}$ tensors form an abelian group with the operation of addition. Similarly the requirements of Axiom (2) in Appendix A are clearly met.

3.23   (b) Prove that a basis for the vector space formed from the set of all $\binom{M}{N}$ tensors for fixed M and N is the set:

$$\{\vec{e}_\alpha \otimes \vec{e}_\beta \otimes \cdots \otimes \vec{e}_\gamma \quad \otimes \quad \tilde{\omega}^\mu \otimes \tilde{\omega}^\nu \otimes \cdots \otimes \tilde{\omega}^\lambda\} \tag{3.57}$$

with M vectors labeled with $\alpha \ldots \gamma$ and N one-forms labeled $\mu \ldots \lambda$.

Solution: This is a nice question because it forces us to think about what we mean by a basis. The answer is a straightforward generalization of the argument for the basis of the $\binom{0}{2}$ tensors starting after Schutz Eq. (3.22) and ending with Eq. (3.26) of §3.4.

The notation is cumbersome because one needs to refer to M superscripts and N subscripts where M and N are arbitrary positive integers. In defining the basis eqn. (3.57) above Schutz has used a series of Greek letters like $\alpha \ldots \gamma$. Here we put subscript indices on the Greek letters $\alpha_1, \alpha_2, \ldots \alpha_M$ to be explicit about how many there are. Remember that each Greek letter index can take on four values, e.g. $\alpha_1 \in \{0, 1, 2, 3\}$ corresponding to the four dimensions.

As in Schutz Eq. (3.23) we write the $\binom{M}{N}$ tensor as a sum of components times the basis that we seek:

$$\mathbf{R} = R^{\alpha_1, \alpha_2, \ldots, \alpha_M}{}_{\beta_1, \beta_2, \ldots, \beta_N} \, \tilde{\omega}_{\alpha_1, \alpha_2, \ldots, \alpha_M}{}^{\beta_1, \beta_2, \ldots, \beta_N}. \tag{3.58}$$

And furthermore, the components correspond to the real values produced by applying the tensor to arguments that are the basis one-forms and basis vectors. So,

$$R^{\alpha_1,\alpha_2,...,\alpha_M}{}_{\beta_1,\beta_2,...,\beta_N} = \mathbf{R}(\tilde{\omega}^{\alpha_1}, \tilde{\omega}^{\alpha_2}, \ldots, \tilde{\omega}^{\alpha_M}; \vec{e}_{\beta_1}, \vec{e}_{\beta_2}, \ldots, \vec{e}_{\beta_N}), \tag{3.59}$$

which is the generalization of the formula given between Schutz Eq. (3.23) and Eq. (3.24). Now, we simply substitute the tensor eqn. (3.58) into eqn. (3.59) to obtain:

$$R^{\mu_1,...,\mu_M}{}_{\nu_1,...,\nu_N} = R^{\alpha_1,...,\alpha_M}{}_{\beta_1,...,\beta_N}\, \tilde{\omega}_{\alpha_1,...,\alpha_M}{}^{\beta_1,...,\beta_N}(\tilde{\omega}^{\mu_1}, \ldots, \tilde{\omega}^{\mu_M}; \vec{e}_{\nu_1}, \ldots, \vec{e}_{\nu_N}). \tag{3.60}$$

This implies the analogue to Schutz Eq. (3.24),

$$\tilde{\omega}_{\alpha_1,...,\alpha_M}{}^{\beta_1,...,\beta_N}(\tilde{\omega}^{\mu_1}, \ldots, \tilde{\omega}^{\mu_M}; \vec{e}_{\nu_1}, \ldots, \vec{e}_{\nu_N}) = \delta_{\alpha_1}{}^{\mu_1} \ldots \delta_{\alpha_M}{}^{\mu_M}\, \delta^{\beta_1}{}_{\nu_1} \ldots \delta^{\beta_N}{}_{\nu_N}. \tag{3.61}$$

Using eqn. (3.91) we identify

$$\delta^{\beta_1}{}_{\nu_1} = \tilde{\omega}^{\beta_1}(\vec{e}_{\nu_1}), \qquad \delta^{\beta_2}{}_{\nu_2} = \tilde{\omega}^{\beta_2}(\vec{e}_{\nu_2}), \qquad \ldots \qquad \delta^{\beta_N}{}_{\nu_N} = \tilde{\omega}^{\beta_N}(\vec{e}_{\nu_N}). \tag{3.62}$$

Based upon the dualism between vectors and one-forms, we identify:

$$\delta_{\alpha_1}{}^{\mu_1} = \vec{e}_{\alpha_1}(\tilde{\omega}^{\mu_1}), \qquad \delta_{\alpha_2}{}^{\mu_2} = \vec{e}_{\alpha_2}(\tilde{\omega}^{\mu_2}), \qquad \ldots \qquad \delta_{\alpha_M}{}^{\mu_M} = \vec{e}_{\alpha_M}(\tilde{\omega}^{\mu_M}).$$

So focusing on just the tensor, i.e. dropping the arguments, we are left with the basis that is the analogue to Schutz Eq. (3.25),

$$\tilde{\omega}_{\alpha_1,\alpha_2,...,\alpha_M}{}^{\beta_1,\beta_2,...,\beta_N} = \vec{e}_{\alpha_1} \otimes \vec{e}_{\alpha_2} \otimes \cdots \vec{e}_{\alpha_M} \otimes \tilde{\omega}^{\beta_1} \otimes \tilde{\omega}^{\beta_2} \cdots \otimes \tilde{\omega}^{\beta_N}, \tag{3.63}$$

consistent with eqn. (3.57).

Note we have introduced the idea of an outer product of N one-forms as a simple extension of the case when $N = 2$ introduced by Schutz at the start of §3.4. That is, the outer product of N one-forms,

$$\tilde{p}^1 \otimes \tilde{p}^2 \cdots \otimes \tilde{p}^N,$$

is simply the tensor that, when supplied with N vector inputs, say $\vec{A}_1, \vec{A}_2, \ldots, \vec{A}_N$, as arguments, produces that number that results from multiplying together each real number that results from applying $\tilde{p}^n$ to vector argument $\vec{A}_n$, i.e.

$$\tilde{p}^1 \otimes \tilde{p}^2 \ldots \otimes \tilde{p}^N(\vec{A}_1, \vec{A}_2, \ldots, \vec{A}_N) = \tilde{p}^1(\vec{A}_1)\tilde{p}^2(\vec{A}_2) \ldots \tilde{p}^N(\vec{A}_N).$$

3.24   (b) For the $\binom{1}{1}$ tensor whose components are $M^{\alpha}{}_{\beta}$, does it make sense to speak of its symmetric and antisymmetric parts? If so, define them. If not, say why [not].

Solution: A $\binom{1}{1}$ tensor can be represented by a matrix, so it is tempting to suppose that it has symmetric and antisymmetric parts. But a tensor is *not just a matrix*; it also has a basis composed of basis one-forms and vectors and this must be taken into account. Symmetry has to do with the interchange of the order of the arguments. For a $\binom{1}{1}$ tensor, one argument is a vector, the other a one-form. So they cannot be interchanged – if you did then the vector argument would not be contracted with the one-form basis and vice versa. For instance, it makes sense to assert

$$A_{\alpha\beta} u^\alpha v^\beta = A_{\beta\alpha} u^\alpha v^\beta,$$

but it does *not* make sense to assert

$$A^\alpha{}_\beta u_\alpha v^\beta = A^\beta{}_\alpha u_\alpha v^\beta.$$

The RHS of the above expression cannot be evaluated.

3.25 Show that if **A** is a $\binom{2}{0}$ tensor and **B** is a $\binom{0}{2}$ tensor, then

$$A^{\alpha\beta} B_{\alpha\beta}$$

is frame invariant, i.e. a scalar.

Solution: The simplest solution is analogous to that for the contraction of a one-form and a vector, given by Schutz in Eq. (3.10). Because we have twice the number of indices, we also have twice the number of Λs. Applying a Λ for each index in $A^{\alpha\beta} B_{\alpha\beta}$ we have:

$$A^{\alpha\beta} B_{\alpha\beta} = \left(A^{\bar\alpha\bar\beta} \Lambda^\alpha{}_{\bar\alpha} \Lambda^\beta{}_{\bar\beta} \right) \left(B_{\bar\mu\bar\nu} \Lambda^{\bar\mu}{}_\alpha \Lambda^{\bar\nu}{}_\beta \right). \qquad \text{used Schutz Eqs. (2.7) and (3.9)}$$

$$(3.64)$$

Deciding which Λ was easy; the correct one has the indices in the right place. For instance, we used eqn. (3.87) to tell us how to transform the lower "covariant" indices on $B_{\bar\mu\bar\nu}$, and Schutz Eq. (2.7) for the upper indices. If you are concerned that Schutz Eq. (2.7) was a transformation from $\mathcal{O}$ to $\overline{\mathcal{O}}$ while on the RHS of eqn. (3.64) we are transforming in the other sense, see the discussion on inverse transformations after Schutz Eq. (2.18). Carrying on,

$$A^{\alpha\beta} B_{\alpha\beta} = A^{\bar\alpha\bar\beta} B_{\bar\mu\bar\nu} \left(\Lambda^\alpha{}_{\bar\alpha} \Lambda^{\bar\mu}{}_\alpha \right) \left(\Lambda^\beta{}_{\bar\beta} \Lambda^{\bar\nu}{}_\beta \right) \qquad \text{rearranged eqn. (3.64) above}$$

$$= A^{\bar\alpha\bar\beta} B_{\bar\mu\bar\nu} \left(\delta_{\bar\alpha}{}^{\bar\mu} \right) \left(\delta_{\bar\beta}{}^{\bar\nu} \right) \qquad \text{sum on } \alpha \text{ and } \beta, \text{ used eqn. (3.23)}$$

$$= A^{\bar\mu\bar\nu} B_{\bar\mu\bar\nu}. \qquad \text{sum on } \bar\alpha \text{ and } \bar\beta$$

$$(3.65)$$

And we see that the contraction of these two tensors in an arbitrary frame $\overline{\mathcal{O}}$ gives the same result as in the original frame $\mathcal{O}$.

See SP3.10 for a related problem.

3.27 (a) Suppose **A** is an antisymmetric $\binom{2}{0}$ tensor. Show that $\{A_{\alpha\beta}\}$, obtained by lowering indices by using the metric tensor, are components of an antisymmetric $\binom{0}{2}$ tensor.

Solution:

$$A_{\alpha\beta} = \eta_{\alpha\mu}\eta_{\beta\nu}A^{\mu\nu}, \qquad \text{used Schutz Eq. (3.56)} \qquad (3.66)$$

So in matrix notation:

$$(A_{\alpha\beta}) = \begin{pmatrix} -1 & 0 & 0 & 0 \\ 0 & 1 & 0 & 0 \\ 0 & 0 & 1 & 0 \\ 0 & 0 & 0 & 1 \end{pmatrix} \begin{pmatrix} 0 & a_{12} & a_{13} & a_{14} \\ -a_{12} & 0 & a_{23} & a_{24} \\ -a_{13} & -a_{23} & 0 & a_{34} \\ -a_{14} & -a_{24} & -a_{34} & 0 \end{pmatrix} \begin{pmatrix} -1 & 0 & 0 & 0 \\ 0 & 1 & 0 & 0 \\ 0 & 0 & 1 & 0 \\ 0 & 0 & 0 & 1 \end{pmatrix}$$

$$= \begin{pmatrix} -1 & 0 & 0 & 0 \\ 0 & 1 & 0 & 0 \\ 0 & 0 & 1 & 0 \\ 0 & 0 & 0 & 1 \end{pmatrix} \begin{pmatrix} 0 & a_{12} & a_{13} & a_{14} \\ a_{12} & 0 & a_{23} & a_{24} \\ a_{13} & -a_{23} & 0 & a_{34} \\ a_{14} & -a_{24} & -a_{34} & 0 \end{pmatrix}$$

$$= \begin{pmatrix} 0 & -a_{12} & -a_{13} & -a_{14} \\ a_{12} & 0 & a_{23} & a_{24} \\ a_{13} & -a_{23} & 0 & a_{34} \\ a_{14} & -a_{24} & -a_{34} & 0 \end{pmatrix}. \qquad (3.67)$$

The signs of the elements in the first row and first column were changed but the matrix is still antisymmetric.

3.27 (b) Suppose $V^{\alpha} = W^{\alpha}$. Prove that $V_{\alpha} = W_{\alpha}$.

Solution: Recall we lower the indices using the metric so that,

$$V_{\alpha} = \eta_{\alpha\beta}V^{\beta}, \qquad W_{\alpha} = \eta_{\alpha\beta}W^{\beta}. \qquad \text{Schutz Eq. (3.39)} \qquad (3.68)$$

Subtracting,

$$V_{\alpha} - W_{\alpha} = \eta_{\alpha\beta}V^{\beta} - \eta_{\alpha\gamma}W^{\gamma}$$
$$= \eta_{\alpha\beta}\left(V^{\beta} - W^{\beta}\right) \qquad \text{relabeled dummy index}$$
$$= 0.$$

This proves the desired result, $V^{\alpha} = W^{\alpha} \implies V_{\alpha} = W_{\alpha}$.

3.29 Prove that tensor differentiation obeys the Leibniz (product) rule:

$$\nabla(\mathbf{A} \otimes \mathbf{B}) = (\nabla\mathbf{A}) \otimes \mathbf{B} + \mathbf{A} \otimes (\nabla\mathbf{B}).$$

[See SP3.15 for a more concrete version of this question whose solution emphasizes the techniques of Schutz §3.8.]

Solution: Let $\binom{m}{n}$ and $\binom{p}{q}$ be the ranks of $\mathbf{A}$ and $\mathbf{B}$. Then $\mathbf{A} \otimes \mathbf{B}$ is a mapping from $(n + q)$ vectors and $(m + p)$ one-forms into the reals. We can write this as:

$$\left(\mathbf{A} \otimes \mathbf{B}\right)(\mathbf{u}, \mathbf{v}) = R, \tag{3.69}$$

where R is a real number, $\mathbf{u}$ is a set of n vectors and m one-forms, and $\mathbf{v}$ is a set of q vectors and p one-forms. Suppose $\mathbf{u}$ and $\mathbf{v}$ are constants; they don't vary over the spacetime. It is easy to take the gradient of R:

$$\nabla R = \nabla\left(\mathbf{A} \otimes \mathbf{B}\right)(\mathbf{u}, \mathbf{v}) = \nabla(\mathbf{A}(\mathbf{u}) \cdot \mathbf{B}(\mathbf{v})). \tag{3.70}$$

On the RHS the "$\cdot$" is just ordinary multiplication between the two real scalars $\mathbf{A}(\mathbf{u})$ and $\mathbf{B}(\mathbf{v})$. So using the ordinary rules of differential calculus we have:

$$\nabla R = \nabla(\mathbf{A}(\mathbf{u}))\, \mathbf{B}(\mathbf{v}) + \mathbf{A}(\mathbf{u})\, \nabla(\mathbf{B}(\mathbf{v})). \tag{3.71}$$

Now we ask what tensor, when applied to the set of constant vectors and one-forms $\mathbf{u}$ and $\mathbf{v}$, gives ∇R? Comparing eqn. (3.70) and eqn. (3.71) above, this tensor must be:

$$\nabla\left(\mathbf{A} \otimes \mathbf{B}\right) = \left((\nabla\mathbf{A}) \otimes \mathbf{B} + \mathbf{A} \otimes (\nabla\mathbf{B})\right). \tag{3.72}$$

This is the unique solution because a tensor is a linear mapping from its arguments into the reals.

3.31 Consider a timelike unit four-vector $\vec{U}$, and the tensor $\mathbf{P}$ whose components are given by

$$P_{\mu\nu} = \eta_{\mu\nu} + U_\mu U_\nu.$$

(a) Show that $\mathbf{P}$ is a projection operator that projects an arbitrary vector $\vec{V}$ into one orthogonal to $\vec{U}$. That is, show that the vector $\vec{V}_\perp$ whose components are

$$V_\perp^\alpha = P^\alpha{}_\beta V^\beta = (\eta^\alpha{}_\beta + U^\alpha U_\beta)V^\beta$$

is

  (i) orthogonal to $\vec{U}$, and
  (ii) unaffected by $\mathbf{P}$:

$$V_{\perp\perp}^\alpha := P^\alpha{}_\beta V_\perp^\beta = V_\perp^\alpha.$$

Solution: Recall that just before Schutz Eq. (2.27) we were told that a timelike unit four-vector has magnitude -1, so

$$\vec{U} \cdot \vec{U} = -1. \tag{3.73}$$

  (i) To show that $\vec{V}_\perp$ is orthogonal to $\vec{U}$ we simply show that their scalar (dot) product must vanish:

$$\eta_{\alpha\mu} V_{\perp}^{\alpha} U^{\mu} = \eta_{\alpha\mu} U^{\mu} (\eta^{\alpha}{}_{\beta} + U^{\alpha} U_{\beta}) V^{\beta} \qquad \text{used Schutz Eq. (3.1)}$$
$$= U_{\alpha} (\eta^{\alpha}{}_{\beta} + U^{\alpha} U_{\beta}) V^{\beta}, \qquad \text{used eqn. (3.68)}$$
$$= U_{\alpha} (\delta^{\alpha}{}_{\beta} + U^{\alpha} U_{\beta}) V^{\beta}, \qquad \text{used Schutz Eq. (3.60)}$$
$$= (U_{\beta} + U_{\alpha} U^{\alpha} U_{\beta}) V^{\beta}, \qquad \text{summed over } \alpha$$
$$= (U_{\beta} - U_{\beta}) V^{\beta}. \qquad \text{used eqn. (3.73)}$$
$$= 0. \qquad (3.74)$$

(ii) Similarly for showing that $\vec{V}_{\perp}$ is unaffected by **P** we contract them, giving

$$P^{\mu}{}_{\alpha} V_{\perp}^{\alpha} = (\eta^{\mu}{}_{\alpha} + U^{\mu} U_{\alpha}) V_{\perp}^{\alpha}$$
$$= (\eta^{\mu}{}_{\alpha} + U^{\mu} U_{\alpha})(\eta^{\alpha}{}_{\beta} + U^{\alpha} U_{\beta}) V^{\beta}$$
$$= (\eta^{\mu}{}_{\alpha} \eta^{\alpha}{}_{\beta} + U^{\mu} U_{\alpha} \eta^{\alpha}{}_{\beta} + \eta^{\mu}{}_{\alpha} U^{\alpha} U_{\beta} + U^{\mu} U_{\alpha} U^{\alpha} U_{\beta}) V^{\beta}$$
$$= (\eta^{\mu}{}_{\beta} + U^{\mu} U_{\beta} + U^{\mu} U_{\beta} + (-1) U^{\mu} U_{\beta}) V^{\beta} \qquad \text{summed over } \alpha$$
$$= (\eta^{\mu}{}_{\beta} + U^{\mu} U_{\beta}) V^{\beta}$$
$$= V_{\perp}^{\mu}. \qquad (3.75)$$

(b) Show that for an arbitrary non-null vector $\vec{q}$, the tensor that projects orthogonally to it has components

$$\eta_{\mu\nu} - \frac{q_{\mu} q_{\nu}}{q^{\alpha} q_{\alpha}},$$

How does this fail for null vectors? How does this relate to the definition of **P**?

Hint: Based upon (a) we interpret "projects orthogonally" as meaning that this tensor converts vectors into one-forms that are orthogonal to $\vec{q}$. (Note that the given tensor produces a one-form from a vector input because it is a $\binom{0}{2}$ tensor.) Much like in (a) we simply apply the given tensor to an arbitrary vector, say $\vec{s}$.

(c) Show that **P** defined above is the metric tensor for vectors perpendicular to $\vec{U}$:

$$\mathbf{P}(\vec{V}_{\perp}, \vec{W}_{\perp}) = \mathbf{g}(\vec{V}_{\perp}, \vec{W}_{\perp}) = \vec{V}_{\perp} \cdot \vec{W}_{\perp}. \qquad (3.76)$$

Solution: The metric tensor plays the role of forming the scalar product between any two vectors,

$$\mathbf{g}(\vec{A}, \vec{B}) = \vec{A} \cdot \vec{B}, \qquad \text{Schutz Eq. (3.3)} \qquad (3.77)$$

so eqn. (3.76) is a special case of this. Our strategy here will be to show that **P** plays this role for vectors orthogonal to $\vec{U}$. We find

$$\mathbf{P}(\vec{V}_\perp, \vec{W}_\perp) = P_{\alpha\beta} \, V_\perp^\alpha \, W_\perp^\beta = (\eta_{\alpha\beta} + U_\alpha \, U_\beta) \, W_\perp^\beta \, V_\perp^\alpha,$$

$$= \eta_{\alpha\beta} \, W_\perp^\beta \, V_\perp^\alpha + U_\alpha V_\perp^\alpha \, U_\beta W_\perp^\beta, \qquad \text{rearranged}$$

$$= \eta_{\alpha\beta} \, W_\perp^\beta \, V_\perp^\alpha, \qquad \text{using (a) (i) above,}$$

$$= \vec{W}_\perp \cdot \vec{V}_\perp, \qquad \text{used Schutz Eq. (3.1)}$$

$$= \mathbf{g}(\vec{V}_\perp, \vec{W}_\perp). \qquad \text{used eqn. (3.76)}$$

3.33   The result of Exercise 3.32(c) establishes that Lorentz transformations form a group, represented by multiplication of their matrices. This is called the *Lorentz group*, denoted by $L(4)$ or $O(1,3)$. [This exercise requires familiarity with the material in Exercise 3.32.]

(a) Find the matrices of the identity element of the Lorentz group and of the element inverse to that whose matrix is implicit in eqn. (1.24).

Solution: The term "identity element" is a very general concept, beyond just group theory (Hassani, 1999). For the Lorentz group the binary operation is matrix multiplication, and we seek the matrix $\mathbf{I}$ such that $\mathbf{I}\,\mathbf{L} = \mathbf{L}$ for all matrices $\mathbf{L}$. Clearly the 4×4 identity matrix $\mathbf{I}$ meets this requirement. Note that $\Lambda(\mathbf{v} = \mathbf{0}) = \mathbf{I}$.

The implicit matrix in eqn. (1.24) was found in Exercise 1.20 to be

$$\begin{pmatrix} \gamma & -v\gamma & 0 & 0 \\ -v\gamma & \gamma & 0 & 0 \\ 0 & 0 & 1 & 0 \\ 0 & 0 & 0 & 1 \end{pmatrix},$$

where $\gamma = 1/\sqrt{1 - v^2}$.

Its inverse is

$$\begin{pmatrix} \gamma & v\gamma & 0 & 0 \\ v\gamma & \gamma & 0 & 0 \\ 0 & 0 & 1 & 0 \\ 0 & 0 & 0 & 1 \end{pmatrix},$$

which is obvious on physical grounds, and can be easily confirmed by multiplication:

$$\begin{pmatrix} \gamma & v\gamma & 0 & 0 \\ v\gamma & \gamma & 0 & 0 \\ 0 & 0 & 1 & 0 \\ 0 & 0 & 0 & 1 \end{pmatrix} \begin{pmatrix} \gamma & -v\gamma & 0 & 0 \\ -v\gamma & \gamma & 0 & 0 \\ 0 & 0 & 1 & 0 \\ 0 & 0 & 0 & 1 \end{pmatrix} = \begin{pmatrix} \gamma^2(1 - v^2) & 0 & 0 & 0 \\ 0 & \gamma^2(1 - v^2) & 0 & 0 \\ 0 & 0 & 1 & 0 \\ 0 & 0 & 0 & 1 \end{pmatrix}$$

$$= \begin{pmatrix} 1 & 0 & 0 & 0 \\ 0 & 1 & 0 & 0 \\ 0 & 0 & 1 & 0 \\ 0 & 0 & 0 & 1 \end{pmatrix}. \qquad (3.78)$$

3.33   (b) Prove that the determinant of any matrix representing a Lorentz transformation is ± 1.

Solution: It is easy to show that the determinant of the Lorentz transformation associated with the "velocity boost" v in the x-direction is $+1$. But what about more general Lorentz transformations, e.g. for velocity components in different directions? In Chapter 1 we did find the most general Lorentz transformation for arbitrarily oriented velocity. But it would be messy to find the determinant. It turns out to be even easier to work with generalized Lorentz transformations defined by

$$\eta_{\bar{\alpha}\bar{\beta}} = \Lambda^{\mu}{}_{\bar{\alpha}}\Lambda^{\nu}{}_{\bar{\beta}}\eta_{\mu\nu}. \qquad \text{Schutz Eq. (3.71)} \qquad (3.79)$$

Let η and L be the matrices associated with tensors $\eta_{\mu\nu}$ and $\Lambda^{\mu}{}_{\bar{\alpha}}$ respectively.

$$\eta = L^T \eta L \qquad \qquad \text{eqn. (3.79) in matrix notation}$$

$$\det(\eta) = \det(L^T \eta L) \qquad \qquad \text{taken determinant of both sides}$$

$$= \det(L^T)\,\det(\eta)\,\det(L) \qquad \text{used (Hassani, 1999, Theorem 3.5.7)}$$

$$-1 = -\det(L)^2 \qquad \qquad \text{used (Hassani, 1999, Theorem 3.5.2)}$$

$$\det(L) = \pm 1. \qquad \qquad (3.80)$$

We have used properties of the determinant of a matrix, in particular $\det(A^T) = \det(A)$ for any (square) matrix (Hassani, 1999, Theorem 3.5.2) and $\det(AB) = \det(A)\det(B)$ for any square matrices A and B (Hassani, 1999, Theorem 3.5.7).

3.33   (c) Prove that those elements whose matrices have determinant $+1$ form a subgroup, while those with -1 do not.

Solution: The axioms of a group were given in Exercise 3.2 above, and can be found in books on mathematical physics (e.g. Schutz, 1980). Here we only have to show that the closure property applies to the subgroup, since all three other properties of groups are automatically satisfied by virtue of the fact that the elements are members of the Lorentz group. By the closure property we mean that if A and B are members of the group, then AB and BA are members of the group too, where multiplication is understood as matrix multiplication. Clearly if $\det(A) = 1$, and $\det(B) = 1$, then $\det(C) = \det(AB) = 1$, and this forms a subgroup. But if $\det(A) = -1$, and $\det(B) = -1$, then $\det(C) = \det(AB) = 1$. Thus C is not a member and the set of matrices with determinant of -1 do not form a subgroup because they fail to meet the closure axiom.

3.33   (d) The three-dimensional orthogonal group $O(3)$ is the analogous group for the metric of three-dimensional Euclidean space. In Exercise 3.20(b), we saw that it was represented by the orthogonal matrices. Show that the orthogonal matrices do form a group, and then show that $O(3)$ is (isomorphic to) a subgroup of $L(4)$.

Solution: The identity matrix is orthogonal and matrix multiplication enjoys associativity and invertibility (Schutz, 1980). We need to show that the set of 3D orthogonal matrices also meets the fourth axiom of a group, i.e. it is closed under the binary operation of matrix multiplication. That is, if A and B are elements of the set, then $C = AB$ is also an element.

$$
\begin{aligned}
C^T C &= (AB)^T (AB) \\
&= (B^T A^T)(A\,B) && \text{used property of transpose} \\
&= B^T\,(A^T A)\,B && \text{used associativity} \\
&= B^T\,(I)\,B && \text{used orthogonality of } A \\
&= B^T B \\
&= I, && \text{used orthogonality of } B
\end{aligned}
$$

which implies the orthogonality of C. Thus the set of 3D orthogonal matrices is closed, and forms a group.

The group $O(3)$ represents all the rotations of the spatial coordinate axes. This is a special case of the general Lorentz transformation wherein the relative velocity of the two coordinate systems is zero $\Lambda(\mathbf{v} = \mathbf{0})$. So we can just set $\mathbf{v} = 0, \gamma(0) = 1$ in the general Lorentz transformation. (See Supplementary Problem SP2.3.) This corresponds to matrices of the form,

$$
(R) = \begin{pmatrix} 1 & 0 & 0 & 0 \\ 0 & & & \\ 0 & & (A) & \\ 0 & & & \end{pmatrix}. \tag{3.81}
$$

To confirm that such transformations with matrices (R) both satisfy eqn. (3.79) and form a one-to-one correspondence with those of an orthogonal matrix, we observe the following. Inserting (R) into eqn. (3.79) gives the condition for orthogonal 3D matrices given in Exercise 3.20:

$$
\delta_{\bar{i}\bar{j}} = \delta_{ij} A_{\bar{i}}{}^{i} A_{\bar{j}}{}^{j}.
$$

This is more transparent in matrix notation:

$$
(\eta) = (R)^T (\eta)(R) = \begin{pmatrix} 1 & 0 & 0 & 0 \\ 0 & & & \\ 0 & & (A)^T & \\ 0 & & & \end{pmatrix} \begin{pmatrix} -1 & 0 & 0 & 0 \\ 0 & 1 & 0 & 0 \\ 0 & 0 & 1 & 0 \\ 0 & 0 & 0 & 1 \end{pmatrix} \begin{pmatrix} 1 & 0 & 0 & 0 \\ 0 & & & \\ 0 & & (A) & \\ 0 & & & \end{pmatrix}.
$$

$$\tag{3.82}$$

Clearly the condition on (A) is just

$$
I = \begin{pmatrix} 1 & 0 & 0 \\ 0 & 1 & 0 \\ 0 & 0 & 1 \end{pmatrix} = (A)^T I (A) = (A)^T (A).
$$

In other words, the transpose of (A) is its inverse, i.e. it must be orthogonal. So matrices of the form (R) in eqn. (3.81) above form a subset of the Lorentz group $L(4)$.

To show that matrices of the form (R) form a subgroup of $L(4)$, we observe that

$$
\begin{pmatrix} 1 & 0 & 0 & 0 \\ 0 & & & \\ 0 & & (A) & \\ 0 & & & \end{pmatrix}
\begin{pmatrix} 1 & 0 & 0 & 0 \\ 0 & & & \\ 0 & & (B) & \\ 0 & & & \end{pmatrix}
=
\begin{pmatrix} 1 & 0 & 0 & 0 \\ 0 & & & \\ 0 & & (C) & \\ 0 & & & \end{pmatrix},
$$

where $(C) = (A)(B)$. The closure property of (R) follows from that of orthogonal matrices.

In summary, the orthogonal 3D matrices form a group that we identify with the set of rotations of the three spatial coordinate axes. For each orthogonal matrix (A) we can form a 4D matrix (R) as shown in eqn. (3.81) above, and we just showed that they form a subgroup of $L(4)$.

3.2 Supplementary problems

SP 3.1 Compare the two equations below for the square of a one-form:

$$\tilde{p}^2 = \vec{p}^2 = \eta_{\alpha\beta}\, p^\alpha p^\beta, \qquad\qquad \text{Schutz Eq. (3.47)}$$

$$\tilde{p}^2 = \eta^{\alpha\beta}\, p_\alpha p_\beta. \qquad\qquad \text{Schutz Eq. (3.50)} \qquad (3.83)$$

It is tempting to generalize the idea that we equate expressions of the form

$$A^\alpha\, B_\alpha = A_\alpha\, B^\alpha.$$

Prove that this is indeed true. That is, prove that when a dummy index is repeated as a superscript and subscript in the same term (implying a sum via the Einstein summation convention), then we can interchange the upper and lower indices. Show that in general, $A^\alpha\, B_\alpha = A_\alpha\, B^\alpha$.

Solution

$$
\begin{aligned}
A^\alpha\, B_\alpha &= \eta^{\alpha\sigma}\, A_\sigma\, B_\alpha & \text{cf. Schutz Eq. (3.58)} \\
&= A_\sigma\, (\eta^{\alpha\sigma}\, B_\alpha) & \text{rearranged order} \\
&= A_\sigma\, B^\sigma & \text{cf. Schutz Eq. (3.58)} \\
&= A_\alpha\, B^\alpha. & \text{relabeled dummy index} \qquad (3.84)
\end{aligned}
$$

SP 3.2   Show that in general, $A^{\alpha\beta} B_{\gamma\alpha} = A_\alpha{}^\beta B_\gamma{}^\alpha$. That is, we can lower a dummy index on one term if we raise this dummy index on the corresponding term.

SP 3.3   Establish how to obtain the components of the addition of two one-forms. That is, given $\tilde{s} = \tilde{p} + \tilde{q}$, find s_α in terms of p_β and q_γ.

Solution

Intuitively one might guess we just add the components, $s_\alpha = p_\alpha + q_\alpha$. Indeed this is so, but to establish this rigorously we start with the definition of addition in eqns. (3.1, 3.2): if $\tilde{s} = \tilde{p} + \tilde{q}$ then $\tilde{s}(\vec{A}) = \tilde{p}(\vec{A}) + \tilde{q}(\vec{A})$ for all $\vec{A}$. Suppose

$$\vec{A} \underset{\mathcal{O}}{\rightarrow} (1,0,0,0).$$

Then

$$\tilde{p}(\vec{A}) = p_\alpha A^\alpha = p_0, \quad \tilde{q}(\vec{A}) = q_\alpha A^\alpha = q_0, \quad \tilde{s}(\vec{A}) = s_\alpha A^\alpha = s_0 = p_0 + q_0.$$

Similarly for $\vec{A} \underset{\mathcal{O}}{\rightarrow} (0,1,0,0)$, etc. This establishes that eqns. (3.1, 3.2) implies that to add two one-forms one just adds the components.

SP 3.4   Derive

$$\tilde{p} \cdot \tilde{q} = -p_0 q_0 + p_1 q_1 + p_2 q_2 + p_3 q_3 \qquad\qquad \text{eqn. (3.44)}$$

from

$$\tilde{p} \cdot \tilde{q} = \frac{1}{2}\left[(\tilde{p} + \tilde{q})^2 - \tilde{p}^2 - \tilde{q}^2\right] \qquad\qquad \text{eqn. (3.45)}$$

$$\tilde{p}^2 = -(p_0)^2 + (p_1)^2 + (p_2)^2 + (p_3)^2. \qquad \text{Schutz Eq. (3.51)}$$

Hint: You can add the two one-forms component-wise, see SP3.3.

SP 3.5   We often encounter a pair of expressions, involving a sum over several terms, that must be equal. And we make the argument that because of some arbitrariness of the quantities involved the equality extends to the individual terms. For example, in the solution to Exercise 3.7 above we relabeled dummy indices, because the equality in eqn. (3.20) had to hold for arbitrary $\tilde{p}$, and then eliminate p_α. Justify this step.

Solution

The sums on the two sides of the equal sign are equal in eqn. (3.20)

$$p_\alpha \tilde{\omega}^\alpha = \Lambda^\beta{}_{\bar{\alpha}} p_\beta \tilde{\omega}^{\bar{\alpha}}, \qquad\qquad \text{see solution to Exercise 3.7} \qquad (3.85)$$

but we require further argumentation to say the individual terms are equal. We note that $\tilde{p}$ is arbitrary, and therefore we can imagine the case where all components of $\tilde{p}$ are zero but one, $\tilde{p} = a\tilde{\omega}^0$ say, and then:

$$a\tilde{\omega}^0 = \Lambda^0{}_{\bar{\alpha}}\, a\tilde{\omega}^{\bar{\alpha}}$$

$$\tilde{\omega}^0 = \Lambda^0{}_{\bar{\alpha}}\, \tilde{\omega}^{\bar{\alpha}}. \tag{3.86}$$

And similarly, we can imagine $\tilde{p} = a\tilde{\omega}^1$, leading to $\tilde{\omega}^1 = \Lambda^1{}_{\bar{\alpha}}\, \tilde{\omega}^{\bar{\alpha}}$. Repeating this argument for all four components we see that indeed it is valid to set $\alpha = \beta$ in eqn. (3.20) and eqn. (3.85) above and then eliminate p_α.

SP 3.6 Suppose B^μ is a set of four numbers and we are told that for arbitrary four-vector $\vec{A}$, $A_\mu B^\mu$ is frame invariant. Prove this implies that B^μ is also a four-vector.

SP 3.7 From the expression for the transformation of one-form components under a change of basis:

$$p_{\bar{\beta}} = \Lambda^\alpha{}_{\bar{\beta}}\, p_\alpha, \qquad\qquad \text{Schutz Eq. (3.9)} \qquad\qquad (3.87)$$

we can just relabel indices and arrive at

$$p_\mu = \Lambda^{\bar{\nu}}{}_\mu\, p_{\bar{\nu}}. \tag{3.88}$$

Let's practice our manipulation skills by showing that one can arrive at eqn. (3.88) starting with its RHS and using eqn. (3.87) and the fact that $\Lambda^\alpha{}_{\bar{\alpha}}$ is the inverse of the Lorentz transformation $\Lambda^{\bar{\alpha}}{}_\beta$, i.e. eqn. (3.23). Also prove eqn. (3.88) explicitly using a procedure analogous to that used to justify Schutz Eq. (3.9), wherein one-forms are considered functions taking vectors as arguments and producing reals.

Solution

We start with the RHS of eqn. (3.88) and work towards arriving at p_μ.

$$p_{\bar{\nu}}\, \Lambda^{\bar{\nu}}{}_\mu = p_\alpha\, \Lambda^\alpha{}_{\bar{\nu}}\Lambda^{\bar{\nu}}{}_\mu \qquad\qquad \text{used eqn. (3.87)}$$

$$= \delta^\alpha{}_\mu\, p_\alpha \qquad\qquad \text{used eqn. (3.23)}$$

$$= p_\mu. \qquad\qquad \text{summed over } \bar{\nu} \tag{3.89}$$

For the second demonstration start with the definition of the LHS of eqn. (3.88):

$$p_\mu = \tilde{p}(\vec{e}_\mu) \qquad\qquad \text{definition eqn. (3.17)}$$

$$= \tilde{p}(\Lambda^{\bar{\beta}}{}_\mu\, \vec{e}_{\bar{\beta}}) \qquad\qquad \text{used eqn. (2.8)}$$

$$= \Lambda^{\bar{\beta}}{}_\mu\, \tilde{p}(\vec{e}_{\bar{\beta}}) \qquad\qquad \text{linear in the arguments}$$

$$= \Lambda^{\bar{\beta}}{}_\mu\, p_{\bar{\beta}}. \qquad\qquad \text{used definition eqn. (3.17)} \tag{3.90}$$

SP 3.8   (a) The term *contraction* was introduced by Schutz as the operation of a one-form on a vector. One will also encounter the phrase *contraction over a pair of indices* designating an operation on a $\binom{n}{m}$ tensor that reduces the ranks to $\binom{n-1}{m-1}$ and is obtained by setting an upper index equal to a lower index and performing the required summation (Wald, 1984; Rindler, 2006; Lawden, 2002; Carroll, 2004; Hobson et al., 2006). For example, for a $\binom{1}{1}$ tensor $R^{\alpha}{}_{\beta}$ the only contraction possible is $R = R^{\alpha}{}_{\alpha}$. Because this amounts to taking the trace of the corresponding matrix, some textbooks refer to the *trace* of a $\binom{0}{2}$ tensor, $R_{\alpha\beta}$, as defined by $R \equiv \eta^{\alpha\sigma} R_{\sigma\alpha} = R^{\alpha}{}_{\alpha}$ (Carroll, 2004). Show that the trace of a $\binom{0}{2}$ tensor is a scalar and therefore invariant under Lorentz transformations. Sometimes the term *Lorentz scalar* is used to emphasize its invariance under Lorentz transformations.

(b) Show that the trace of an antisymmetric tensor $\binom{0}{2}$ tensor, say $F_{\alpha\beta}$, vanishes.

SP 3.9   (a) Suppose we are told that the basis vector $\vec{e}_t \underset{\mathcal{O}'}{\rightarrow} (1, 1, 0, 0)$ in some coordinate system $\mathcal{O}'$. Without knowing the other basis vectors $\{\vec{e}_x, \vec{e}_y, \vec{e}_z\}$, is it possible to find the dual basis $\tilde{\omega}^t$ using

$$\tilde{\omega}^{\alpha}(\vec{e}_{\beta}) = \delta^{\alpha}{}_{\beta}? \qquad\qquad \text{Schutz Eq. (3.12)} \qquad\qquad (3.91)$$

> **Solution**
>
> No, given $\vec{e}_t$ alone we do not have enough information to find its dual basis one-form $\tilde{\omega}^t$. We must solve a system of equations
>
> $$\tilde{\omega}^t(\vec{e}_{\beta}) = \delta^t{}_{\beta}, \Rightarrow \begin{cases} \tilde{\omega}^t(\vec{e}_t) = 1 = \tilde{\omega}^t_{t'} + \tilde{\omega}^t_{x'}, \\ \tilde{\omega}^t(\vec{e}_x) = 0, \\ \tilde{\omega}^t(\vec{e}_y) = 0, \\ \tilde{\omega}^t(\vec{e}_z) = 0. \end{cases}$$
>
> The first equation has two unknowns, $\tilde{\omega}^t_{t'}$ and $\tilde{\omega}^t_{x'}$, and cannot be solved without further information. And the other three equations cannot help because they all involve the unknown basis vectors.

(b) Suppose in addition we are told that $\vec{e}_x \underset{\mathcal{O}'}{\rightarrow} (-1, 1, 0, 0)$, $\vec{e}_y \underset{\mathcal{O}'}{\rightarrow} (0, 0, 1, 0)$, and $\vec{e}_z \underset{\mathcal{O}'}{\rightarrow} (0, 0, 0, 1)$. Find the dual basis one-form $\tilde{\omega}^t$.

> **Solution**
>
> Now we have enough information to find its dual basis one-form $\tilde{\omega}^t$. We must solve a system of equations
>
> $$\tilde{\omega}^t(\vec{e}_{\beta}) = \delta^t{}_{\beta}, \Rightarrow \begin{cases} \tilde{\omega}^t(\vec{e}_t) = 1 = \tilde{\omega}^t_{t'} + \tilde{\omega}^t_{x'}, \\ \tilde{\omega}^t(\vec{e}_x) = 0 = -\tilde{\omega}^t_{t'} + \tilde{\omega}^t_{x'}, \\ \tilde{\omega}^t(\vec{e}_y) = 0 = \tilde{\omega}^t_{y'}, \\ \tilde{\omega}^t(\vec{e}_z) = 0 = \tilde{\omega}^t_{z'}. \end{cases}$$

The complete system of equations above can be solved immediately (in this case by inspection), giving

$$\tilde{\omega}' \underset{\mathcal{O}'}{\rightarrow} \left(\frac{1}{2}, \frac{1}{2}, 0, 0\right).$$

SP 3.10 Show that if $\mathbf{A}$ is a $\binom{2}{0}$ tensor and $\mathbf{B}$ is a $\binom{0}{2}$ tensor, then

$$\mathbf{A}(\mathbf{B}) = A^{\alpha\beta} B_{\alpha\beta}$$

by writing $\mathbf{A}$ and $\mathbf{B}$ in terms of their components and bases and then performing the contraction, justifying each step. Give a brief argument why $A^{\alpha\beta} B_{\alpha\beta}$ should be Lorentz-invariant, as we confirmed in Exercise 3.25. See also SP3.11 and SP3.12 below.

SP 3.11 Show that a $\binom{2}{0}$ tensor $\mathbf{A}$ is Lorentz frame invariant by transforming the components and bases as described in Schutz §3.6.

Solution

We can demonstrate the frame-invariance of $\mathbf{A}$ as follows:

$$\mathbf{A} = A^{\bar{\alpha}\bar{\beta}} \vec{e}_{\bar{\alpha}} \otimes \vec{e}_{\bar{\beta}} \qquad\qquad \text{basis described in §3.6}$$

$$= \left(A^{\alpha\beta} \Lambda^{\bar{\alpha}}{}_{\alpha} \Lambda^{\bar{\beta}}{}_{\beta}\right) \vec{e}_{\bar{\alpha}} \otimes \vec{e}_{\bar{\beta}} \qquad\qquad \text{one } \Lambda \text{ for each index, cf. Schutz Eq. (3.55)}$$

$$= \left(A^{\alpha\beta} \Lambda^{\bar{\alpha}}{}_{\alpha} \Lambda^{\bar{\beta}}{}_{\beta}\right) \left(\Lambda^{\mu}{}_{\bar{\alpha}} \vec{e}_{\mu}\right) \otimes \left(\Lambda^{\nu}{}_{\bar{\beta}} \vec{e}_{\nu}\right) \qquad \text{transform basis too, Schutz Eq. (2.14)}$$

$$= A^{\alpha\beta} \left(\Lambda^{\bar{\alpha}}{}_{\alpha} \Lambda^{\bar{\beta}}{}_{\beta} \Lambda^{\mu}{}_{\bar{\alpha}} \Lambda^{\nu}{}_{\bar{\beta}}\right) \vec{e}_{\mu} \otimes \vec{e}_{\nu} \qquad\qquad \text{rearranged}$$

$$= A^{\alpha\beta} \delta_{\alpha}{}^{\mu} \delta_{\beta}{}^{\nu} \vec{e}_{\mu} \otimes \vec{e}_{\nu} \qquad\qquad \text{used eqn. (3.23)}$$

$$= A^{\alpha\beta} \vec{e}_{\alpha} \otimes \vec{e}_{\beta} = \mathbf{A}. \qquad\qquad \text{summed over } \mu \text{ and } \nu$$
$$\text{(3.92)}$$

SP 3.12 In Exercise 3.25 we showed that if $\mathbf{A}$ is a $\binom{2}{0}$ tensor and $\mathbf{B}$ is a $\binom{0}{2}$ tensor, then

$$A^{\alpha\beta} B_{\alpha\beta}$$

is frame invariant, i.e. a scalar. The same tensor contraction can also be written as $\mathbf{A}(\mathbf{B})$, including the bases:

$$\mathbf{A}(\mathbf{B}) = A^{\alpha\beta} \vec{e}_{\alpha} \otimes \vec{e}_{\beta} (B_{\mu\nu} \tilde{\omega}^{\mu} \otimes \tilde{\omega}^{\nu}).$$

Show, by also transforming the bases, that indeed all is consistent; $\mathbf{A}(\mathbf{B})$ is Lorentz-frame invariant too.

SP 3.13 Recall that the symmetry of a tensor was defined in Schutz §3.4 based upon the result of interchanging pairs of arguments. So of course transforming the components from one reference frame to another via a Lorentz transformation does not change the symmetry. Prove this for a $\binom{2}{0}$ tensor.

Solution

Suppose $F^{\alpha\beta}$ is a symmetric $\binom{2}{0}$ tensor. Under a Lorentz transformation, the components of **F** become

$$
\begin{aligned}
F^{\alpha'\beta'} &= \Lambda^{\alpha'}{}_{\mu}\Lambda^{\beta'}{}_{\nu}F^{\mu\nu}, \\
&= \Lambda^{\alpha'}{}_{\mu}\Lambda^{\beta'}{}_{\nu}F^{\nu\mu}, && \text{because } \mathbf{F} \text{ is symmetric} \\
&= \Lambda^{\beta'}{}_{\nu}\Lambda^{\alpha'}{}_{\mu}F^{\nu\mu}, && \text{rearranged} \\
&= F^{\beta'\alpha'}. && \text{contracted over } \mu \text{ and } \nu \qquad (3.93)
\end{aligned}
$$

If **F** were an antisymmetric $\binom{2}{0}$ tensor we would have found $F^{\alpha'\beta'} = -F^{\beta'\alpha'}$.

SP 3.14   Recall in the solution to Exercise 2.7(a) we claimed that $(\vec{e}_{\alpha})^{\beta}$ was a $\binom{1}{1}$ tensor. Prove this by verifying it transforms accordingly under the Lorentz transformation.

SP 3.15   Assume that **A** and **B** are $\binom{1}{1}$ tensors. As in Exercise 3.29, prove that tensor differentiation obeys the Leibniz (product) rule:

$$
\nabla(\mathbf{A} \otimes \mathbf{B}) = (\nabla\mathbf{A}) \otimes \mathbf{B} + \mathbf{A} \otimes (\nabla\mathbf{B}).
$$

Unlike in the solution of Exercise 3.29, use the procedures learned in Schutz §3.8.

Solution

Write the tensors in terms of their bases:

$$
\begin{aligned}
\mathbf{A} &= A_{\beta}{}^{\alpha}\, \tilde{\omega}^{\beta} \otimes \vec{e}_{\alpha}, && \text{Schutz Eq. 3.61} \\
\mathbf{B} &= B_{\beta}{}^{\alpha}\, \tilde{\omega}^{\beta} \otimes \vec{e}_{\alpha}. && (3.94)
\end{aligned}
$$

We then have

$$
\begin{aligned}
\mathbf{A} \otimes \mathbf{B} &= A_{\beta}{}^{\alpha}\, \tilde{\omega}^{\beta} \otimes \vec{e}_{\alpha} \otimes (B_{\nu}{}^{\mu}\, \tilde{\omega}^{\nu} \otimes \vec{e}_{\mu}) \\
&= A_{\beta}{}^{\alpha}\, B_{\nu}{}^{\mu}\, \tilde{\omega}^{\beta} \otimes \vec{e}_{\alpha} \otimes \tilde{\omega}^{\nu} \otimes \vec{e}_{\mu}. && (3.95)
\end{aligned}
$$

Redo what Schutz did in §3.8. Differentiate $\mathbf{A} \otimes \mathbf{B}$ with respect to proper time, as in Schutz Eq. (3.63):

$$
\frac{\mathrm{d}(\mathbf{A} \otimes \mathbf{B})}{\mathrm{d}\tau} = \frac{\mathrm{d}(A_{\beta}{}^{\alpha}\, B_{\nu}{}^{\mu})}{\mathrm{d}\tau}\, \tilde{\omega}^{\beta} \otimes \vec{e}_{\alpha} \otimes \tilde{\omega}^{\nu} \otimes \vec{e}_{\mu}, \qquad (3.96)
$$

where use have assumed that the basis one-forms and basis vectors are uniform in spacetime. (In curved spacetime this will no longer be the case!) And

$$\frac{\mathrm{d}(A_\beta{}^\alpha B_\nu{}^\mu)}{\mathrm{d}\tau}$$

is just the ordinary derivative of this function $(A_\beta{}^\alpha B_\nu{}^\mu)$ along the world line. So we can use the ordinary product rule of differential calculus:

$$\frac{\mathrm{d}(A_\beta{}^\alpha B_\nu{}^\mu)}{\mathrm{d}\tau} = \left(\frac{\mathrm{d}A_\beta{}^\alpha}{\mathrm{d}\tau}\right) B_\nu{}^\mu + A_\beta{}^\alpha \left(\frac{\mathrm{d}B_\nu{}^\mu}{\mathrm{d}\tau}\right). \tag{3.97}$$

Now just substitute this into eqn. (3.96):

$$\frac{\mathrm{d}(\mathbf{A} \otimes \mathbf{B})}{\mathrm{d}\tau} = \left[\left(\frac{\mathrm{d}A_\beta{}^\alpha}{\mathrm{d}\tau}\right) B_\nu{}^\mu + A_\beta{}^\alpha \left(\frac{\mathrm{d}B_\nu{}^\mu}{\mathrm{d}\tau}\right)\right] \tilde{\omega}^\beta \otimes \vec{e}_\alpha \otimes \tilde{\omega}^\nu \otimes \vec{e}_\mu,$$

$$\nabla(\mathbf{A} \otimes \mathbf{B})(\vec{U}) = \nabla\mathbf{A}(\vec{U}) \otimes \mathbf{B} + \mathbf{A} \otimes \nabla\mathbf{B}(\vec{U}). \tag{3.98}$$

In the final step we identified the LHS with the directional derivative using Schutz Eqs. (3.65) and (3.66). If we wrote out $\vec{U} = U^\gamma \vec{e}_\gamma$ in the last line of eqn. (3.98) we would find that it cancels leaving:

$$\nabla(\mathbf{A} \otimes \mathbf{B}) = (\nabla\mathbf{A}) \otimes \mathbf{B} + \mathbf{A} \otimes (\nabla\mathbf{B}). \tag{3.99}$$

4 Perfect fluids in special relativity

> Our relativistic approach has unified these two notions [number density and three-vector flux] into a single, frame-independent four-vector. This is progress in our thinking, of the most fundamental sort: the union of apparently disparate notions into a single coherent one.
>
> Bernard Schutz, §4.2

Modern introductory general relativity books at the advanced undergraduate level tend to cover mostly the same material: special relativity, tensor calculus, curvature and the Riemann tensor, Einstein's field equations, Schwarzschild solution, gravity waves, and cosmology. But they do not necessarily cover fluid mechanics, the material of this chapter. This chapter has come in lieu of one on electricity and magnetism. Either of these topics would give us practice working with tensors and developing frame-invariant equations as required to understand the development of the Einstein field equations. Schutz's choice here is especially useful because fluid mechanics is of more general relevance to astrophysical applications of general relativity. The last exercise of Schutz §4.10, Exercise 4.25, is on electricity and magnetism. Several texts (Rindler, 2006; Lawden, 2002; Hobson et al., 2006) have a full chapter on electricity and magnetism with the aim of preparing the student for the development of the Einstein field equations (but without a fluids chapter).

4.1 Exercises

4.1 Comment on whether the continuum approximation is likely to apply to the following physical systems:

(a) Planetary motions in the solar system.

Solution: The continuum hypothesis does not apply because there are only eight or nine planets, and they are well separated and have different orbits, periods, velocities, etc.

(b) A lava flow from a volcano.

Solution: A lava flow from a volcano is a heterogeneous mixture of minerals with different melting temperatures that starts out mostly liquid but gradually the solid component increases with the growth of crystals of minerals with higher temperature of solidification. Flow regimes range from Newtonian fluid turbulence to creeping plastic flow as the lava cools to solid rock (Griffiths, 2000). The continuum approximation applies to most of a high temperature lava flow because the molten rock flows like a liquid and fluid "parcels" can be described that are much bigger than the solid crystals and gas bubbles such that the mean properties (like temperature, density, vicosity) over the fluid parcel vary gradually between adjacent fluid parcels.

(c) Traffic on a major road at rush hour.

Solution: Traffic on a major road at rush hour is likely to be well suited to the continuum approximation if one considers scales much larger than an individual car so there are many cars in the element, but small enough that speed and direction of the cars in one element is roughly constant. At rush hour it is more likely to have bumper to bumper traffic which would force the cars in a vicinity to travel at the same speed.

(d) Traffic at an intersection controlled by stop signs for each incoming road.

Solution: Traffic at an intersection with stop signs is likely to be *not* well suited to the continuum approximation. The stop signs ensure that the cars in an element will have different speeds. There is no near-uniform element.

(e) Plasma dynamics.

Solution: Plasma dynamics often is well suited to the continuum approximation unless the plasma is extremely rarefied. In the latter case it might not be possible to describe fluid parcels large enough for there to be sufficient collisions to bring the gas molecules and ions into statistical equilibrium, so that parcel-mean properties such as temperature, density, etc. are well-defined, yet small enough that these mean quantities vary smoothly. The relevant dimensionless quantity is the Knudsen number Kd defined as the ratio of mean free path of the gas particles to the local length scale of the bulk properties. When $Kd \gtrsim 1$ statistical mechanics rather than continuum mechanics must be used.

4.3   (a) Describe how the Galilean concept of momentum is frame dependent in a manner
             in which the relativistic concept is not.

Solution: Galilean momentum **p** of a particle of mass m is the ordinary three-vector
velocity **v** times the (frame-independent) m:

$$\mathbf{p} = m\mathbf{v}. \tag{4.1}$$

The velocity depends very much on the frame. It is *not just* the components that
change with reference frame, but the *vector itself* that changes.

   In contrast with Galilean momentum, the relativistic momentum, $\vec{p}$, is a four-vector,
created by the scalar rest mass m times the four-velocity $\vec{U}$:

$$\vec{p} = m\vec{U}. \qquad\text{Schutz Eq. (2.19)} \tag{4.2}$$

The rest mass is, by definition, frame-invariant (it's a Lorentz scalar). The four-
velocity, while its components do depend upon reference frame, is a geometric
quantity; it is frame-invariant as discussed by Schutz in Chapter 2. Note for instance
that the magnitude of the four-velocity is always,

$$\vec{U} \cdot \vec{U} = U_\alpha U^\alpha = -1. \qquad\text{eqn. (2.40)} \tag{4.3}$$

In this sense Galilean momentum is frame-dependent in a manner that relativistic
momentum is not.

4.3   (b) How is this possible, since the relativistic definition is nearly the same as the
             Galilean one for small velocities? (Define a *Galilean* four-momentum vector.)

Solution: How is this possible? We were able to describe points or events in a way
that was independent of the uniform motion of the reference frame by introducing
four dimensional spacetime called Minkowski spacetime. With time one of our
independent variables locating events in spacetime, a uniformly moving observer is
analogous to a rotation of the $x-y$ coordinates of the Euclidean plane. This analogy
is explored in our Table 4.1 herein where we deliberately limited the consideration to
two dimensions to emphasize the role of the time coordinate. (In fact one can derive
the Lorentz transformation from this analogy (Hassani, 2008).)

   But in contrast to the four-momentum, Galilean momentum is a three-vector. As a
three-vector it lacks the time component necessary to handle the case of a reference
frame that changes position with time relative to another frame.

   To help make the connection with Euclidean space and the Galilean transformation,
consider the "Galilean four-momentum" that would look like the regular three-
momentum but with a time component,

$$\vec{p}_G \underset{\mathcal{O}}{\rightarrow} m(1, v^x, v^y, v^z). \tag{4.4}$$

Table 4.1 Analogy between Lorentz transformation and rotations	
Rotations in Euclidean plane	Lorentz transformation (for boost in x-direction)
$\begin{pmatrix} x' \\ y' \end{pmatrix} = \begin{pmatrix} \cos\theta & \sin\theta \\ -\sin\theta & \cos\theta \end{pmatrix} \begin{pmatrix} x \\ y \end{pmatrix}$	$\begin{pmatrix} \bar{t} \\ \bar{x} \end{pmatrix} = \begin{pmatrix} \cosh V & -\sinh V \\ -\sinh V & \cosh V \end{pmatrix} \begin{pmatrix} t \\ x \end{pmatrix}$
θ is rotation angle about z-axis	$V = \operatorname{arctanh}(v)$ is velocity parameter, see Exercises 1.18–1.20.
$x'^2 + y'^2 = x^2 + y^2$	$-\bar{t}^2 + \bar{x}^2 = -t^2 + x^2$

where (v^x, v^y, v^z) is the ordinary three-velocity in frame $\mathcal{O}$. In the small velocity limit, $|\mathbf{v}| \ll 1$, note that the true four-momentum $\vec{p}$ approaches this "Galilean four-momentum" because $\gamma(|\mathbf{v}|) \to 1$. Furthermore the Lorentz transformation approaches the Galilean transformation when $|\mathbf{v}| \ll 1$, see SP1.12.

4.5 Complete the proof that

$$\mathbf{T}(\tilde{d}x^\alpha, \tilde{d}x^\beta) = T^{\alpha\beta} := \left\{ \begin{array}{l} \text{flux of } \alpha-\text{momentum across} \\ \text{a surface of constant } x^\beta \end{array} \right\} \quad \text{Schutz Eq. (4.14)}$$

(4.5)

defines a tensor by arguing that it must be linear in both its arguments.

Solution: Of course we require this flux to be proportional to the area of the surface of constant x^β; that's what we mean by a flux – it's the rate of stuff flowing across a surface per unit area of the surface. This ensures that the expression for the stress–energy tensor, $\mathbf{T}(\tilde{d}x^\alpha, \tilde{d}x^\beta)$, is linear in the second argument $\tilde{d}x^\beta$. What about the first argument? The α-component of the four-momentum is

$$p^\alpha = \langle \tilde{d}x^\alpha, \vec{p} \rangle = \vec{p}(\tilde{d}x^\alpha), \qquad \text{notation of Schutz Eq. (3.54)} \quad (4.6)$$

which is linear in $\tilde{d}x^\alpha$ according to the properties of one-forms and vectors. So we've shown that $\mathbf{T}(\tilde{d}x^\alpha, \tilde{d}x^\beta)$ must be linear in both its arguments: $\tilde{d}x^\alpha$ and $\tilde{d}x^\beta$.

4.7 Derive the expressions for the components of the stress–energy tensor of dust:

$$T^{\bar{0}\bar{0}} = \rho U^{\bar{0}} U^{\bar{0}} = \rho/(1-v^2) \qquad T^{\bar{0}\bar{i}} = \rho U^{\bar{0}} U^{\bar{i}} = \rho v^i/(1-v^2)$$

$$T^{\bar{i}\bar{j}} = \rho U^{\bar{i}} U^{\bar{j}} = \rho v^i v^j/(1-v^2) \quad T^{\bar{i}\bar{0}} = \rho U^{\bar{0}} U^{\bar{i}} = \rho v^i/(1-v^2) \quad \text{Schutz Eq. (4.21)}$$

(4.7)

Solution: The terms in eqn. (4.7) follow immediately from the expression for the stress–energy tensor for dust $T^{\alpha\beta} = \rho U^\alpha U^\beta$, (relabeling indices so we are in the $\overline{\mathcal{O}}$

frame) and using the expression for the four-velocity in frame $\overline{\mathcal{O}}$:

$$\vec{U} \underset{\overline{\mathcal{O}}}{\rightarrow} (\gamma, v^x \gamma, v^y \gamma, v^z \gamma), \qquad \text{used eqn. (2.39)} \qquad (4.8)$$

where $\gamma = \frac{1}{\sqrt{1-v^2}}$, and $v^2 = \sum_i v^i v^i$.

4.9   Show that the divergence of the stress–energy tensor,

$$T^{\alpha\beta}{}_{,\beta} = 0, \qquad \text{Schutz Eq. (4.34)} \qquad (4.9)$$

when α is any spatial index, is just Newton's second law.

Solution: Set the free index $\alpha = i$ to make explicit the choice to focus on the spatial components of the first index; recall the convention $i \in \{1, 2, 3\}$. When $\beta = 0$, the LHS of eqn. (4.9) above gives the following term:

$$T^{i0}{}_{,0} = \frac{\partial T^{i0}}{\partial x^0} = \frac{\partial T^{i0}}{\partial t}, \qquad (4.10)$$

which we interpret as the time rate of change of the i-direction momentum density; recall the definition in Schutz Eq. (4.17), which is quite general. The remaining terms, obtained from the LHS of eqn. (4.9) when $\beta = j$,

$$T^{ij}{}_{,j} = \frac{\partial T^{ij}}{\partial x^j},$$

we interpret as the (spatial) divergence of the flux of i-direction momentum, using Schutz Eq. (4.18). This is a way of writing Newton's second law, because

$$\frac{\partial T^{i0}}{\partial t} = -\frac{\partial T^{ij}}{\partial x^j}, \qquad (4.11)$$

expresses the time rate of increase in the i-direction momentum density equals the convergence of the flux of i-direction momentum. While eqn. (4.11) is a relativistic equation, each term has a classical counterpart in Newton's second law. This can be made more transparent by considering the Newtonian limit of eqn. (4.11) applied to a cubical fluid element of volume l^3. Then T^{0i} cooresponds to the fluid momentum density and

$$\frac{\partial T^{i0}}{\partial t} \implies \frac{\partial \rho v^i}{\partial t} = \frac{1}{l^3} \frac{\partial m v^i}{\partial t}, \qquad (4.12)$$

where on the RHS we have used the standard symbols of classical fluid mechanics: $\rho = m/l^3$ is the mass density, v^i is the fluid three-velocity in the x^i-direction. Furthermore, T^{ij} is the force due to surface stresses on a fluid element in the direction x^i per unit area normal to x^j. To find the net force due to surface stresses, place the cubical fluid element with one corner at the origin of a Carstesian coordinate system, as in Schutz Fig. 4.8. The net (interfacial) force in the x^i-direction is

$$F^i = l^2 \left[T^{ix}(x=0) - T^{ix}(x=l) + T^{iy}(y=0) - T^{iy}(y=l) \right.$$

$$\left. + T^{iz}(z=0) - T^{iz}(z=l) \right]$$

$$= -l^3 \left[\frac{T^{ix}(x=l) - T^{ix}(x=0)}{l} + \frac{T^{iy}(y=l) - T^{iy}(y=0)}{l} \right.$$

$$\left. + \frac{T^{iz}(z=l) - T^{iz}(z=0)}{l} \right]$$

$$= -l^3 T^{ij}{}_{,j}. \tag{4.13}$$

In the final step we took the limit $l \to 0$. Combining eqn. (4.12) and eqn. (4.13) and cancelling the l^3 we obtain

$$\frac{\partial T^{i0}}{\partial t} + \frac{\partial T^{ij}}{\partial x^j} = 0 \implies \frac{\partial m v^i}{\partial t} = F^i. \tag{4.14}$$

The equation on the right of eqn. (4.14) states that the rate of change of momentum in the x^i-direction equals the sum of the (interfacial) forces in this direction. In Exercise 7.5(a) one derives the full (inviscid) Navier–Stokes equations from the Newtonian limit of the generalization of eqn. (4.11) to curved spacetime.

4.12   Derive

$$T^{\alpha\beta} = (\rho + p)U^\alpha U^\beta + p\eta^{\alpha\beta} \qquad \text{Schutz Eq. (4.37)} \tag{4.15}$$

from,

$$(T^{\alpha\beta}) = \begin{pmatrix} \rho & 0 & 0 & 0 \\ 0 & p & 0 & 0 \\ 0 & 0 & p & 0 \\ 0 & 0 & 0 & p \end{pmatrix}. \qquad \text{Schutz Eq. (4.36)} \tag{4.16}$$

Solution: This exercise provides practice using the powerful technique of developing a general tensorial equation in a convenient reference frame. In this case we work in the MCRF of the local fluid parcel showing the correspondance between the terms in eqn. (4.15) and eqn. (4.16). Step through the possible combinations of indices $\{\alpha\beta\}$ systematically. Schutz has already addressed $\alpha = \beta = 0$. Using the convention $i \in \{1, 2, 3\}$, consider the three terms T^{0i}:

$$\begin{aligned} T^{0i} &= (\rho + p)U^0 U^i + p\eta^{0i} & \text{start with eqn. (4.15)} \\ &= p\eta^{0i} & \text{in MCRF, } U^i = 0 \\ &= 0, & \eta^{\mu\nu} \text{ is diagonal} \end{aligned} \tag{4.17}$$

which confirms agreement between eqn. (4.15) and the final three columns of the first row in eqn. (4.16). Then by symmetry, $T^{i0} = T^{0i} = 0$, as required by the final three rows of the first column of eqn. (4.16). Finally,

$$T^{ij} = (\rho + p)U^i U^j + p\eta^{ij} = p\eta^{ij} = p\delta^{ij} \qquad (4.18)$$

because in the MCRF $\vec{U}$ has nil spatial components. This confirms the agreement between eqn. (4.15) and the lower-right three rows and columns of eqn. (4.16). Now we have a tensorial equation, eqn. (4.15) for the stress–energy tensor of a perfect fluid, which is valid in all reference frames related by Lorentz transformations.

4.14 Argue that

$$n\, U^\beta U_\alpha \left(\frac{\rho + p}{n} U^\alpha \right)_{,\beta} + p_{,\beta}\, \eta^{\alpha\beta} U_\alpha = 0 \qquad \text{Schutz Eq. (4.46)} \qquad (4.19)$$

is the time component of

$$n\, U^\beta \left(\frac{\rho + p}{n} U^\alpha \right)_{,\beta} + p_{,\beta}\, \eta^{\alpha\beta} = 0 \qquad \text{Schutz Eq. (4.45)} \qquad (4.20)$$

in the MCRF.

Solution: First consider what is the time component of eqn. (4.20). Although there are two indices in eqn. (4.20), the β is just a dummy index (it appears both upstairs and downstairs implying a sum over β). The time component is when $\alpha = 0$. However it would be reckless to simply set $\alpha = 0$ in eqn. (4.20) because U^α appears inside the scope of the derivative $\partial/\partial x^\beta$, and we would be unjustifiably setting $U^0{}_{,\beta} = 0$. See SP4.3.

On the other hand, we can argue in the other direction, starting with eqn. (4.19) and recognizing it as the contraction of $-U_\alpha$ with eqn. (4.20). In the MCRF, $\vec{U} = \vec{e}_t$ so lowering the index we have $-U_0 = -\eta_{0\alpha} U^\alpha = 1$.

There's a subtle point here involving the order of steps. First we contracted a tensorial equation, eqn. (4.20), with a one-form field, U_α. Contraction is a valid tensorial operation, producing a new tensorial equation from an old. And then we evaluated the new equation at a given point in a given reference frame, recognizing the result as just the time component of the original.

Finally we emphasize that this worked in SR but will not always work in GR. For a reference frame in GR where $U^\alpha \xrightarrow[\text{MCRF}]{} (1, 0, 0, 0)$ it is not always the case that $U_\alpha \xrightarrow[\text{MCRF}]{} (-1, 0, 0, 0)$! Indeed eqn. (12.90) of SP12.13 provides a concrete example.

4.16 In the MCRF, the components of the four-velocity vanish, $U^i = 0$. Why can we not assume the gradient vanishes, $U^i{}_{,\beta} = 0$?

Solution: The analogous statement in 3D space is also true. In fluid mechanics for instance, one can alway transform the equations into a frame momentarily co-moving with the local fluid velocity, but that does not mean the velocity gradient will be zero. The three-velocity in fluid mechanics, and four-velocity in SR, can depend upon space, and spacetime, so that adjacent fluid elements have different MCRFs.

4.18 Sharpen the discussion at the end of Schutz §4.6 by showing that $-\nabla p$ is actually the net force per unit volume on the fluid element in the MCRF.

Solution: In the MCRF we can use the same argument as in classical fluid mechanics. Imagine a cube with one corner at the origin, with sides parallel to the Cartesian coordinate axes, and of volume $\delta x\, \delta y\, \delta z$. Without loss of generality let the pressure gradient be in the y-direction. The pressure force on the face at $y = 0$ is $\delta x\, \delta z\, p(y = 0)$, while the pressure force on the face at $y = \delta y$ is $-\delta x\, \delta z\, p(y = \delta y)$. So the pressure gradient force per unit volume is

$$\mathrm{PGF} = -[p(y = \delta y) - p(y = 0)]\, \frac{\delta x\, \delta z}{\delta x\, \delta y\, \delta z}$$

$$= -\left[\frac{p(y = \delta y) - p(y = 0)}{\delta y}\right]. \tag{4.21}$$

Taking the limit $\delta x \to 0$, $\delta y \to 0$, $\delta z \to 0$,

$$\mathrm{PGF} = -\frac{\partial p}{\partial y}. \tag{4.22}$$

4.20 (a) Show that if particles are not conserved but are generated locally at a rate ε particles per unit volume per unit time in the MCRF, then the conservation law becomes:

$$N^{\alpha}{}_{,\alpha} = \varepsilon \tag{4.23}$$

instead of

$$N^{\alpha}{}_{,\alpha} = (nU^{\alpha})_{\alpha} = 0. \qquad\qquad \text{Schutz Eq. (4.35)} \qquad (4.24)$$

Solution: We must essentially derive eqn. (4.24) but including the source term. We were told just before Schutz Eq. (4.35) that the procedure was the same as for Schutz Eq. (4.34), see p. 98. Consider a fluid element as described in Schutz Fig. 4.8 and bottom of p. 98. The number density is n in the MCRF, and is γn in a reference frame moving at speed v relative to the fluid element because of Lorentz contraction, with $\gamma = 1/\sqrt{1 - v^2}$, see Schutz Eq. (4.2). The four-velocity of the fluid is, by definition, $\vec{U} = \vec{e}_0$ in the MCRF and the time component is in general $U^0 = \gamma$. Thus we can write the number of particles in an element of volume l^3 as

$$l^3 n\gamma = n\, l^3\, U^0.$$

The rate of flow (or flux of) particles across the $x = 0$ surface (surface 4 in Schutz Fig. 4.8) is $l^2 n U^x(x = 0)$. (This may seem strange because we know that $U^x = 0$ in the MCRF, but soon we're going to take a derivative that will not be zero – recall Exercise 4.16 above.) The flux of particles across the parallel surface at $x = l$, i.e. surface 2 in Schutz Fig. 4.8, is $l^2 n U^x(x = l)$. Similarly, in the y-direction and z-direction the net inflow of particles is $l^2 n U^y(y = 0) - l^2 n U^y(y = l)$ and $l^2 n U^z(z = 0) - l^2 n U^z(z = l)$ respectively. These net inflow terms increase the particle density in the fluid element at a rate

$$\frac{\partial n l^3 U^0}{\partial t} = l^2[(nU^x)(x = 0) - (nU^x)(x = l) + (nU^y)(y = 0) - (nU^y)(y = l)$$
$$+ (nU^z)(z = 0) - (nU^z)(z = l)] + \cdots \text{other terms}. \tag{4.25}$$

There are other terms contributing now, unlike in deriving Schutz Eq. (4.35), because there is also a source term giving,

$$\frac{\partial n l^3 U^0}{\partial t} = l^2[(nU^x)(x = 0) - (nU^x)(x = l) + (nU^y)(y = 0) - (nU^y)(y = l)$$
$$+ (nU^z)(z = 0) - (nU^z)(z = l)] + l^3 \varepsilon. \tag{4.26}$$

Note that this relation should be frame-invariant because n is obviously frame invariant and $\vec{U}$ is also frame invariant.

Note: ε is frame-invariant! Recall ε is the rate of generation of particles per unit volume per unit time in the MCRF. In another reference frame, there is a factor of γ to account for the fact that the volume will be smaller, thus tending to increase the generation rate, but the time will be slower by a factor $1/\gamma$. In short, time dilation cancels length contraction.

And we can pull l^3 out of the derivative because it is a specified constant, and then divide both sides by l^3:

$$\frac{\partial n U^0}{\partial t} = -\frac{(nU^x)(x = l) - (nU^x)(x = 0)}{l} - \frac{(nU^y)(y = l) - (nU^y)(y = 0)}{l}$$
$$- \frac{(nU^z)(z = l) - (nU^z)(z = 0)}{l} + \varepsilon. \tag{4.27}$$

In the limit $l \to 0$,

$$\frac{\partial n U^0}{\partial t} = -\frac{\partial n U^x}{\partial x} - \frac{\partial n U^y}{\partial y} - \frac{\partial n U^z}{\partial z} + \varepsilon. \tag{4.28}$$

Or

$$N^\alpha{}_{,\alpha} = \varepsilon.$$

4.20   (b) Generalize (a) to show that if the energy and momentum of a body are not conserved (e.g. because it interacts with other systems), then there is a non-zero relativistic force four-vector F^α defined by

$$T^{\alpha\beta}{}_{,\beta} = F^\alpha. \tag{4.29}$$

Interpret the components of F^α in the MCRF.

Solution: This problem is easier than Exercise 20(a) because we follow the derivation of Schutz Eq. (4.34) on pp. 98 and 99. Recall

$$\frac{\partial T^{00}}{\partial t} = -T^{0i}{}_{,i}, \qquad\qquad \text{Schutz Eq. (4.31)} \tag{4.30}$$

where the LHS is the time rate of change of energy per unit volume, and the RHS is the net influx of energy per unit time per unit volume. Thus for non-conservative systems we must add an F^0 to the RHS of eqn. (4.30), representing the net rate of energy forcing (or supply of energy from external sources and not associated with fluxes across the boundaries of the fluid element) per unit time per unit volume. So eqn. (4.30) becomes:

$$\frac{\partial T^{00}}{\partial t} = -T^{0i}{}_{,i} + F^0. \tag{4.31}$$

Interestingly, unlike the source of particles ε we encountered in Exercise 4.20(a), F^0 will be frame-dependent since it is a component of a four-vector. For suppose it is associated with particles being generated at a rate ε. We argued that ε was frame-invariant in Exercise 20(a). But the total energy of each particle is $m\gamma$, where m is the rest mass. So the energy source F^0 will increase as one changes to a reference frame moving relative to the fluid element.

Similarly for the other components. Consider

$$\frac{\partial T^{x0}}{\partial t} = -T^{xi}{}_{,i}. \tag{4.32}$$

The LHS of eqn. (4.32) is the time rate of change of x-direction momentum per unit volume, and the RHS in the net influx of x-direction momentum per unit time per unit volume. Thus we must add any external forces and eqn. (4.32) becomes

$$\frac{\partial T^{x0}}{\partial t} = -T^{xi}{}_{,i} + F^x, \tag{4.33}$$

where F^x is the external force on the fluid element per unit volume in the MCRF. Similarly for the other two spatial components. Combining results eqn. (4.31) and eqn. (4.33) gives the equation we were to derive, eqn. (4.29).

4.22 Many physical systems may be idealized as collections of non-colliding particles (for example, black-body radiation, rarified plasmas, galaxies, and globular clusters). By assuming that such a system has a random distribution of velocities at every point, with no bias in any direction in the MCRF, prove that the stress-energy tensor is that of a perfect fluid. If all particles have the same speed v and mass m, express p and ρ as functions of m, v, and n. Show that a photon gas has $p = \frac{1}{3}\rho$.

Solution: Here we must simply argue that heat conduction and viscosity are zero, so that the conditions of a perfect fluid are met. Heat conduction and momentum diffusion (viscosity) result from the net transfer of energy and momentum, respectively, due to particle motions. In classical fluid dynamics this results from the fluxes associated with molecular motions having a preferred direction due to temperature gradients or momentum gradients. But here we are assuming a priori that there is no preferred direction in the MCRF (this must be the same MCRF for the entire system so that there are no gradients). So there can be no net transfer of energy or momentum due to the motion of the particles; hence there is no heat transfer by conduction or momentum transfer by viscosity. With the conditions of a perfect fluid being met, the argument of Schutz §4.6 applies for the form of the stress-energy tensor.[1]

If all particles have the same speed v and rest mass m then the energy of each particle is $E = \gamma m$, with $\gamma = (1 - v^2)^{-1/2}$. (This is just Einstein's famous equation with $c = 1$ and it follows from the definition of particle energy given by Schutz in §2.4: $E := p^0 = mU^0 = m\gamma$.) The energy density is

$$\rho = En = nm\gamma, \tag{4.34}$$

that is simply the energy per particle times the number density, which is n in the MCRF.

The pressure requires more effort to derive but it is quite instructive. We can argue intuitively as follows (embellishing and generalizing the argument of Batchelor (1969, §1.7) to the SR). Imagine a surface orthogonal to the z-axis. The pressure is the normal stress (the force per unit area) and in the ideal fluid case arises only from the flux of momentum across this surface. The particle's relativistic momentum in the z-direction, say at an angle θ to the direction of particle motion, is $P^z = |P^i| \cos\theta$. For a particle of mass m this is just $|P^i| = m\gamma(v)v$. For a photon $|P^i| = E$, in order that its mass is zero, cf. eqn. (2.31). Associated with this massive particle or photon is a momentum flux in the z-direction $P^z v^z = |P^i| v \cos^2\theta$. We need to sum this momentum flux over all n particles per unit volume, taking account of their random direction. Suppose a particle is at the origin. Because the direction of a given particle's trajectory is random, the probability of leaving through a given piece of the unit sphere centered at the origin is proportional to the area of the piece. Dividing by the area of the sphere we get the

---

[1] The condition of no bias in any direction of the particle velocities (in the MCRF) is a statistical condition, applying in mean. But if the velocities are truly randomly distributed, then we should expect random fluctuations about the mean, implying a random heat conduction and viscosity effect, albeit with a time mean of zero. In classical atmosphere/ocean fluid dynamics this is only recently being considered under the name of "stochastic parameterization" (Mana and Zanna, 2014).

solid angle. The solid angle of the strip of width $d\theta$ making an angle θ to the z-axis is $2\pi r \sin(\theta)\, r d\theta /4\pi r^2 = \sin(\theta)\, d\theta/2$. So the expected (or mean) value of the momentum flux of one particle in the z-direction is obtained by integrating over the sphere[2]

$$|P^i|v \frac{\int_0^\pi \sin(\theta)\cos^2(\theta) d\theta}{2} = \frac{|P^i|v}{2}\left(\frac{-1}{3}\right)[\cos^3(\theta)]_0^\pi = \frac{1}{3}|P^i|v. \qquad (4.35)$$

By isotropy there is nothing special about the z-direction so eqn. (4.35) applies for any direction. The pressure p is the total momentum flux and is simply n times the momentum flux per particle,

$$p = \frac{1}{3}n|P^i|v = \frac{1}{3}\rho v^2. \qquad (4.36)$$

We used eqn. (4.34) above for ρ in the final step in eqn. (4.36) obtaining a single expression valid for either massive particles or photons (in the latter case of course $v = 1$). This is the equation of state of a photon gas or an isotropic homogeneous collection of particles of uniform speed v.

4.24 Astronomical observations of the brightness of objects are measurements of the flux of radiation T^{0i} from the object at Earth. This problem calculates how that flux depends on the relative velocity of the object and Earth.

(a) Show that, in the rest frame $\mathcal{O}$ of a star of constant luminosity L (total energy radiated per second), the stress–energy tensor of the radiation from the star at the event $(t, x, 0, 0)$ has components $T^{00} = T^{0x} = T^{x0} = T^{xx} = L/(4\pi x^2)$. The star sits at the origin.

Solution: A sphere of radius x centered on the star at the origin has radiation flowing out of it at a rate of L. Assume the star emits radiation isotropically in a reference frame stationary with respect to the star so the energy flux will be evenly distributed over the surface of the sphere. The surface area of the sphere is $4\pi x^2$. Thus the flux per unit time per unit area is of magnitude $L/(4\pi x^2)$ everywhere on the surface. And in particular at event $(t, x, 0, 0)$ the energy is flowing in the x-direction so $T^{0x} = L/(4\pi x^2)$. By the symmetry properties of $\mathbf{T}$ (see discussion in Schutz §4.5), we know that in general $T^{0x} = T^{x0} = L/(4\pi x^2)$.

Now consider the energy density. In time period δt the energy flow out of the sphere will be $L\delta t$ and this energy will fill a spherical shell of volume $(4\pi x^2)\, c\, \delta t =$

---

[2] Don't make the mistake I initially did of only integrating over half the sphere thinking that only half the particles contribute to the momentum flux in a given direction. This is wrong because the momentum flux goes like the square of the velocity – a particle moving in the negative z-direction carries negative momentum in the negative z-direction and thus results in a positive momentum flux! This was clarified in Exercise 4.21(c).

$(4\pi x^2)\,\delta t$, since $c = 1$ in geometric units. Thus the energy density at a distance of x from the origin will be $T^{00} = L\,\delta t/(4\pi x^2\,\delta t) = L/(4\pi x^2)$.

Finally, the radiation flows radially which, at the event $(t, x, 0, 0)$, is in the x-direction. The energy flux in this direction is photon number flux, say F_p, times the energy per photon, $E = h\nu$, giving $T^{0x} = F_p h\nu$. And the momentum flux will be the photon number flux times the momentum per photon, h/λ (where λ is the wavelength),

$$T^{xx} = F_p \frac{h}{\lambda} = F_p h \frac{\nu}{c} = T^{0x}$$

because again $c = 1$ in geometric units, cf. Schutz §2.7.

4.24 (b) Let $\vec{X}$ be the null vector that separates the events of emission and reception of the radiation. Show that $\vec{X} \underset{\mathcal{O}}{\to} (x, x, 0, 0)$ for radiation observed at the event $(x, x, 0, 0)$. Show that the stress–energy tensor of (a) has the frame-invariant form

$$\mathbf{T} = \frac{L}{4\pi} \frac{\vec{X} \otimes \vec{X}}{(\vec{U}_s \cdot \vec{X})^4},$$

where $\vec{U}_s$ is the star's four-velocity, $\vec{U}_s \underset{\mathcal{O}}{\to} (1, 0, 0, 0)$.

Solution: Radiation is emitted at the spatial origin at some time t, say event $\mathcal{E} \underset{\mathcal{O}}{\to} (t, 0, 0, 0)$ and is absorbed at event $\mathcal{A} \underset{\mathcal{O}}{\to} (x, x, 0, 0)$, see fig. 4.1. The light must of course travel at the speed of light, and so a photon traveling along the x-axis will be located at event

$$\vec{X}(t) \underset{\mathcal{O}}{\to} (t, \pm t, 0, 0).$$

This is clearly a null vector: $\vec{X} \cdot \vec{X} = -t^2 + (\pm t)^2 = 0$. At the absorption event $\vec{X}(t = x) \equiv \vec{X} \underset{\mathcal{O}}{\to} (x, x, 0, 0)$.

To show that the stress–energy tensor has the given frame-invariant form, we must show that it is indeed Lorentz frame-invariant, and that it reproduces the results of (a) in the MCRF. Vectors are Lorentz frame-invariant by construction, and so their outer product will form a $\binom{2}{0}$ tensor that is also frame-invariant. The radiation emitted per second will depend upon reference frame but the luminosity L is that measured in the rest frame[3] so is a scalar by definition. Finally the denominator has an inner product of two four-vectors, and is also frame-invariant. So the given expression for $\mathbf{T}$ is frame-invariant.

---

[3] This is implicit in the definition given in stardard textbooks, e.g. Weinberg (2008, Eq. (1.3.3)) or Sparke and Gallagher III (2007, Eq. (1.1)).

In the MCRF it is easy to verify that the expression given produces the results of (a):

$$T^{00} = \frac{L}{4\pi} \frac{X^0 X^0}{(U^\alpha X_\alpha)^4} = \frac{L}{4\pi} \frac{x \cdot x}{(-1 \cdot x)^4} = \frac{L}{4\pi x^2}. \tag{4.37}$$

Similarly

$$T^{0x} = \frac{L}{4\pi} \frac{X^0 X^1}{(U^\alpha X_\alpha)^4} = \frac{L}{4\pi} \frac{x \cdot x}{(-1 \cdot x)^4} = \frac{L}{4\pi x^2}, \tag{4.38}$$

and

$$T^{xx} = \frac{L}{4\pi} \frac{X^1 X^1}{(U^\alpha X_\alpha)^4} = \frac{L}{4\pi} \frac{x \cdot x}{(-1 \cdot x)^4} = \frac{L}{4\pi x^2}. \tag{4.39}$$

Furthermore,

$$T^{0y} = \frac{L}{4\pi} \frac{X^0 X^2}{(U^\alpha X_\alpha)^4} = \frac{L}{4\pi} \frac{x \cdot 0}{(-1 \cdot x)^4} = 0, \tag{4.40}$$

and similarly for the remaining terms: $T^{0z} = T^{xy} = T^{xz} = T^{yz} = 0$.

4.24 (c) Let the Earth-bound observer $\overline{\mathcal{O}}$, traveling with speed v away from the star in the x-direction, measure the same radiation, again with the star on the $\bar{x}$-axis. Let $\vec{X} \underset{\overline{\mathcal{O}}}{\to} (R, R, 0, 0)$ and find R as a function of x. Express $T^{\bar{0}\bar{x}}$ in terms of R. Explain why R and $T^{\bar{0}\bar{x}}$ depend as they do on v.

Solution:
First transform $\vec{X}$ from the reference frame of the star to that of Earth:

$$\vec{X} \underset{\overline{\mathcal{O}}}{\to} (\gamma(1-v)x, \gamma(1-v)x, 0, 0) \equiv (R, R, 0, 0), \qquad \text{used eqn. (1.34)} \tag{4.41}$$

where $\gamma = 1/\sqrt{1 - v^2}$. This determines R:

$$R = \gamma(1-v)x = \sqrt{\frac{1-v}{1+v}}\, x. \tag{4.42}$$

By the so-called reciprocity principle,[4] the star moves away from Earth with speed v in the $-\bar{x}$-direction. So the four-velocity of the star in the Earth's frame of reference is

$$\vec{U}_s \underset{\overline{\mathcal{O}}}{\to} (\gamma, -v\gamma, 0, 0), \qquad \text{used eqn. (2.39)} \tag{4.43}$$

---

[4] Rindler discusses reciprocity briefly (Rindler, 2006, §2.5) while Berzi and Gorini (1969) analyze it in detail.

which gives for $\vec{U}_s \cdot \vec{X} = -R\gamma(1+v)$. Using eqn. (4.42) one can show that this also equals $\vec{U}_s \cdot \vec{X} = -x$, in agreement with the same calculation in the reference frame of the star, cf. the calculation in Exercise 4.24(b). Using the frame-invariant expression for the stress–energy tensor we find in the Earth's frame of reference:

$$
T^{\bar{0}\bar{x}} = \frac{L}{4\pi} \frac{X^{\bar{0}} X^{\bar{1}}}{(U^\alpha X_\alpha)^4} = \frac{L}{4\pi} \frac{R^2}{x^4} = \frac{L}{4\pi R^2} \left(\frac{R}{x}\right)^4
$$
$$
= \frac{L}{4\pi R^2} \frac{(1-v)^2}{(1+v)^2}.
\tag{4.44}
$$

Interpretation There are two steps to understand the relation eqn. (4.42) between the distance the photon has traveled from the star to the Earth measured in the star's frame, x, and measured in Earth's frame, R. First a bit of algebra gives that the distance of the Earth from the star at the point of emission in the star's frame, x-coordinate of $\mathcal{B}$ in fig. 4.1 is $(1-v)x$. Next we recognize this as the Lorentz contraction of the proper length R in the Earth's frame, giving eqn. (4.42). Consult fig. 4.1 for a more detailed explanation. The $\mathcal{O}$ frame is that of the star, while the $\overline{\mathcal{O}}$ frame is that of the Earth. Let their origins coincide at event $\mathcal{E}$, the point of emission of the photon that will be detected on Earth. The origin of $\overline{\mathcal{O}}$ moves at speed v along the x-axis, and the Earth travels parallel to this, starting at event $\mathcal{B}$ on the x-axis. To find the x-coordinate of event $\mathcal{B}$ we need the time of travel of the photon. The photon is absorbed at event $\mathcal{A}$ with coordinates in $\mathcal{O}$ of $(x, x, 0, 0)$; that is $t = x$ because the

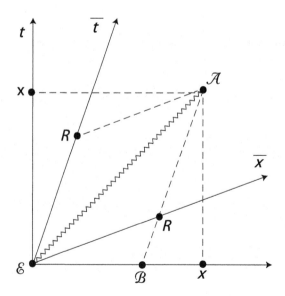

The photon travels from the origin $\mathcal{E}$ along the zigzag line and is absorbed at event $\mathcal{A}$ with coordinates $(x, x, 0, 0)$ in frame $\mathcal{O}$. Meanwhile the Earth travels parallel to the $\bar{t}$-axis (it is stationary in the $\overline{\mathcal{O}}$ frame) from event $\mathcal{B}$ and intersects the photon at event $\mathcal{A}$ with coordinates $(R, R, 0, 0)$ in frame $\overline{\mathcal{O}}$.

speed of light is unity. During this time period x the Earth has moved $\Delta x = vt = vx$ along the x-axis. Solving for the position at $t = 0$ we find event $\mathcal{B}$ has coordinates $(0, (1-v)x, 0, 0)$ in $\mathcal{O}$. This distance $(1-v)x$ is the initial distance between the Earth and the origin of $\mathcal{O}$. But from the Earth's point of view, its distance from the origin of $\overline{\mathcal{O}}$ remains R, i.e. a "rod" of proper length R in the frame moving with the Earth. This "rod" is Lorentz contracted in the frame $\mathcal{O}$ to the length $(1-v)x$, cf. Schutz Fig. 1.13.

$$(1-v)x = \frac{R}{\gamma(v)}, \qquad \text{so} \qquad \frac{R}{x} = \gamma(1-v).$$

That was actually the "easy part." Now we have to understand eqn. (4.44) for the stress–energy tensor in the Earth's frame! First an interpretation inspired by cosmology.[5] From eqn. (4.44) we have

$$T^{\bar{0}\bar{x}} = \frac{L}{4\pi R^2}\left(\frac{R}{x}\right)^4. \tag{4.45}$$

In the $\bar{\mathcal{O}}$ reference frame the length scales have been contracted by a factor $R/x = \gamma(1-v)$. The energy density, and hence the flux, will be exaggerated by this factor to the fourth power R^4/x^4 for two reasons: the electromagnetic radiation energy is contained within a correspondingly smaller volume $(R/x)^3$ and the wavelength of the photons has been contracted by the factor R/x.

Here is an alternative interpretation inspired by Weinberg's explanation of luminosity distance in an expanding universe (Weinberg, 2008, §1.4). The *frequency* of the radiation received by the moving star will be redshifted by the relativistic Doppler effect,

$$\frac{\bar{\nu}}{\nu} = (1-v)\gamma = \left(\frac{1-v}{1+v}\right)^{\frac{1}{2}}. \qquad \text{used Schutz Eq. (2.39)} \tag{4.46}$$

Likewise the *rate* of receiving photons has been reduced by the same factor. If you happen to be Weinberg this is obvious. But if not obvious, suppose the star emits $\mathcal{N}$ photons per second. For now forget about their frequency and focus just on the rate of emission $\mathcal{N}$. The arrival of photons is a periodic signal of sorts and it too is Doppler shifted by exactly the same relativistic Doppler factor $(1-v)\gamma$ and for the same reasons. So the emission rate observed in frame $\overline{\mathcal{O}}$ will be reduced to $\bar{\mathcal{N}} = \mathcal{N}(1-v)\gamma$. Thus the apparent luminosity observed in $\overline{\mathcal{O}}$ will be reduced by the Doppler shift factor squared:

$$\bar{L} = \bar{\mathcal{N}}h\bar{\nu} = \mathcal{N}h\nu[(1-v)\gamma]^2 = L\left(\frac{1-v}{1+v}\right).$$

---

[5] We thank Professor Richard Tweed for this interpretation.

Based on the above argument

$$T^{\bar{0}\bar{x}} = \frac{L}{4\pi x^2}\left(\frac{1-v}{1+v}\right) = \frac{L}{4\pi x^2}\left(\frac{R}{x}\right)^2 = T^{0x}\left(\frac{R}{x}\right)^2,$$

consistent with eqn. (4.45) above.

4.25 *Electromagnetism in SR.* (This exercise is suitable only for students who have already encountered Maxwell's equations in some form.) Maxwell's equations for the electric and magnetic fields in vacuum, **E** and **B**, in three-vector notation are

$$\nabla \times \mathbf{B} - \frac{\partial \mathbf{E}}{\partial t} = 4\pi\mathbf{J},$$

$$\nabla \times \mathbf{E} + \frac{\partial \mathbf{B}}{\partial t} = 0,$$

$$\nabla \cdot \mathbf{E} = 4\pi\rho,$$

$$\nabla \cdot \mathbf{B} = 0, \qquad\qquad \text{Schutz Eq. (4.59)}$$

in units where $\mu_0 = \epsilon_0 = c = 1$. (Here ρ is the density of electric charge and **J** the current density.)

(a) An *antisymmetric* $\binom{2}{0}$ tensor **F** can be defined on spacetime by the equations $F^{0i} = E^i (i = 1, 2, 3), F^{xy} = B^z, F^{yz} = B^x, F^{zx} = B^y$. Find from this definition all other components $F^{\mu\nu}$ in this frame and write them down in a matrix.

Solution: Because $F^{\alpha\beta}$ is antisymmetric it immediately follows that the diagonal elements are zero, for zero is the only real number for which $F^{\alpha\alpha} = -F^{\alpha\alpha}$. The lower diagonal elements follow from −ve the given upper diagonal ones.

$$(F^{\mu\nu}) = \begin{pmatrix} 0 & E^x & E^y & E^z \\ -E^x & 0 & B^z & -B^y \\ -E^y & -B^z & 0 & B^x \\ -E^z & B^y & -B^x & 0 \end{pmatrix}. \tag{4.47}$$

4.25 (b) A rotation by an angle θ about the z-axis is one kind of Lorentz transformation, with the matrix

$$(\Lambda^{\beta'}{}_{\alpha}) = \begin{pmatrix} 1 & 0 & 0 & 0 \\ 0 & \cos\theta & -\sin\theta & 0 \\ 0 & \sin\theta & \cos\theta & 0 \\ 0 & 0 & 0 & 1 \end{pmatrix}. \tag{4.48}$$

Show that the new components of **F**,

$$F^{\alpha'\beta'} = \Lambda^{\alpha'}{}_\mu \Lambda^{\beta'}{}_\nu F^{\mu\nu}, \tag{4.49}$$

define new electric and magnetic three-vector components (by the rule given in (a)) that are just the same as the components of the old **E** and **B** in the rotated three-space. (This shows that a spatial rotation of **F** makes a spatial rotation of **E** and **B**.)

Solution: The transformation eqn. (4.48) corresponds to an *active* transformation (reference frame stays the same but the tensors change, see SP5.8), which is potentially confusing since the Lorentz transformation introduced in Schutz Chapter 1 is a passive transformation (tensors remain the same, the reference frame changes). If this is confusing, just think of eqn. (4.48) as corresponding to a rotation of the coordinate axes of $-\theta$ about the z-axis.

We simply go through the computations of $F^{\alpha'\beta'}$. There are a few shortcuts. First the antisymmetry of $F^{\alpha'\beta'}$ follows from the antisymmetry of $F^{\mu\nu}$. (This is obvious because the antisymmetry is defined independent of reference frame $\mathbf{F}(\vec{A}, \vec{B}) = -\mathbf{F}(\vec{B}, \vec{A})$, see also SP3.13.) So there are only the four "rules" of (a) to check. For example, we must check $F^{0'i'} = E^{i'}$, where $E^{i'}$ should be the E^i vector in the rotated frame $\mathcal{O}'$:

$$\begin{aligned}
F^{\alpha'\beta'} &= \Lambda^{\alpha'}{}_\mu \Lambda^{\beta'}{}_\nu F^{\mu\nu}, && \text{in general} \\
F^{0'i'} &= \Lambda^{0'}{}_\mu \Lambda^{i'}{}_\nu F^{\mu\nu}, && \text{used eqn. (4.49)} \\
&= \Lambda^{i'}{}_\nu F^{0\nu}, && \text{summed over } \mu \\
&= \Lambda^{i'}{}_i F^{0i}, && \text{because } F^{00} = 0 \\
&= \Lambda^{i'}{}_i E^i, && \text{interpreted } F^{0i} \\
&= E^{i'}. && (4.50)
\end{aligned}$$

The last step follows from the fact that here $\Lambda^{i'}{}_i$ is a rotation about the z-axis. In particular,

$$F^{0'x'} = \cos\theta\, E^x - \sin\theta\, E^y, \quad F^{0'y'} = \sin\theta\, E^x + \cos\theta\, E^y, \quad F^{0'z'} = E^z. \tag{4.51}$$

We have recovered the first rule now in the $\mathcal{O}'$ frame.

Now check the second rule: $F^{x'y'} = B^{z'}$:

$$\begin{aligned}
F^{x'y'} &= \Lambda^{x'}{}_\mu \Lambda^{y'}{}_\nu F^{\mu\nu}, && \text{used eqn. (4.49)} \\
&= (\cos\theta\, F^{x\nu} - \sin\theta\, F^{y\nu})\Lambda^{y'}{}_\nu, && \text{summed over } \mu \\
&= \cos\theta\, F^{xy} \cos\theta - \sin\theta\, F^{yx} \sin\theta, && \text{summed over } \nu \\
&= F^{xy}, && \text{used antisymmetry and trig} \\
&= B^z = B^{z'}. && \text{interpreted } F^{xy} = B^z \quad (4.52)
\end{aligned}$$

The final equality again follows from the fact that our transformation here is a rotation about the z-axis. We have recovered the second rule in the $\mathcal{O}'$ frame.

Check the final two rules:

$$
\begin{aligned}
F^{y'z'} &= \Lambda^{y'}{}_{\mu}\Lambda^{z'}{}_{\nu}F^{\mu\nu} & \text{used eqn. (4.49)} \\
&= \Lambda^{y'}{}_{\mu}F^{\mu z} & \text{summed over } \nu \\
&= \sin\theta\, F^{xz} + \cos\theta\, F^{yz} & \text{summed over } \mu \\
&= \sin\theta(-B^{y}) + \cos\theta\, B^{x} & \text{interpreted } \mathbf{F} \\
&= B^{x'}. & (4.53)
\end{aligned}
$$

The final equality again follows from recognizing the transformation as an active rotation about the z-axis (or passive rotation of the axes). We have recovered the third rule now in the $\mathcal{O}'$ frame.

Finally

$$
\begin{aligned}
F^{z'x'} &= \Lambda^{z'}{}_{\mu}\Lambda^{x'}{}_{\nu}F^{\mu\nu}, & \text{used eqn. (4.49)} \\
&= \Lambda^{x'}{}_{\nu}F^{z\nu}, & \text{summed over } \mu \\
&= \cos\theta\, F^{zx} - \sin\theta\, F^{zy} & \text{summed over } \nu \\
&= \cos\theta\, F^{zx} + \sin\theta\, F^{yz} & \text{used antisymmetry} \\
&= \cos\theta\, B^{y} + \sin\theta\, B^{x} & \text{interpreted } \mathbf{F} \\
&= B^{y'}. & (4.54)
\end{aligned}
$$

The final equality again follows from recognizing the transformation as an active rotation about the z-axis (or passive rotation of the axes). We have recovered the final rule now in the $\mathcal{O}'$ frame.

4.25 (c) Define the current four-vector $\vec{J}$ by $J^{0} = \rho$, $J^{i} = (\mathbf{J})^{i}$, and show that two of Maxwell's equations are just

$$
F^{\mu\nu}{}_{,\nu} = 4\pi J^{\mu}. \qquad\qquad \text{Schutz Eq. (4.60)} \qquad (4.55)
$$

[Note we corrected a typo in the original question: $J^{i} = (\vec{J})^{i} \implies J^{i} = (\mathbf{J})^{i}$.]

Solution: Of course the third Maxwell Equation $\nabla \cdot \mathbf{E} = 4\pi\rho$ follows immediately from setting $\mu = 0$. The other one is more subtle. Inspection of Maxwell's equations reveals it should be the first one since that is the only one in which $\mathbf{J} = J^{i}$ appears explicitly.

$$
\nabla \times \mathbf{B} - \frac{\partial \mathbf{E}}{\partial t} = 4\pi\mathbf{J}.
$$

Then obviously to get the RHS correct we set $\mu = i$ in eqn. (4.55):

$$F^{i\nu}{}_{,\nu} = 4\pi J^i$$

$$F^{i0}{}_{,0} + F^{ij}{}_{,j} = 4\pi J^i$$

$$-\frac{\partial E^i}{\partial t} + F^{ij}{}_{,j} = 4\pi J^i. \qquad\qquad \text{interpreted } F^{i0} \text{ as } -E^i.$$

We've adopted J^i as the notation for both the four-vector and the corresponding three-vector components: $(\mathbf{J})^i \equiv J^i$. To show that $F^{ij}{}_{,j}$ corresponds to the curl of $\mathbf{B}$ we simply expand the expression and use the definition of $\mathbf{F}$ in Exercise 4.25(a):

$$-\frac{\partial E^x}{\partial t} + F^{xj}{}_{,j} = 4\pi J^x$$

$$-\frac{\partial E^x}{\partial t} + F^{xy}{}_{,y} + F^{xz}{}_{,z} = 4\pi J^x$$

$$-\frac{\partial E^x}{\partial t} + B^z{}_{,y} - B^y{}_{,z} = 4\pi J^x$$

$$-\frac{\partial E^x}{\partial t} + (\nabla \times \mathbf{B})^x = 4\pi J^x. \qquad (4.56)$$

The y and z components work out in a similar manner.

4.2 Supplementary problems

SP 4.1   The expression for the number flux across a surface of constant $\bar{x}$,

$$(\text{flux})^{\bar{x}} = \frac{n v^{\bar{x}}}{\sqrt{1 - v^2}}, \qquad\qquad \text{Schutz Eq. (4.3)} \qquad (4.57)$$

looks similar to the expression just prior to it,

$$(\text{flux})^{\bar{x}} = \frac{n v}{\sqrt{1 - v^2}} \qquad (4.58)$$

which is the same quantity when $v^{\bar{y}} = 0$. Are they *exactly* the same? If not, what is the difference?

Solution

Of course the two equations are manifestly different because the numerator of eqn. (4.57) has $v^{\bar{x}}$ in the place of v in eqn. (4.58). But the equations also differ because in eqn. (4.57) the v in the denominator is

$$v^2 = (v^{\bar{x}})^2 + (v^{\bar{y}})^2.$$

SP 4.2   The derivation for the number flux,

$$\vec{N} \underset{\mathcal{O}}{\to} \left(\frac{n}{\sqrt{1 - v^2}}, \frac{n v^{\bar{x}}}{\sqrt{1 - v^2}}, \frac{n v^{\bar{y}}}{\sqrt{1 - v^2}}, \frac{n v^{\bar{z}}}{\sqrt{1 - v^2}} \right), \qquad \text{Schutz Eq. (4.5)}$$

used the fact that

$$\vec{U} \underset{\mathcal{O}}{\to} \left(\frac{1}{\sqrt{1 - v^2}}, \frac{v^{\bar{x}}}{\sqrt{1 - v^2}}, \frac{v^{\bar{y}}}{\sqrt{1 - v^2}}, \frac{v^{\bar{z}}}{\sqrt{1 - v^2}} \right).$$

Where did this come from? Show that this is the general expression for the four-velocity.

Hint

You can skip this problem if you've done Exercise 2.15!

SP 4.3   In the solution to Exercise 4.14, to obtain the time component of eqn. (4.20), why did we not just immediately set $U^0 = 1$? That is, what is wrong with the following argument?

$$0 = n U^{\beta} \left(\frac{\rho + p}{n} U^0 \right)_{,\beta} + p_{,\beta} \, \eta^{0\beta} \qquad \qquad \text{set } \alpha = 0$$

$$= n U^{\beta} \left(\frac{\rho + p}{n} \right)_{,\beta} + p_{,\beta} \, \eta^{0\beta}, \qquad \text{since } \vec{U} = \vec{e}_0 \text{ in the MCRF}$$

$$= n U^{\beta} \left(\frac{\rho + p}{n} \right)_{,\beta} - \frac{\partial p}{\partial t}. \qquad \text{used Schutz Eq. (3.44)}$$

Solution

Although $\vec{U} = \vec{e}_0$ in MCRF, so that $U^0 = 1$ *locally*, it does necessarily hold globally or even in a finite region. So we cannot immediately conclude that the gradient of the time component is zero.

SP 4.4   Confirm your answer to Exercise 4.24(c) by calculating the stress–energy tensor in the reference frame of the Earth, $T^{\bar{\alpha}\bar{\beta}}$, using the Lorentz transformation from the reference frame $\mathcal{O}$ of the star.

SP 4.5   More advanced textbooks introduce the so-called *energy conditions*, which take the form of restrictions on the stress–energy tensor. They are meant to embody restrictions on the properties we can reasonably expect for matter. One of these is called the *weak energy condition*, and can be written

$$T_{\alpha\beta} U^{\alpha} U^{\beta} \geq 0, \qquad \text{(Poisson, 2004, Eq. 2.6)} \qquad (4.59)$$

where $T_{\alpha\beta}$ is the stress–energy tensor and U^α is the four-velocity of any observer. Argue that this represents the requirement that observers always measure non-negative energy density. See Poisson (2004, §2.1) for a more complete introduction.

Solution

The argument here is similar to that in Exercise 4.14. At any event of interest we can choose a reference frame co-moving with the observer. Then $\vec{U} = \vec{e}_t$ and we have $U^\alpha \xrightarrow[\text{MCRF}]{} (1, 0, 0, 0)$. But then eqn. (4.59) reduces to

$$T_{00} \underset{\text{MCRF}}{\geq} 0 \qquad \text{the inequality holds locally in the MCRF}$$

$$\rho \underset{\text{MCRF}}{\geq} 0 \qquad\qquad \text{used definition of } T_{00}$$

$$\rho \geq 0. \qquad \text{eqn. (4.59) is a valid tensor equation} \qquad (4.60)$$

The last line deserves elobartion. The eqn. (4.59) is the contraction of a $\binom{0}{2}$ tensor with two four-vectors, producing a (Lorentz invariant) scalar. Because it is true in the MCRF, it must be true in *all* legitimate reference frames.

Preface to curvature

The successful operation of GPS [navigational system] can be taken to be a very accurate verification of the [gravitational] redshift. This experimental verification of the redshift is comforting from the point of view of energy conservation. But it is the death-blow to our chances of finding a simple, special-relativistic theory of gravity . . .

Bernard Schutz, §5.1

5.1 Exercises

5.1 Recall the argument that led to,

$$\frac{E'}{E} = \frac{hv'}{hv} = \frac{m}{m + m\,g\,H + O(\mathbf{v}^4)} = 1 - g\,H + O(\mathbf{v}^4) \quad \text{Schutz Eq. (5.1)} \quad (5.1)$$

where E is the total energy of a massive particle at the surface of the Earth, E' is the energy of a photon at the top of the tower, H is the height of the tower,[1] while h is Planck's constant, v' and v are photon frequency at the top and bottom of the tower, and $\mathbf{v}$ is the three-velocity of the particle at ground level. Repeat this argument under more realistic assumptions: suppose a fraction ε of the kinetic energy of the mass at the bottom can be converted into a photon and sent back up, the remaining energy staying at ground level in a useful form. Devise a perpetual motion engine if eqn. (5.1) is violated.

Solution: Taken literally, ε is the fraction of the *kinetic energy* of the mass at the bottom, not the fraction of the total energy. The "remaining energy" then means the rest mass energy plus $(1 - \varepsilon)$ of the kinetic energy. Although the calculations would simplify if instead we considered ε the fraction of the *total energy*, we will take the literal interpretation. Conceptually it does not really matter since we are just supposed to see that the Einstein thought experiment carries forward the same message even when inefficiencies are introduced.

Introduce an index i to keep track of the iterations of the mass falling and photon propagating to the top of the tower. Say we start with mass m_0 at the top, it falls gaining kinetic energy $m_0 g H / c^2 = m_0 g H$ where the constants are in geometric units

[1] Changed from h in original text to avoid confusion with Planck's constant.

so that $c = 1$ and gH is dimensionless. For Earth conditions of course $gH \ll 1$. Of this kinetic energy only a fraction ε is available for generating the photon at the bottom of the tower, $\varepsilon m_0 gH = 2\pi \hbar \nu_0$, while the remaining energy is accumulated, apparently in useful form, at the base of the tower:

$$\tilde{m}_0 = [(1 - \varepsilon)gH + 1]m_0. \qquad (5.2)$$

The key assumption is that the radiation is *unaffected* by the gravitational field (in violation of eqn. (5.1)), yielding a photon at the top of the tower of the same energy as at the bottom: $2\pi \hbar \nu_0$. Now this is converted into mass at the top of the tower, $m_1 = 2\pi \hbar \nu_0 = m_0 \varepsilon gH$. And the process repeats. Upon falling to the base of the tower this will yield kinetic energy: $m_1 gH = m_0 \varepsilon (gH)^2$, of which the fraction ε is available for the second photon: $\varepsilon m_1 gH = 2\pi \hbar \nu_1 = m_0 (\varepsilon gH)^2$. The remaining energy accumulates at the bottom:

$$\tilde{m}_1 = [(1 - \varepsilon)gH + 1]m_1 = [(1 - \varepsilon)gH + 1](\varepsilon gH)m_0, \qquad (5.3)$$

giving us $(\tilde{m}_0 + \tilde{m}_1)$ useful energy at the bottom after two iterations. The kinetic energy at the top of the tower is again taken as the total energy in the photon at the base of the tower, yielding a new mass at the top of the tower: $m_2 = 2\pi \hbar \nu_1 = m_0 (\varepsilon gH)^2$. At the bottom we generate another photon, $2\pi \hbar \nu_2 = \varepsilon m_2 gH = m_0 (\varepsilon gH)^3$, and accumulate more mass,

$$\tilde{m}_2 = [(1 - \varepsilon)gH + 1]m_2 = [(1 - \varepsilon)gH + 1](\varepsilon gH)^2 m_0. \qquad (5.4)$$

After three iterations we have accumulated $(\tilde{m}_0 + \tilde{m}_1 + \tilde{m}_3)$ useful energy at the bottom. The process will repeat indefinitely. Inspection of eqns. (5.2)–(5.4) above for $\tilde{m}_i$ reveals that after $n + 1$ iterations we have accumulated a geometric series of mass at the base of the tower:

$$\sum_{i=0}^{n} \tilde{m}_i = \sum_{i=0}^{n} ar^i,$$

with $a = [(1 - \varepsilon)gH + 1]m_0$ and $r = (\varepsilon gH)$. As $n \to \infty$, the accumulated mass approaches

$$M = \frac{a}{1 - r} = \frac{[(1 - \varepsilon)gH + 1]m_0}{1 - \varepsilon gH} = m_0 \left(1 + \frac{gH}{1 - \varepsilon gH}\right),$$

see Boas (1983, Eq. (1.8)) for the sum of an infinite geometric series. Assuming Earth-like values, $gH \ll 1$, and we can solve for the mass gained for free, ΔM:

$$\Delta M \equiv M - m_0(1 + gH) = m_0 gH[\varepsilon gH + O((\varepsilon gH)^2)]. \quad \text{used eqn. (B.4)} \quad (5.5)$$

The accumulated mass is not much more than the starting mass so the process is not an efficient way to create energy. However, we gained something for nothing and generated an infinite process. Clearly something is wrong, and in particular, it was the violation of eqn. (5.1) describing the gravitational redshift. Einstein's simple thought experiment is robust to the inclusion of inefficiencies.

A very clear presentation of the gravitational redshift was written by Earman and Glymour (1980), and includes an entertaining discussion of the historical confusion surrounding the search for a clear derivation.

5.3   (a) Show that the coordinate transformation $(x, y) \rightarrow (\xi, \eta)$ with $\xi = x$ and $\eta = 1$ violates,

$$\det \begin{pmatrix} \frac{\partial \xi}{\partial x} & \frac{\partial \xi}{\partial y} \\ \frac{\partial \eta}{\partial x} & \frac{\partial \eta}{\partial y} \end{pmatrix} \neq 0. \qquad\qquad \text{Schutz Eq. (5.6)} \qquad\qquad (5.6)$$

Solution: Consider coordinate transformation $(x, y) \rightarrow (\xi, \eta)$ with $\xi = x$ and $\eta = 1$. Note that $\partial \eta / \partial x = 0$ and $\partial \eta / \partial y = 0$. This violates eqn. (5.6), implying that this coordinate transformation is not good, or degenerate. In fact this same example was worked out in Schutz §5.2, complete with an example of a distinct pair of points (x, y) points having the same (ξ, η) coordinates.

5.3   (b) Are the following coordinates transformations good ones? Compute the Jacobian and list any points where the transformations fail.

(i) $\xi = (x^2 + y^2)^{1/2}, \eta = \arctan(y/x)$.

Solution: This is of course Schutz Eq. (5.3), the polar coordinate transformation. Note that traditionally one clarifies the ambiguity in arctan by specifying that the range of $\arctan(y/x)$ as $-\pi/2 < \eta < \pi/2$, and for $x < 0$ we have $\eta = \arctan(y/x) + \pi$. The Jacobian is the determinant of the transformation matrix,

$$\begin{pmatrix} \frac{\partial \xi}{\partial x} & \frac{\partial \xi}{\partial y} \\ \frac{\partial \eta}{\partial x} & \frac{\partial \eta}{\partial y} \end{pmatrix} = \begin{pmatrix} \frac{x}{\sqrt{x^2+y^2}} & \frac{y}{\sqrt{x^2+y^2}} \\ \frac{-y}{x^2+y^2} & \frac{x}{x^2+y^2} \end{pmatrix}.$$

The determinant is $1/\sqrt{x^2 + y^2}$ so the only problem is at the origin, where $r = 0$ and derivatives above are undefined.

(ii) $\xi = \ln(x), \eta = y$.

Solution: The transformation matrix is

$$\begin{pmatrix} \frac{\partial \xi}{\partial x} & \frac{\partial \xi}{\partial y} \\ \frac{\partial \eta}{\partial x} & \frac{\partial \eta}{\partial y} \end{pmatrix} = \begin{pmatrix} \frac{1}{x} & 0 \\ 0 & 1 \end{pmatrix}.$$

The determinant is $\frac{1}{x}$ so again the only problems are along $x = 0$ and of course for all $x \leq 0$ where $\xi = \ln(x)$ is undefined.

(iii) $\xi = \arctan(y/x), \eta = (x^2 + y^2)^{-1/2}$.

Solution: This is related to the polar coordinate transformation, so again one should avoid the ambiguity in arctan by specifying that the range of $\arctan(y/x)$ as $-\pi/2 < \eta < \pi/2$, and for $x < 0$ use $\eta = \arctan(y/x) + \pi$.

$$\begin{pmatrix} \frac{\partial \xi}{\partial x} & \frac{\partial \xi}{\partial y} \\ \frac{\partial \eta}{\partial x} & \frac{\partial \eta}{\partial y} \end{pmatrix} = \begin{pmatrix} \frac{-y}{x^2+y^2} & \frac{x}{x^2+y^2} \\ \frac{-x}{(x^2+y^2)^{3/2}} & \frac{-y}{(x^2+y^2)^{3/2}} \end{pmatrix}.$$

The determinant is $1/(x^2 + y^2)^{3/2}$ so the only problem is at the origin, where the derivatives above are undefined.

5.5   Sketch the following curves. Which have the same paths? Find also their tangent vectors where the parameter equals zero.

Solution: Although the computations in this exercise are simple it is still quite instructive. The tangent vector to a curve with coordinates (x, y) given by the functions $x(\lambda)$ and $y(\lambda)$ of the real parameter λ is

$$\vec{V} = \frac{dx(\lambda)}{d\lambda} \vec{e}_x + \frac{dy(\lambda)}{d\lambda} \vec{e}_y. \tag{5.7}$$

If one likes this approach, one might like the introductory textbook by Faber (1983). The plots are in fig. 5.1 and were produced by the accompanying Maple™ worksheet.

(a)

$$x = \sin \lambda, \quad y = \cos \lambda. \tag{5.8}$$

Solution: This is a unit circle centered at the origin. When $\lambda = 0$ the tangent vector is at $(0, 1)$, points in the x-direction and has unit length.

(b)

$$x = \cos(2\pi t^2), \quad y = \sin(2\pi t^2 + \pi). \tag{5.9}$$

Solution: This is also a unit circle centered at the origin and has the same path as in (a). The tangent vector is a bit subtle. Differentiating the coordinates of the curve with respect to the parameter t:

$$\dot{x} = -\sin(2\pi t^2)4\pi t, \quad \dot{y} = \cos(2\pi t^2 + \pi)4\pi t. \tag{5.10}$$

When $t = 0$ the tangent vector is the zero vector.

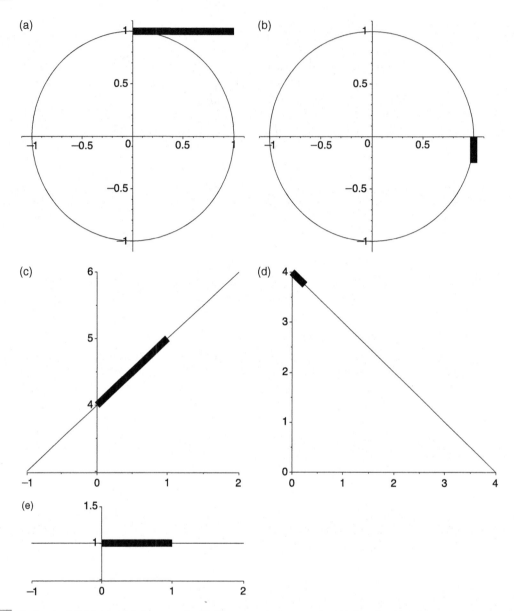

Figure 5.1 For Exercise 5.5(a) through (e). Curves are the thin lines, tangent vectors are the thick lines. Note that the tangent vectors in (b) and (d) are not well-defined at the given parameter value (little stubs have been drawn to show where the tangent vector would go if the curve were differently parameterized).

(c)

$$x = s, \quad y = s + 4. \tag{5.11}$$

Solution: The path is a straight line with slope 1 and y-intercept at $y = 4$. The tangent vector is uniform $(1, 1)$.

(d)

$$x = s^2, \quad y = -(s-2)(s+2) = 4 - s^2. \tag{5.12}$$

Solution: The path is a straight line with slope -1 and y-intercept at $y = 4$, but now we are restricted to $x \geq 0$. The tangent vector is not uniform but depends upon s:

$$\dot{x} = 2s, \quad \dot{y} = -2s. \tag{5.13}$$

As in (b) the tangent vector is the zero vector at $s = 0$.

(e)

$$x = \mu, \quad y = 1. \tag{5.14}$$

Solution: The path is a horizontal straight line at $y = 1$. The tangent vector is uniform:

$$\dot{x} = 1, \quad \dot{y} = 0. \tag{5.15}$$

5.7 Calculate all elements of the transformation matrices $\Lambda^{\alpha'}{}_{\beta}$ and $\Lambda^{\mu}{}_{\nu'}$ for the transformation from Cartesian (x, y) – the unprimed indices – to polar (r, θ) – the primed indices.

Solution: Start with $\Lambda^{\mu}{}_{\nu'}$. Using Schutz Eq. (5.3) for the cooridnates (x, y) in terms of polar coordinate variables (r, θ), we calculate the terms of the transformation given in Schutz Eq. (5.13):

$$(\Lambda^{\mu}{}_{\nu'}) = \begin{pmatrix} \frac{\partial x}{\partial r} & \frac{\partial x}{\partial \theta} \\ \frac{\partial y}{\partial r} & \frac{\partial y}{\partial \theta} \end{pmatrix} = \begin{pmatrix} \cos\theta & -r\sin\theta \\ \sin\theta & r\cos\theta \end{pmatrix}. \tag{5.16}$$

Slightly more awkward is $\Lambda^{\alpha'}{}_{\beta}$. The simplest solution is to invert the matrix in eqn. (5.16) above using eqn. (B.1). To calculate it directly use Schutz Eq. (5.8) with $\xi = r$ and $\eta = \theta$. Note for the second row we must differentiate arctan:

$$\frac{\partial \eta}{\partial x} = \frac{\partial \theta}{\partial x} = \frac{\partial \arctan(y/x)}{\partial x}. \tag{5.17}$$

It's simpler to write $\tan\theta = y/x$ and differentiate both sides with respect to x, then solve for $\partial\theta/\partial x$. Finally we arrive at:

$$(\Lambda^{\alpha'}{}_{\beta}) = \begin{pmatrix} \frac{x}{\sqrt{x^2+y^2}} & \frac{y}{\sqrt{x^2+y^2}} \\ \frac{-y}{x^2+y^2} & \frac{x}{x^2+y^2} \end{pmatrix} = \begin{pmatrix} \cos\theta & \sin\theta \\ \frac{-\sin\theta}{r} & \frac{\cos\theta}{r} \end{pmatrix}, \tag{5.18}$$

where in the second equality we changed to polar coordinates using the definition in Schutz Eq. (5.3).

As a check we can multiple the two transformations together to confirm that they are indeed a pair of inverses. Indeed we find their product gives the identity matrix,

$$(\Lambda^{\alpha'}{}_{\beta})(\Lambda^{\beta}{}_{\gamma'}) = \begin{pmatrix} \cos\theta & \sin\theta \\ -\frac{\sin\theta}{r} & \frac{\cos\theta}{r} \end{pmatrix} \begin{pmatrix} \cos\theta & -r\sin\theta \\ \sin\theta & r\cos\theta \end{pmatrix} = \begin{pmatrix} 1 & 0 \\ 0 & 1 \end{pmatrix}. \tag{5.19}$$

5.9 Draw a diagram similar to Schutz Fig. 5.6 to explain Schutz Eq. (5.38a, b),

$$\frac{\partial \vec{e}_{\theta}}{\partial r} = \frac{1}{r}\vec{e}_{\theta}, \qquad\qquad \text{Schutz Eq. (5.38a)} \tag{5.20}$$

$$\frac{\partial \vec{e}_{\theta}}{\partial \theta} = -r\vec{e}_r. \qquad\qquad \text{Schutz Eq. (5.38b)} \tag{5.21}$$

Solution: In eqn. (5.20) we see that changing r does not change the direction of the polar coordinate basis vectors. But $\vec{e}_{\theta}$ does change in magnitude since it must increase in length as one moves further from the origin, albeit more slowly the farther one is from the origin, see fig. 5.2(a) herein.

Changing θ on the other hand does change the orientation of the basis vectors. Increasing θ when one is in the first quadrant for instance results in $\vec{e}_{\theta}$ pointing more toward the $-x$-direction, see fig. 5.2(b) herein.

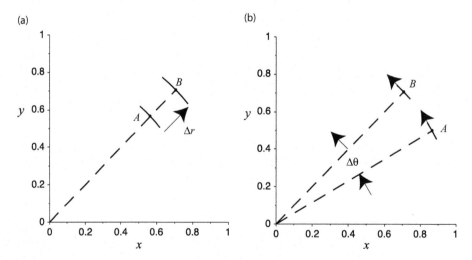

(a)

(b)

Figure 5.2 (a) Moving from point A to B we see the basis vector $\vec{e}_{\theta}$ increases in length but does not change direction. For larger r the relative change is smaller. (b) Moving from point A to B we see the basis vector $\vec{e}_{\theta}$ changes direction, in this case pointing more toward the $-x$-direction, but does not change in length. Plots were partly generated with Maple™, see accompanying worksheet.

5.11 (Uses the result of Exercises 5.7 and 5.8.) For the vector field $\vec{V}$ whose Cartesian components are $(x^2 + 3y, y^2 + 3x)$, compute:

(a) $V^\alpha{}_{,\beta}$ in Cartesian coordinates.

Solution:

$$V^\alpha{}_{,\beta} \underset{\text{Car}}{\rightarrow} \begin{pmatrix} \frac{\partial V^x}{\partial x} & \frac{\partial V^x}{\partial y} \\ \frac{\partial V^y}{\partial x} & \frac{\partial V^y}{\partial y} \end{pmatrix} = \begin{pmatrix} 2x & 3 \\ 3 & 2y \end{pmatrix} = \begin{pmatrix} 2r\cos\theta & 3 \\ 3 & 2r\sin\theta \end{pmatrix}. \tag{5.22}$$

The tensor components on the far RHS of eqn. (5.22) are still in the Cartesian coordinate basis. We've simply expressed those Cartesian components using polar coordinates for comparison with later results.

5.11 (b) Find the transformation of $V^\alpha{}_{,\beta}$ to polar coordinates.

Solution: Transformation of vectors between different coordinates was explained in Schutz §2.2 and §5.2. This exercise then is very important for clarifying how to transform tensors. The key ingredient is that one needs to apply a transformation matrix for each index or rank of the tensor – Schutz Eq. (5.8) for the superscript indices (what other books called the contravariant components) and Schutz Eq. (5.13) for the subscript indices (what other books call the covariant components).

$$V^{\mu'}{}_{;\nu'} = \Lambda^{\mu'}{}_\alpha \, V^\alpha{}_{,\beta} \, \Lambda^\beta{}_{\nu'}$$

$$= \begin{pmatrix} \cos\theta & \sin\theta \\ -\frac{\sin\theta}{r} & \frac{\cos\theta}{r} \end{pmatrix} \begin{pmatrix} 2r\cos\theta & 3 \\ 3 & 2r\sin\theta \end{pmatrix} \begin{pmatrix} \cos\theta & -r\sin\theta \\ \sin\theta & r\cos\theta \end{pmatrix} = \begin{pmatrix} A & B \\ C & D \end{pmatrix}, \tag{5.23}$$

where

$$A = 2r(\cos^3\theta + \sin^3\theta) + 3\sin 2\theta,$$

$$B = 2r^2(-\cos^2\theta\sin\theta + \sin^2\theta\cos\theta) + 3r\cos 2\theta,$$

$$C = 2(-\cos^2\theta\sin\theta + \sin^2\theta\cos\theta) + \frac{3}{r}\cos 2\theta,$$

$$D = 2r(\cos^2\theta\sin\theta + \sin^2\theta\cos\theta) - 3\sin 2\theta. \tag{5.24}$$

5.11 (c) Find the components $V^{\mu'}{}_{;\nu'}$ directly in polar coordinates using the Christoffel symbols,

$$(1) \ \partial\vec{e}_r/\partial r = 0 \qquad\qquad \Longrightarrow \ \Gamma^\mu{}_{rr} = 0 \qquad \text{for all } \mu,$$

$$(2) \ \partial\vec{e}_r/\partial\theta = \frac{1}{r}\vec{e}_\theta \qquad \Longrightarrow \ \Gamma^r{}_{r\theta} = 0, \quad \Gamma^\theta{}_{r\theta} = \frac{1}{r},$$

(3) $\partial \vec{e}_\theta / \partial r = \dfrac{1}{r}\vec{e}_\theta$ $\quad\Longrightarrow\quad$ $\Gamma^r{}_{\theta r} = 0, \quad \Gamma^\theta{}_{\theta r} = \dfrac{1}{r},$

(4) $\partial \vec{e}_\theta / \partial \theta = -r\vec{e}_r$ $\quad\Longrightarrow\quad$ $\Gamma^r{}_{\theta\theta} = -r, \quad \Gamma^\theta{}_{\theta\theta} = 0, \quad$ Schutz Eq. (5.45)

$$(5.25)$$

in the formula for covariant derivative:

$$V^\alpha{}_{;\beta} := V^\alpha{}_{,\beta} + V^\mu \Gamma^\alpha{}_{\mu\beta}. \qquad\qquad \text{Schutz Eq. (5.50)} \qquad (5.26)$$

Solution: The velocity field in polar coordinates was found in Exercise 5.8(a):

$$\vec{V} \underset{\text{Pol}}{\to} \begin{pmatrix} r^2\,(\cos^3\theta + \sin^3\theta) + 3\,r\,\sin 2\theta \\ -r\,\cos^2\theta\,\sin\theta + r\,\sin^2\theta\,\cos\theta + 3\,\cos 2\theta \end{pmatrix}. \qquad (5.27)$$

From eqn. (5.26) above the velocity gradient has two parts. The first part, due to the gradient of the components:

$$V^{\mu'}{}_{,\nu'} = \begin{pmatrix} \frac{\partial V^r}{\partial r} & \frac{\partial V^r}{\partial \theta} \\ \frac{\partial V^\theta}{\partial r} & \frac{\partial V^\theta}{\partial \theta} \end{pmatrix} = \begin{pmatrix} A & E \\ F & G \end{pmatrix}, \qquad (5.28)$$

where

$A = 2r(\cos^3\theta + \sin^3\theta) + 3\sin 2\theta$ $\qquad\qquad$ as in (b) above

$E = 3r^2(-\cos^2\theta\,\sin\theta + \sin^2\theta\,\cos\theta) + 6r\,\cos 2\theta$

$F = (-\cos^2\theta\,\sin\theta + \sin^2\theta\,\cos\theta)$

$G = r(2\sin^2\theta\,\cos\theta + 2\cos^2\theta\,\sin\theta - \cos^3\theta - \sin^3\theta) - 6\sin 2\theta.$ $\qquad (5.29)$

The second part is due to the gradient in basis vectors. Using eqn. (5.25) for the Christoffel symbols:

$$V^{\gamma'}\Gamma^{\mu'}{}_{\gamma'\nu'} = \begin{pmatrix} 0 & V^\theta\Gamma^r{}_{\theta\theta} \\ V^\theta\Gamma^\theta{}_{\theta r} & V^r\Gamma^\theta{}_{r\theta} \end{pmatrix}$$

$$= \begin{pmatrix} 0 & r^2(\cos^2\theta\,\sin\theta - \sin^2\theta\,\cos\theta) - 3r\,\cos 2\theta \\ (-\cos^2\theta\,\sin\theta + \sin^2\theta\,\cos\theta) + \frac{3\cos 2\theta}{r} & r\,(\cos^3\theta + \sin^3\theta) + 3\,\sin 2\theta \end{pmatrix}. $$

$$(5.30)$$

Adding these two parts eqn. (5.28) and eqn. (5.30) we obtain

$$V^{\mu'}{}_{;\nu'} \underset{\text{Pol}}{\to}$$

$$\begin{pmatrix} 2r(\cos^3\theta + \sin^3\theta) + 3\sin 2\theta & 2r^2(-\cos^2\theta\,\sin\theta + \sin^2\theta\,\cos\theta) + 3r\,\cos 2\theta \\ 2(-\cos^2\theta\,\sin\theta + \sin^2\theta\,\cos\theta) + \frac{3\cos 2\theta}{r} & r(2\sin^2\theta\,\cos\theta + 2\cos^2\theta\,\sin\theta) - 3\sin 2\theta \end{pmatrix}.$$

$$(5.31)$$

And of course it agrees with the much simpler calculation in Cartesian coordinates, transformed to polar eqn. (5.23).

5.11 (d) The divergence $V^\alpha{}_{,\alpha}$ using your results in (a).

Solution: The divergence should be reference frame-independent and is the trace of the matrix of the covariant derivative of the vector. Using the covariant derivative in Cartesian coordinates from Exercise 5.11(a) above:

$$V^{\alpha}_{,\alpha}{}_{\text{car}} = 2r(\cos\theta + \sin\theta). \tag{5.32}$$

5.11   (e) The divergence using results from (b) or (c).

Solution: Using results from Exercise 5.11(c) we contract the indices in eqn. (5.31) above, setting $\mu' = \nu' = \alpha$ and summing :

$$V^{\alpha}_{;\alpha} = 2r(\cos^3\theta + \sin^3\theta + \cos^2\theta\sin\theta + \sin^2\theta\cos\theta),$$
$$= 2r[\cos^2\theta(\cos\theta + \sin\theta) + \sin^2\theta(\sin\theta + \cos\theta)],$$
$$= 2r(\cos\theta + \sin\theta). \tag{5.33}$$

And of course this agrees with the result eqn. (5.32) obtained in Exercise 5.11(d).

5.11   (f) The divergence using

$$V^{\alpha}_{;\alpha} = \frac{1}{r}\frac{\partial}{\partial r}(rV^r) + \frac{\partial}{\partial\theta}V^{\theta} \qquad \text{Schutz Eq. (5.56)} \tag{5.34}$$

directly.

Solution: The vector field $\vec{V}$ in polar coordinates eqn. (5.27) from Exercise 5.8(a) was:

$$\vec{V} \underset{\text{Pol}}{\to} \begin{pmatrix} r^2(\cos^3\theta + \sin^3\theta) + 6r\sin\theta\cos\theta \\ -r\cos^2\theta\sin\theta + r\sin^2\theta\cos\theta + 3(\cos^2\theta - \sin^2\theta) \end{pmatrix}, \tag{5.35}$$

where we have simplified using trignometric identities. So applying eqn. (5.34) we get

$$\nabla \cdot \vec{V} = \frac{1}{r}\frac{\partial}{\partial r}\left[rV^r\right] + \frac{\partial}{\partial\theta}V^{\theta}$$
$$= \frac{1}{r}\frac{\partial}{\partial r}\left[r(r^2(\cos^3\theta + \sin^3\theta) + 6r\sin\theta\cos\theta)\right]$$
$$\quad + \frac{\partial}{\partial\theta}[-r\cos^2\theta\sin\theta + r\sin^2\theta\cos\theta + 3(\cos^2\theta - \sin^2\theta)]$$
$$= \left[3(r(\cos^3\theta + \sin^3\theta) + 12\sin\theta\cos\theta)\right]$$
$$\quad + [2r\cos\theta\sin^2\theta - r\cos^3\theta + 2r\sin\theta\cos^2\theta - r\sin^3\theta - 6\sin(2\theta)]$$
$$= 2r(\cos\theta + \sin\theta). \tag{5.36}$$

And of course this also agrees with the result eqn. (5.32) obtained in Exercise 5.11(d) above.

5.13   Show that one could have obtained the results in Exercise 5.12(b) by lowering the index using the metric,

$$p_{\mu';\nu'} = g_{\mu'\sigma'} V^{\sigma'}_{\;;\nu'}. \tag{5.37}$$

Here $\vec{V}$ is the vector field from Exercise 5.11 and $\tilde{p}$ from Exercise 5.12 is its corresponding one-form.

Solution: Recall from Schutz Eq. (5.31) that the 2D Euclidean space metric in polar coordinates is $g_{rr} = 1, g_{r\theta} = g_{\theta r} = 0, g_{\theta\theta} = r^2$. In Exercise 5.12(b) we found

$$p_{r;r} = \Lambda^{\alpha}_{\;r} \Lambda^{\beta}_{\;r} p_{\alpha,\beta} = 2r(\cos^3\theta + \sin^3\theta) + 6\cos\theta\,\sin\theta = V^r_{\;;r}. \tag{5.38}$$

While using the metric to lower the index of $V^{\sigma'}_{\;;r}$ we find:

$$p_{r;r} = g_{r\sigma'} V^{\sigma'}_{\;;r} = g_{rr} V^r_{\;;r} = V^r_{\;;r}. \tag{5.39}$$

In Exercise 5.12(b) we found

$$p_{\theta;r} = \Lambda^{\alpha}_{\;\theta} \Lambda^{\beta}_{\;r} p_{\alpha,\beta} = r^2\{2[-\cos^2\theta\,\sin\theta + \sin^2\theta\,\cos\theta] + \frac{3}{r}\cos(2\theta)\}$$
$$= r^2 V^{\theta}_{\;;r}. \tag{5.40}$$

While using the metric to lower the index of $V^{\sigma'}_{\;;r}$ we find:

$$p_{\theta;r} = g_{\theta\sigma'} V^{\sigma'}_{\;;r} = g_{\theta\theta} V^{\theta}_{\;;r} = r^2\, V^{\theta}_{\;;r}. \tag{5.41}$$

In Exercise 5.12(b) we found $p_{r;\theta} = p_{\theta;r}$ (a result particular to the special one-form given). And using the metric to lower the index of $V^{\sigma'}_{\;;\theta}$ we find:

$$p_{r;\theta} = g_{r\sigma'} V^{\sigma'}_{\;;\theta} = g_{rr} V^r_{\;;\theta} = V^r_{\;;\theta}$$
$$= r^2 V^{\theta}_{\;;r} = p_{\theta;r}. \qquad\qquad \text{used result eqn. (5.24)} \tag{5.42}$$

Finally in Exercise 5.12(b) we found

$$p_{\theta;\theta} = 2r^3[\cos^2\theta\,\sin\theta + \sin^2\theta\,\cos\theta] - 3r^2\sin(2\theta)$$
$$= r^2 V^{\theta}_{\;;\theta}. \tag{5.43}$$

While using the metric to lower the index of $V^{\sigma'}_{\;;\theta}$ we find:

$$p_{\theta;\theta} = g_{\theta\sigma'} V^{\sigma'}_{\;;\theta} = g_{\theta\theta} V^{\theta}_{\;;\theta} = r^2\, V^{\theta}_{\;;\theta}. \tag{5.44}$$

In summary, all four components of $p_{\mu';\nu'}$ found here using eqn. (5.37) agree with the results of Exercise 5.12(b).

5.15 For the vector whose polar components are $(V_r = 1, V_\theta = 0)$, compute in polars all components of the second covariant derivative $V^\alpha{}_{;\mu;\nu}$. Hint: to find the second derivative, treat the first derivative $V^\alpha{}_{;\mu}$ as any tensor: Schutz Eq. (5.66),

$$\nabla_\beta B^\mu{}_\nu = B^\mu{}_{\nu,\beta} + B^\alpha{}_\nu \Gamma^\mu{}_{\alpha\beta} - B^\mu{}_\alpha \Gamma^\alpha{}_{\nu\beta}.$$

Solution: In principle this is quite straightforward, but there are several places one might slip up. First it is a good idea to write down the general expression, and then substitute the given vector field. Write

$$T^\alpha{}_\mu \equiv V^\alpha{}_{;\mu} = V^\alpha{}_{,\mu} + V^\sigma \, \Gamma^\alpha{}_{\sigma\mu}. \qquad \text{used Schutz's 2nd Eq. (5.64)} \qquad (5.45)$$

Do not substitute the given vector at this point because we are still going to take another derivative:

$$V^\alpha{}_{;\mu;\nu} \equiv T^\alpha{}_{\mu;\nu} = \left(T^\alpha{}_{\mu,\nu}\right) + [T^\sigma{}_\mu \, \Gamma^\alpha{}_{\sigma\nu}] - \{T^\alpha{}_\sigma \, \Gamma^\sigma{}_{\mu\nu}\}. \quad \text{used Schutz Eq. (5.66)}$$
$$(5.46)$$

(To help you debug, we have split eqn. (5.46) into parts enclosed in the three types of brackets that will be referred to below.) Now it is straightforward substitution. The problem simplifies tremendously because there are only three non-zero components of the Christoffel symbol, see eqn. (5.25). The three parts of eqn. (5.46) making the contributions indicated by the respective types of brackets:

$$V^\theta{}_{;\theta;r} = \left(-\frac{1}{r^2}\right) + \left[\frac{1}{r^2}\right] - \left\{\frac{1}{r^2}\right\} = -\frac{1}{r^2},$$

$$V^\theta{}_{;r;\theta} = \left\{-\frac{1}{r^2}\right\} = -\frac{1}{r^2}, \qquad\qquad V^r{}_{;\theta;\theta} = [-1] = -1. \qquad (5.47)$$

All other components are zero. The above solution was verified with Maple™ in the accompanying worksheet.

5.17 Discover how each expression $V^\beta{}_{,\alpha}$ and $V^\mu \Gamma^\beta{}_{\mu\alpha}$ separately transforms under a change of coordinates; for $\Gamma^\beta{}_{\mu\alpha}$, begin with the definition

$$\frac{\partial \vec{e}_\alpha}{\partial x^\beta} = \Gamma^\mu{}_{\alpha\beta} \, \vec{e}_\mu. \qquad\qquad \text{Schutz Eq. (5.44)} \qquad (5.48)$$

Show that neither is the standard tensor law, but that their sum does obey the standard law.

Hint: See derivation by Carroll (2004, §3.2), his Eq. (3.10). SP5.4 and SP5.5 were created as an alternative to this problem. They carry the same message in a straightforward way that follows naturally from what we did in Chapter 2 for vectors. See also SP5.11.

5.19 Verify that the calculation from Schutz Eq. (5.81) to Schutz Eq. (5.84), when repeated
for $\tilde{d}r$ and $\tilde{d}\theta$, shows them to be a coordinate basis.

Solution: We simply repeat the argument but instead of substituting Schutz Eq. (5.77)
we use $\tilde{d}r$ and $\tilde{d}\theta$ as the one-form basis under investigation. We find Schutz Eq. (5.81)
changes to:

$$\tilde{d}r = \cos\theta\,\tilde{d}x + \sin\theta\,\tilde{d}y, \qquad\qquad \text{used Schutz Eq. (5.27)}$$

$$\tilde{d}\theta = -\frac{\sin\theta}{r}\,\tilde{d}x + \frac{\cos\theta}{r}\tilde{d}y. \qquad\qquad \text{used Schutz Eq. (5.26)} \qquad (5.49)$$

We must show that both coordinate functions $\xi(x,y)$ and $\eta(x,y)$ exist. So now instead
of Schutz Eq. (5.82) we get

$$\frac{\partial\eta}{\partial x} = \frac{-1}{r}\sin\theta, \qquad \frac{\partial\eta}{\partial y} = \frac{1}{r}\cos\theta. \qquad (5.50)$$

And instead of Schutz Eq. (5.83) we have factors of $1/r$ on both sides,

$$\frac{\partial^2\eta}{\partial y\partial x} = \frac{\partial}{\partial y}\left(\frac{-1}{r}\sin\theta\right) = \frac{\partial}{\partial y}\left(\frac{-y}{x^2+y^2}\right)$$

$$= -\frac{1}{x^2+y^2} + \frac{2y^2}{(x^2+y^2)^2} \qquad\qquad \text{chain rule}$$

$$= -\frac{x^2+y^2}{(x^2+y^2)^2} + \frac{2y^2}{(x^2+y^2)^2} = \frac{-x^2+y^2}{(x^2+y^2)^2}. \qquad (5.51)$$

On the other hand,

$$\frac{\partial^2\eta}{\partial x\partial y} = \frac{\partial}{\partial x}\left(\frac{1}{r}\cos\theta\right) = \frac{\partial}{\partial x}\left(\frac{x}{x^2+y^2}\right) = \frac{1}{x^2+y^2} - \frac{2x^2}{(x^2+y^2)^2} = \frac{-x^2+y^2}{(x^2+y^2)^2} = \frac{\partial^2\eta}{\partial y\partial x},$$

verifying that $\eta(x,y)$ exists. Repeating the above argument we find that for $\xi(x,y)$ to
exist we require:

$$\frac{\partial\xi}{\partial x} = \cos\theta, \qquad \frac{\partial\xi}{\partial y} = \sin\theta. \qquad (5.52)$$

Find the common mixed partial derivative,

$$\frac{\partial^2\xi}{\partial y\partial x} = \frac{\partial}{\partial y}(\cos\theta) = \frac{\partial}{\partial y}\left(\frac{x}{\sqrt{x^2+y^2}}\right) = -\frac{2xy}{(x^2+y^2)^{3/2}}. \qquad (5.53)$$

And

$$\frac{\partial^2\xi}{\partial x\partial y} = \frac{\partial}{\partial x}(\sin\theta) = \frac{\partial}{\partial x}\left(\frac{y}{\sqrt{x^2+y^2}}\right) = -\frac{2xy}{(x^2+y^2)^{3/2}} = \frac{\partial^2\xi}{\partial y\partial x}, \qquad (5.54)$$

verifying that $\xi(x,y)$ indeed exists too. Thus the basis $\tilde{d}r$ and $\tilde{d}\theta$ are consistent with
a coordinate basis.

5.21   Consider the $x-t$ plane of an inertial observer in SR. A certain uniformly accelerated observer wishes to set up an orthonormal coordinate system. By Exercise 2.21 his world line is:

$$t(\lambda) = a \sinh\lambda, \qquad x(\lambda) = a \cosh\lambda, \qquad \text{Schutz Eq. (5.96)} \qquad (5.55)$$

where a is a constant and $a\lambda$ is his proper time (clock time on his wrist watch).

(a) Show that the spacelike line described by eqn. (5.55) with a as the variable parameter and λ fixed is orthogonal to his world line where they intersect. Changing λ in eqn. (5.55) then generates a *family* of such lines.

Solution: We need to show that the dot product of the tangent vectors to the two curves vanishes. Let $\vec{A}$ be the tangent to the curve with λ fixed and varying a. Applying eqn. (5.7) for the tangent vector to the curve parameterized by a gives

$$\vec{A} = \vec{e}_0 \frac{\partial t(a,\lambda)}{\partial a} + \vec{e}_x \frac{\partial x(a,\lambda)}{\partial a} = \vec{e}_0 \sinh(\lambda) + \vec{e}_x \cosh(\lambda). \qquad (5.56)$$

We can verify that $\vec{A}$ is spacelike ($|\vec{A}| > 0$, see Schutz §2.5) providing a check on our computations:

$$\vec{A} \cdot \vec{A} = \sinh^2\lambda\, \vec{e}_t \cdot \vec{e}_t + \cosh^2\lambda\, \vec{e}_x \cdot \vec{e}_x + 2\sinh\lambda\,\cosh\lambda\, \vec{e}_t \cdot \vec{e}_x = 1.$$

(Recall that $\vec{e}_\alpha \cdot \vec{e}_\beta = \eta_{\alpha\beta}$, Schutz Eq. (2.27).)

Let $\vec{B}$ be the tangent to the curve with a fixed and varying λ (i.e. the world lines of the uniformly accelerated observer):

$$\vec{B} = \vec{e}_0 \frac{\partial t(a,\lambda)}{\partial \lambda} + \vec{e}_x \frac{\partial x(a,\lambda)}{\partial \lambda} = \vec{e}_0\, a \cosh(\lambda) + \vec{e}_x\, a \sinh(\lambda). \qquad (5.57)$$

Substituting eqn. (5.56) and eqn. (5.57) for the two tangent vectors it is easy to show that they are orthogonal:

$$\vec{A} \cdot \vec{B} = A^\alpha B^\beta \eta_{\alpha\beta} = -a \sinh(\lambda)\,\cosh(\lambda) + a \sinh(\lambda)\,\cosh(\lambda) = 0. \qquad (5.58)$$

5.21   (b) Show that eqn. (5.55) defines a transformation from coordinates (t,x) to coordinates (λ, a) that form an orthogonal coordinate system. Draw these coordinates and show they only cover half of the original $t-x$ plane. Show that the coordinates are bad on the lines $|x| = |t|$, so they really cover two disjoint quadrants.

Solution: We must first check that the transformation in question is regular (non-zero Jacobian, see Schutz §5.2). Denoting by x^α the (t,x) coordinates and $x^{\bar{\alpha}}$ for (λ, a) we find the transformation matrix

$$\Lambda^\alpha{}_{\bar{\alpha}} = \begin{pmatrix} \frac{\partial t}{\partial \lambda} & \frac{\partial t}{\partial a} \\ \frac{\partial x}{\partial \lambda} & \frac{\partial x}{\partial a} \end{pmatrix} = \begin{pmatrix} a\cosh(\lambda) & \sinh(\lambda) \\ a\sinh(\lambda) & \cosh(\lambda) \end{pmatrix}. \qquad (5.59)$$

And the Jacobian, i.e. the determinant of this transformation matrix, is

$$\det(\Lambda^{\alpha}_{\;\tilde{\alpha}}) = a\cosh^2\lambda - a\sinh^2\lambda = a,$$

which for $a \neq 0$ is a legitimate (regular) transformation.

Is (λ, a) an orthogonal coordinate system? If so then the coordinate basis vectors are orthogonal, $\vec{e}_{\lambda} \cdot \vec{e}_a = 0$. The coordinate basis vectors can be found from the transformation $\Lambda^{\alpha}_{\;\tilde{\alpha}}$ applied to the pseudo-Cartesian basis. $\Lambda^{\alpha}_{\;\tilde{\alpha}}\vec{e}_{\alpha} = \vec{e}_{\tilde{\alpha}}$ giving

$$\vec{e}_a = \frac{\partial t}{\partial a}\vec{e}_t + \frac{\partial x}{\partial a}\vec{e}_x, \qquad\qquad \vec{e}_{\lambda} = \frac{\partial t}{\partial \lambda}\vec{e}_t + \frac{\partial x}{\partial \lambda}\vec{e}_x. \qquad (5.60)$$

But comparing eqn. (5.60) with eqn. (5.56) and eqn. (5.57) we note that these computations correspond to what we did in Exercise 5.21(a) to find the two tangent vectors $\vec{A}$ and $\vec{B}$. In fact now that we have established that (λ, a) are legitimate coordinates we recognize the two families of curves considered in (a), i.e. curves parameterized by a with fixed λ and vice versa, as the *coordinate curves* of these coordinates. *The tangent vectors of coordinate curves are, in fact, the coordinate basis vectors* (e.g. Hobson et al., 2006, Eq. (3.3)). So we have already found the new basis vectors in terms of the old:

$$\vec{e}_a = \vec{A}, \quad \vec{e}_{\lambda} = \vec{B}, \qquad\qquad (5.61)$$

and already shown they were orthogonal in Exercise 5.21(a).

"Plotting the coordinates" might sound vague, but certainly the easiest thing to do is plot the coordinate curves in the $t-x$ plane, see fig. 5.3 herein. Consider first the coordinate curves with fixed λ and variable a. From eqn. (5.55) these are straight lines with slope $\tanh(\lambda)$. Now consider the coordinate curves with fixed a and variable λ. These are hyperbolae. For $a = 1$ and $\lambda \ll -1$, then

$$t(1, \lambda) = 1 \cdot \sinh(\lambda) = \frac{\exp(\lambda) - \exp(-\lambda)}{2} \approx -\frac{\exp(-\lambda)}{2},$$

$$x(1, \lambda) = 1 \cdot \cosh(\lambda) = \frac{\exp(\lambda) + \exp(-\lambda)}{2} \approx +\frac{\exp(-\lambda)}{2},$$

so the curve approaches the straight line $t = -x$ in the 4th quadrant in the limit $\lambda \to -\infty$. And the family of curves approach this same limit regardless of the value of a as long as it is finite. (We will discuss $a < 0$ in a minute. For now think of $a > 0$.) In fig. 5.3(a) the $a = 0.1$ curve is indistinquishable from a straight line. But changing the axes range so that we focus near the origin, fig. 5.3(b), the hyperbolic nature of the $a = 0.1$ curve is revealed. When λ passes through 0 we have $t = 0$ and $x = a$. So for various fixed $a > 0$ we have a family of curves in the 4th quadrant ($t \leq 0$ and $x > 0$).

For $\lambda > 0$ the curves in the 1st quadrant are the reflection about the x-axis of the curves we just described in the 4th quadrant. And all curves with $a > 0$ approach $t = x$ as $\lambda \to +\infty$.

For $a < 0$ we have a family of curves in the 2nd and 3rd quadrants that are the reflection about the t-axis of the curves we just described in the 1st and 4th quadrants respectively.

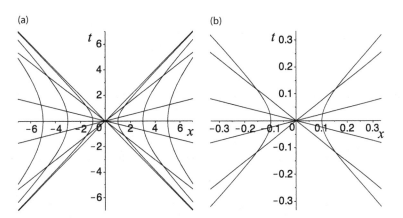

(a) Coordinate curves for eqn. (5.55) with $a = \pm(0.1, 1, 3, 5)$ and for $\lambda = \pm(0.25, 1)$. (b) Same as (a) but focused on the origin region. Plots were generated with Maple^{tm} see accompanying worksheet.

> The coordinates are "bad" along the lines $t = x$ and $t = -x$ in the sense that there are no real values of a and λ that correspond to these lines. Furthermore from the definitions of the hyperbolic functions it is immediately clear that $\cosh \lambda > \sinh \lambda$ for all real λ. So the regions with $|t| \geq |x|$ are "unreachable" by these coordinates. In other words, the regions $t \geq |x|$ and $t \leq -|x|$ are not parameterized by (λ, a).

5.21   (c) Find the metric tensor and all the Christoffel symbols in this coordinate system. This observer will do a perfectly good job, provided that he always uses Christoffel symbols appropriately and sticks to events in his quadrant. In this sense, SR admits accelerated observers. The right-hand quadrant in these coordinates is sometimes called Rindler space, and the boundary lines $x = \pm t$ bear some resemblance to the black-hole horizons we will study later.

> Solution: There are two ways to obtain the metric tensor here. We know the (Minkowski spacetime) metric in $\eta_{\alpha\beta}$ in $t - x$ coordinates. So we can transform this metric to $\lambda - a$ coordinates as we did in §5.2:
>
> $$g_{\bar{\alpha}\bar{\beta}} = \Lambda^{\alpha}{}_{\bar{\alpha}} \Lambda^{\beta}{}_{\bar{\beta}} \, g_{\alpha\beta} = \frac{\partial x^{\alpha}}{\partial x^{\bar{\alpha}}} \frac{\partial x^{\beta}}{\partial x^{\bar{\beta}}} \, g_{\alpha\beta}. \tag{5.62}$$
>
> On the other hand we know the basis vectors from which we can calculate all components of the metric tensor in these coordinates using
>
> $$g_{\bar{\alpha}\bar{\beta}} = \vec{e}_{\bar{\alpha}} \cdot \vec{e}_{\bar{\beta}}. \qquad \text{Schutz Eq. (5.30)} \tag{5.63}$$
>
> In fact we showed in Exercise 5.21(a) that $\vec{A} \cdot \vec{A} = 1$ and $\vec{A} \cdot \vec{B} = 0$, which, using the identification eqn. (5.61), imply $g_{aa} = 1$ and $g_{a\lambda} = g_{\lambda a} = 0$ respectively.

For variety we'll obtain $g_{\lambda\lambda}$ via eqn. (5.62):

$$g_{\lambda\lambda} = \frac{\partial x^\alpha}{\partial\lambda}\frac{\partial x^\beta}{\partial\lambda}\,g_{\alpha\beta} = \frac{\partial t}{\partial\lambda}\frac{\partial t}{\partial\lambda}\,\eta_{00} + \frac{\partial x}{\partial\lambda}\frac{\partial x}{\partial\lambda}\,\eta_{11}$$

$$= \left(\frac{\partial a\sinh\lambda}{\partial\lambda}\right)^2(-1) + \left(\frac{\partial a\cosh\lambda}{\partial\lambda}\right)^2(+1)$$

$$= (a\cosh\lambda)^2\,(-1) + (a\sinh\lambda)^2\,(+1) = -a^2. \tag{5.64}$$

There are only $2\times 3 = 6$ components of the Christoffel symbol to compute. We use

$$\Gamma^\alpha{}_{\mu\nu} = \frac{1}{2}g^{\alpha\sigma}\left(g_{\sigma\mu,\nu} + g_{\sigma\nu,\mu} - g_{\mu\nu,\sigma}\right) \qquad \text{Schutz Eq. (5.75)} \tag{5.65}$$

since we are in a coordinate basis. And we will need the inverse metric tensor:

$$(g^{\bar\alpha\bar\beta}) = \begin{pmatrix} -a^2 & 0 \\ 0 & 1 \end{pmatrix}^{-1} = \begin{pmatrix} -a^{-2} & 0 \\ 0 & 1 \end{pmatrix}. \tag{5.66}$$

The metric depends only upon a since no where does λ appear in our components of **g** above. So we immediately conclude

$$\Gamma^\lambda{}_{\lambda\lambda} = 0. \tag{5.67}$$

$$\Gamma^\lambda{}_{\lambda a} = \frac{1}{2}g^{\lambda\sigma}\left(g_{\sigma\lambda,a} + g_{\sigma a,\lambda} - g_{\lambda a,\sigma}\right)$$

$$= \frac{1}{2}g^{\lambda\lambda}\left(g_{\lambda\lambda,a} + g_{\lambda a,\lambda} - g_{\lambda a,\lambda}\right) \qquad\qquad \text{diagonal metric}$$

$$= \frac{1}{2}g^{\lambda\lambda}\left(g_{\lambda\lambda,a}\right) = \frac{1}{2}\left(\frac{-1}{a^2}\right)\frac{\partial(-a^2)}{\partial a} = \frac{1}{a}$$

$$= \Gamma^\lambda{}_{a\lambda}. \qquad\qquad\qquad\qquad\qquad \text{used Schutz Eq. (5.74)} \tag{5.68}$$

$$\Gamma^\lambda{}_{aa} = \frac{1}{2}g^{\lambda\sigma}\left(g_{\sigma a,a} + g_{\sigma a,a} - g_{aa,\sigma}\right)$$

$$= \frac{1}{2}g^{\lambda\lambda}\left(-g_{aa,\lambda}\right) = 0. \qquad\qquad\qquad \text{diagonal metric} \tag{5.69}$$

$$\Gamma^a{}_{\lambda\lambda} = \frac{1}{2}g^{a\sigma}\left(g_{\sigma\lambda,\lambda} + g_{\sigma\lambda,\lambda} - g_{\lambda\lambda,\sigma}\right)$$

$$= \frac{1}{2}g^{aa}\left(-g_{\lambda\lambda,a}\right) \qquad\qquad\qquad \text{diagonal metric}$$

$$= \frac{1}{2}(+1)\frac{\partial(a^2)}{\partial a} = a. \tag{5.70}$$

$$\Gamma^a{}_{\lambda a} = \frac{1}{2}g^{a\sigma}\left(g_{\sigma\lambda,a} + g_{\sigma a,\lambda} - g_{\lambda a,\sigma}\right)$$

$$= \frac{1}{2}g^{aa}\left(+g_{aa,\lambda}\right) = 0. \qquad\qquad\qquad \text{diagonal metric} \tag{5.71}$$

$$\Gamma^a{}_{aa} = \frac{1}{2} g^{a\sigma} (g_{\sigma a,a} + g_{\sigma a,a} - g_{aa,\sigma})$$

$$= \frac{1}{2} g^{aa} (g_{aa,a} + g_{aa,a} - g_{aa,a}) = 0. \qquad \text{diagonal metric} \qquad (5.72)$$

5.2 Supplementary problems

SP 5.1   Schutz Eq. (5.28b) states that the magnitude of the one-form basis

$$|\tilde{d}\theta| = \frac{1}{r}$$

while Eq. (5.28a) implies that

$$|\vec{e}_\theta| = r.$$

Does this contradict Schutz Eq. (3.47) wherein the magnitude of a one-form was stated to be the same as its associated vector?

Hint

The answer is of course "no." Working through Exercise 3.34 and/or reviewing Schutz §3.3 should help.

Solution

No, there is no contradiction. The one-form bases are not simply the associated one-forms of the vector bases. This was stated explicitly by Schutz in §3.3, p. 61. See also the next supplementary problem SP5.2.

SP 5.2   Write the identity matrix as the product of two transformations and take the partial derivative $\partial/\partial x^\mu$ to show that

$$\Lambda^\alpha{}_{\alpha',\mu} \Lambda^{\alpha'}{}_\beta + \Lambda^\alpha{}_{\alpha'} \Lambda^{\alpha'}{}_{\beta,\mu} = 0. \qquad (5.73)$$

SP 5.3   Where did the equation

$$(\vec{d}\phi)^\alpha = g^{\alpha\beta} \phi_{,\beta} \qquad \text{Schutz Eq. (5.35)} \qquad (5.74)$$

come from? It looks like Schutz Eq. (3.43), $A^\alpha := \eta^{\alpha\beta} A_\beta$, that applied when the metric was that of Minkowski space in pseudo-Cartesian coordinates. Argue that Schutz Eq. (3.43) is more general and applies in general curvilinear coordinates.

Solution

Recall from Schutz §3.5 that the metric plays the role, among other things, of a mapping between vectors and associated one-forms. For example:

$$\mathbf{g}(\vec{V},\ ) := \tilde{V} \qquad \text{Schutz Eq. (3.37)}$$

$$g_{\alpha\beta} V^{\alpha} = V_{\beta}. \qquad \text{component form} \qquad (5.75)$$

The point is that eqn. (5.75) applies not just for the Minkowski metric in pseudo-Cartesian coordinates, but also for the general curvilinear coordinates used in Chapter 5 and also in curved spacetime where the metric is no longer that of Minkowski spacetime. In fact, this will be stated explicitly by Schutz in §6.2. Applying the inverse metric to both sides of eqn. (5.75) we find

$$g^{\beta\mu} g_{\alpha\beta} V^{\alpha} = \delta^{\mu}{}_{\alpha} V^{\alpha} = g^{\beta\mu} V_{\beta}$$

$$V^{\mu} = g^{\beta\mu} V_{\beta}, \qquad (5.76)$$

which has the same form as the equation we were to justify, eqn. (5.74) above. This should be viewed within the general perspective that tensors are mappings from vectors and one-forms into the reals, cf. Exercise 3.17(b). The one-form associated with a given vector and the vector associated with a given one-form can be viewed as being defined by eqn. (5.75) and eqn. (5.76) respectively.

SP 5.4 Multiply

$$\frac{\partial \vec{V}}{\partial x^{\beta}} = \frac{\partial V^{\alpha}}{\partial x^{\beta}} \vec{e}_{\alpha} + V^{\alpha} \frac{\partial \vec{e}_{\alpha}}{\partial x^{\beta}} \qquad \text{Schutz Eq. (5.43)}^{\cdot} \qquad (5.77)$$

by the one-form basis $\tilde{\omega}^{\beta} = \tilde{d}x^{\beta}$ so the basis is explicit in the covariant derivative tensor field. Then show that the result transforms as we would hope, that is like a $\binom{1}{1}$ tensor. This supplementary problem is meant to be easier than Exercise 5.17 but to carry the same message. See also SP5.11.

Hint

Go through Exercise 3.5 again and remind yourself how we showed that $\langle \vec{A}, \tilde{p} \rangle$ was invariant under a change of coordinates. Furthermore, you will need the result eqn. (5.73) from SP5.2.

SP 5.5 The Christoffel symbol is sometimes called the affine connection. It is one of many possible connections that can be used to form a covariant derivative of tensors on a manifold but it is the only one that is *compatible with the metric tensor* in that everywhere $\nabla_{\mu} g_{\alpha\beta} = 0$ and that is *torsion free* in the sense now explained. (In this question we replace the usual semicolon notation with the ∇_{μ} to emphasize that we are considering other possible definitions of the covariant derivative.) Let $S^{\alpha}{}_{\mu\nu}$ be a general (not necessarily affine) connection coefficient. One can prove (Wald, 1984) that in order to obey the Leibniz product rule of differentiation it must be such that the covariant derivative of a given vector can be written in the form of the partial derivative plus the sum of a linear combination of the given vector:

$$\nabla_{\alpha} V^{\beta} = V^{\beta}{}_{,\alpha} + S^{\beta}{}_{\sigma\alpha} V^{\sigma}. \tag{5.78}$$

Prove that the *torsion tensor* defined as $T^{\alpha}{}_{\mu\nu} \equiv \frac{1}{2}(S^{\alpha}{}_{\mu\nu} - S^{\alpha}{}_{\nu\mu})$ is a $\binom{1}{2}$ rank tensor. A connection that is symmetric in the lower indices has $T^{\alpha}{}_{\mu\nu} = 0$, and is said to be torsion-free.

Solution

The covariant derivative of a vector is a tensor; that is what we mean by "covariant derivative." Thus $\nabla_{\alpha} V^{\beta}$ defined via eqn. (5.78) is a tensor. Furthermore, $\tilde{\nabla}_{\alpha} V^{\beta}$ defined by inverting the two lower indices in $S^{\beta}{}_{\sigma\alpha}$ will also be a tensor. The difference of two tensors is also a tensor:

$$\nabla_{\alpha} V^{\beta} - \tilde{\nabla}_{\alpha} V^{\beta} = V^{\beta}{}_{,\alpha} + S^{\beta}{}_{\sigma\alpha} V^{\sigma} - V^{\beta}{}_{,\alpha} - S^{\beta}{}_{\alpha\sigma} V^{\sigma} = 2T^{\beta}{}_{\sigma\alpha} V^{\sigma}. \tag{5.79}$$

The RHS of eqn. (5.79) must be a tensor for any vector V^{σ}, which implies that $T^{\beta}{}_{\sigma\alpha}$ is a $\binom{1}{2}$ tensor. (This solution is based upon (Carroll, 2004, §3.2).)

SP 5.6   The determinant of the metric $g(x^{\mu}) \equiv \det(g_{\mu\nu})$ is, of course, a kind of scalar but it is *not invariant* under a general coordinate transformation $x^{\mu'}(x^{\mu})$. So the determinant of the metric is a different sort of scalar than, say, the rest mass. Show that,

$$g(x^{\mu'}) = \det\left(\frac{\partial x^{\alpha'}}{\partial x^{\alpha}}\right)^{-2} g(x^{\mu}). \tag{5.80}$$

For this reason you might suspect that scalars like g have a different name, and indeed g is sometimes called a scalar density of weight -2. In fact similar considerations apply to tensors, and give rise to the terminology *tensor density* (Carroll, 2004, §2.8).

SP 5.7   Simply by summing over dummy indices it is easy to verify that the Kronecker delta transforms under a general coordinate transformation $x^{\mu'}(x^{\mu})$ like a rank-2 tensor:

$$\delta^{\mu'}{}_{\nu'} = \delta^{\mu}{}_{\nu} \frac{\partial x^{\mu'}}{\partial x^{\mu}} \frac{\partial x^{\nu}}{\partial x^{\nu'}}. \tag{5.81}$$

Derive this from the chain rule of differential calculus.

Solution

For non-degenerate coordinate transformation $x^{\mu'}(x^{\mu})$ we have $\det(\partial x^{\mu'}/\partial x^{\mu}) \neq 0$. Because a matrix is singular if and only if its determinant is zero, we know that the inverse transformation exists $x^\mu/\partial x^{\mu'}$. So we can write:

$$\delta^{\mu'}_{\nu'} = \frac{\partial}{\partial x^{\nu'}} x^{\mu'} \qquad\qquad x^{\mu'} \text{ is independent of } x^{\nu'} \text{ for } \mu' \neq \nu'$$

$$= \frac{\partial x^\mu}{\partial x^{\nu'}} \frac{\partial}{\partial x^\mu} x^{\mu'} \qquad\qquad \text{chain rule, non-degenerate transformation}$$

$$= \frac{\partial x^{\mu'}}{\partial x^\mu} \delta^\mu_{\ \nu} \frac{\partial x^\nu}{\partial x^{\nu'}}, \qquad\qquad \text{rearranged, introduced } \delta^\mu_{\ \nu} \qquad (5.82)$$

in agreement with eqn. (5.81) above.

SP 5.8 Consider a Cartesian coordinate system $\mathcal{S}$ in the plane with coordinate basis $\vec{e}_x$ and $\vec{e}_y$. Now suppose we rotate the coordinates axes by an angle θ to obtain a new coordinate system $\mathcal{S}'$ with coordinate $x'(x, y)$, $y'(x, y)$ and new basis vectors $\vec{e}_{x'}$ and $\vec{e}_{y'}$.

(a) Write the coordinate functions $x'(x, y)$, $y'(x, y)$ as a matrix equation with parameter θ.
(b) For a vector $\vec{A} = A^i \vec{e}_i = A^x \vec{e}_x + A^y \vec{e}_y$, find its components $A^{\alpha'}$ in $\mathcal{S}'$. This is a passive coordinate transformation; the same vector has different components in the new coordinate system.
(c) Find the new basis vectors, $\vec{e}_{x'}$ and $\vec{e}_{y'}$, expressed as linear combinations of the old basis vectors. Hint: These have undergone an active (rotation) transformation.
(d) Convince yourself that $A^{\alpha'} \vec{e}_{\alpha'} = A^\alpha \vec{e}_\alpha$.

SP 5.9 Suppose the acceleration due to gravity was exactly constant near the surface of the Earth, between the base and top of the tower in the Pound–Rebka–Snider experiment. Making *no further approximations*, use the equivalence principle to show that the redshift formula is

$$\frac{\nu_{\text{top}}}{\nu_{\text{bottom}}} = 1 - gh.$$

Hint

Use a free-falling reference frame instantaneously at rest with the tower when the photon is emitted from the base of the tower. In this frame, the tower exhibits uniform acceleration upwards and the results of Exercise 2.19 can be used. See Price (1974) for the solution.

SP 5.10 Reconsider the construction of Schutz Fig. 5.2 where two light pulses are sent vertically from the base to the top of a tower. Treat the gravitational acceleration g as uniform. Construct the Minkowski spacetime diagram by analyzing the experiment from a free-falling reference frame. Trace the world line of the first wave crest from emission at the bottom at event $\mathcal{A}$ to reception at the top at event $\mathcal{B}$. The second wave crest is emitted at the bottom at event $\mathcal{C}$ and received at the top at event $\mathcal{D}$.

(a) What are the shapes of the world lines of the emitter between $\mathcal{A}$ and $\mathcal{C}$, and the receiver between $\mathcal{B}$ and $\mathcal{D}$?

(b) What are the shapes of the world lines of the first and second wave crest?

(c) Show that the length of $\mathcal{BD}$ is $(1 + gh)$ times the length of $\mathcal{AC}$, where h is the height of the tower.

SP 5.11 Find the transformation relations for the Christoffel symbols (also called the *affine connection*) $\Gamma^{\mu}{}_{\alpha\beta}$ defined by

$$\frac{\partial \vec{e}_{\alpha}}{\partial x^{\beta}} = \Gamma^{\mu}{}_{\alpha\beta} \vec{e}_{\mu}. \qquad \text{Schutz Eq. (5.44)} \qquad (5.83)$$

Thus show that the Christoffel symbols are not a tensor. (See also Exercise 5.17 above for a related exercise.)

Solution

Starting with the definition above, we replace the basis vectors $\vec{e}_{\alpha}$ with their representation in another coordinate basis $\vec{e}_{\alpha'}$ linked to coordinate system $x^{\alpha'}(x^{\alpha})$:

$$\vec{e}_{\alpha} = \frac{\partial x^{\alpha'}}{\partial x^{\alpha}} \vec{e}_{\alpha'}, \qquad (5.84)$$

see Schutz Eqs. (5.21)–(5.24). Substituting into eqn. (5.83) above we obtain

$$\Gamma^{\mu}{}_{\alpha\beta} \vec{e}_{\mu} = \frac{\partial}{\partial x^{\beta}} \left(\frac{\partial x^{\alpha'}}{\partial x^{\alpha}} \vec{e}_{\alpha'} \right)$$

$$= \frac{\partial^2 x^{\alpha'}}{\partial x^{\alpha} \partial x^{\beta}} \vec{e}_{\alpha'} + \frac{\partial x^{\alpha'}}{\partial x^{\alpha}} \frac{\partial}{\partial x^{\beta}} \vec{e}_{\alpha'} \qquad \text{used product rule}$$

$$= \frac{\partial^2 x^{\alpha'}}{\partial x^{\alpha} \partial x^{\beta}} \vec{e}_{\alpha'} + \frac{\partial x^{\alpha'}}{\partial x^{\alpha}} \frac{\partial x^{\beta'}}{\partial x^{\beta}} \frac{\partial}{\partial x^{\beta'}} \vec{e}_{\alpha'} \qquad \text{used chain rule}$$

$$= \frac{\partial^2 x^{\mu'}}{\partial x^{\alpha} \partial x^{\beta}} \vec{e}_{\mu'} + \frac{\partial x^{\alpha'}}{\partial x^{\alpha}} \frac{\partial x^{\beta'}}{\partial x^{\beta}} \Gamma^{\mu'}{}_{\alpha'\beta'} \vec{e}_{\mu'} \qquad \text{relabeled, used definition}$$

$$= \left(\frac{\partial^2 x^{\mu'}}{\partial x^{\alpha} \partial x^{\beta}} + \frac{\partial x^{\alpha'}}{\partial x^{\alpha}} \frac{\partial x^{\beta'}}{\partial x^{\beta}} \Gamma^{\mu'}{}_{\alpha'\beta'} \right) \frac{\partial x^{\mu}}{\partial x^{\mu'}} \vec{e}_{\mu} \qquad \text{factored, used eqn. (5.84) above}$$

$$\Gamma^{\mu}{}_{\alpha\beta} = \frac{\partial^2 x^{\mu'}}{\partial x^{\alpha} \partial x^{\beta}} \frac{\partial x^{\mu}}{\partial x^{\mu'}} + \frac{\partial x^{\alpha'}}{\partial x^{\alpha}} \frac{\partial x^{\beta'}}{\partial x^{\beta}} \frac{\partial x^{\mu}}{\partial x^{\mu'}} \Gamma^{\mu'}{}_{\alpha'\beta'}. \qquad (5.85)$$

The first term on the RHS of eqn. (5.85) above precludes the tensorial nature of the Christoffel symbols.

We can obtain the Christoffel symbols in the prime frame by multiplying both sides of eqn. (5.85) by

$$\frac{\partial x^{\nu'}}{\partial x^{\mu}} \frac{\partial x^{\beta}}{\partial x^{\lambda'}} \frac{\partial x^{\alpha}}{\partial x^{\sigma'}}.$$

Then summing over repeated indices and relabelling we find:

$$\Gamma^{\mu'}{}_{\alpha'\beta'} = \frac{\partial x^{\mu'}}{\partial x^\mu}\frac{\partial x^\alpha}{\partial x^{\alpha'}}\frac{\partial x^\beta}{\partial x^{\beta'}}\Gamma^\mu{}_{\alpha\beta} - \frac{\partial^2 x^{\mu'}}{\partial x^\alpha \partial x^\beta}\frac{\partial x^\alpha}{\partial x^{\alpha'}}\frac{\partial x^\beta}{\partial x^{\beta'}}, \tag{5.86}$$

which agrees with Poisson (2004, Eq. (1.6)) derived in another fashion (Poisson, 2004, §1.2).

SP 5.12   Show that the Christoffel symbols $\Gamma^\alpha{}_{\mu\nu}$ for any distinct triplet $\alpha \neq \mu \neq \nu \neq \alpha$ must all vanish for a diagonal metric.

SP 5.13   Many GR textbooks use the notation ∂_α or $\partial/\partial x^\alpha$ as the coordinate basis vectors rather than $\vec{e}_\alpha$ (e.g. Misner et al., 1973; Carroll, 2004; Wald, 1984). This problem aims to explain why. In the beginning of §5.2 Schutz introduced the modern view of a vector as the thing that, given $\phi(x, y)$, produces $\mathrm{d}\phi/\mathrm{d}s$ via

$$\frac{\mathrm{d}\phi}{\mathrm{d}s} = \langle \tilde{\mathrm{d}}\phi, \vec{V} \rangle. \qquad\qquad \text{Schutz Eq. (5.19)} \tag{5.87}$$

The components of $\vec{V}$ were given as $(\mathrm{d}\xi/\mathrm{d}s, \mathrm{d}\eta/\mathrm{d}s)$. But what were the basis vectors? The key to answer this question is the following statement by Schutz: *"This leads to the most modern view, that the tangent vector to the curve should be* called $\mathrm{d}/\mathrm{d}s$*."* Starting from eqn. (5.87) and using the definition of the one-form $\tilde{\mathrm{d}}\phi \rightarrow (\partial\phi/\partial\xi, \partial\phi/\partial\eta)$, argue for the funny-looking basis vectors ∂_α.

Solution

We simply drop the ϕ from eqn. (5.87) and recognize the LHS as the modern view of a vector, $\mathrm{d}/\mathrm{d}s$. But then removing ϕ from the RHS the components of the one-form becomes $(\partial/\partial\xi, \partial/\partial\eta)$:

$$\frac{\mathrm{d}}{\mathrm{d}s} = \left\langle \left(\frac{\partial}{\partial\xi}, \frac{\partial}{\partial\eta}\right), \left(\frac{\mathrm{d}\xi}{\mathrm{d}s}, \frac{\mathrm{d}\eta}{\mathrm{d}s}\right) \right\rangle = \frac{\mathrm{d}\xi}{\mathrm{d}s}\frac{\partial}{\partial\xi} + \frac{\mathrm{d}\eta}{\mathrm{d}s}\frac{\partial}{\partial\eta}. \tag{5.88}$$

The LHS of eqn. (5.88) is a vector, a geometric object, and the RHS is the contraction of the components with $(\partial/\partial\xi, \partial/\partial\eta)$, so the latter must play the role of the vector basis. Generalizing to an arbitrary coordinate system, x^α, we expand the vector $\mathrm{d}/\mathrm{d}s$ as:

$$\frac{\mathrm{d}}{\mathrm{d}s} = \frac{\mathrm{d}x^\alpha}{\mathrm{d}s}\frac{\partial}{\partial x^\alpha} = \frac{\mathrm{d}x^\alpha}{\mathrm{d}s}\partial_\alpha, \tag{5.89}$$

where the second equality just changes the notation for the partial derivative.

6 Curved manifolds

There is an even more striking illustration of the curvature of the sphere. Consider, first, flat space. In Fig. 6.2 a closed path in flat space is drawn, and, starting at A, at each point a vector is drawn parallel to the one at the previous point. This construction is carried around the loop from A to B to C and back to A. The vector finally drawn at A is, of course, parallel to the original one. A completely different thing happens on the sphere! ... the vector field has rotated $90°$ in this construction! Despite the fact that each vector is drawn parallel to its neighbor, the closed loop has caused a discrepancy. ... it must be an effect of the sphere's curvature. This result has radical implications ...

Bernard Schutz, §6.4

6.1 Exercises

6.1 Decide if the following sets are manifolds and say why. If there are exceptional points at which the sets are not manifolds, give them:

> Hint: Schutz has given only an intuitive explanation of manifolds in §6.1, where we are told that a manifold is a space that can be continuously parameterized; that there is a smooth, invertible mapping from points of the manifold to a Euclidean space of the same dimension. This idea is made more rigorous via the mathematics of differential geometry as explained in more advanced GR textbooks (Carroll, 2004; Wald, 1984). An intuitive explanation is all that is required here. If you are struggling with Exercises 1 and 2, it should help to learn the basics of the more rigorous description of a manifold found in differential geometry books (e.g. Faber, 1983, for math students but accessible to physicists) or mathematical physics books (e.g. Schutz, 1980; Hassani, 1999).

6.1 (a) Phase space of Hamiltonian mechanics, the space of the canonical coordinates and momenta p_i and q^i;

> Solution: Yes of course the phase space of Hamiltonian mechanics is a manifold. Suppose there is just one particle, so we have just two canonical coordinates p and q. We could associate p with the x-coordinate and q with the y-coordinate via the

mapping $p = x$ and $q = y$. We can argue that for a physical system the generalized momentum and position must be well-behaved real-valued quantities, $p, q \in \mathbb{R}$. So the canonical coordinates can be plotted on the Euclidean plane. For n particles on a line the index runs from $i = 1, 2, \ldots n$ and we will have $2n$ coordinates describing the system. We could map the canonical coordinates into $\mathbb{R}^{2n}$, the Euclidean space with $2n$ dimensions via the mapping $p_i = X^i$ and $q^i = X^{i+n}$.

6.1   (b) The interior of a circle of unit radius in two-dimensional Euclidean space.

Solution: Yes the interior of a circle in two-dimensional Euclidean space is a manifold. The mapping simply requires restricting the coordinates to be on the interior of the circle in question since the points are already in Euclidean space. Note that there are some subtleties associated with including boundaries (Carroll, 2004) that Schutz is avoiding by specifying the *interior* of the circle.

6.1   (c) The set of permutations of n objects.

Solution: This is not a manifold. Permutations are orderings. For example consider three objects, say a spoon, a fork, and a knife. We can order them by specifying which object goes first, and which second and third. Give the objects more generic names, say X^1, X^2, X^3 respectively. For a given permutation, we can associate values to each name by assigning its cardinal number in the permutation, for example $X^1 = 1$, $X^2 = 2, X^3 = 3$ indicating that the spoon is first, fork second, and knife third; $X^1 = 3, X^2 = 2, X^3 = 1$ is the permutation with the knife first and spoon last. For any ordered triplet of numbers from the set $\{1,2,3\}$, with no repeats, we can identify a unique permutation of the objects. And for any permutation we can describe it with an ordered triplet of numbers. So we have an invertible mapping from the permutations of three objects into a *subset* of 3D Euclidean space. But of course the permutations are mapped to discrete points in Euclidean space; we cannot move continuously from one allowable point in the Euclidean space to the next so this is not a manifold.

6.1   (d) The subset of Euclidean space of two dimensions (coordinates x and y) that is a solution of

$$xy \left(x^2 + y^2 - 1\right) = 0.$$

Solution: The solution of this equation is the union of the x-axis, the y-axis, and the unit circle centered at the origin. Locally the solution set appears one-dimensional

(1D) because one can parameterize it with a single parameter $\theta \in \mathbb{R}$. The union of x- and y-axes and the unit circle, excluding the intersection points, forms a 1D manifold. The points of intersection between the circle and one of the axes are 'exceptional points' that are not part of the manifold because they cannot be parameterized with a single parameter.

In more advanced textbooks for physicists (Carroll, 2004; Wald, 1984; Hawking and Ellis, 1973) or introductions for mathematicians (Faber, 1983) one learns that the manifold must be covered by an atlas consisting of charts that map open subsets of the manifold into open subsets of $\mathbb{R}^n$. These charts form the coordinate systems and can be constructed in different ways. To see how this works we spell out these charts for this exercise. For example, one could form the set of 12 charts $\psi_n : (x, y) \to \theta$ below. Use the standard assumption that $-\pi/2 < \arctan(x) < \pi/2$ for all $x \in \mathbb{R}$. On the unit circle ($x^2 + y^2 = 1$) one has:

$\theta = \psi_1(x, y) = \arctan(y/x)$, $y > 0$ and $x > 0$, (first quadrant),

$\theta = \psi_2(x, y) = \pi + \arctan(y/x)$, $y > 0$ and $x < 0$, (second quadrant),

$\theta = \psi_3(x, y) = -\pi + \arctan(y/x)$, $y < 0$ and $x < 0$, (third quadrant),

$\theta = \psi_4(x, y) = \arctan(y/x)$, $y < 0$ and $x > 0$, (fourth quadrant).

The π was necessary to map the second and third quadrants to the angle $\pi/2 < \theta < \pi$ and $-\pi < \theta < -\pi/2$ respectively. On the y-axis, ($x = 0$), one has four regions to map separately:

$$\theta = \psi_5(0, y) = y, \qquad\qquad -\infty < y < -1, \qquad (6.1)$$

$$\theta = \psi_6(0, y) = y, \qquad\qquad -1 < y < 0, \qquad (6.2)$$

$$\theta = \psi_7(0, y) = y, \qquad\qquad 0 < y < 1, \qquad (6.3)$$

$$\theta = \psi_8(0, y) = y, \qquad\qquad 1 < y < \infty. \qquad (6.4)$$

Similarly on the x-axis, one has four regions to map separately:

$$\theta = \psi_9(x, 0) = x, \qquad\qquad -\infty < x < -1, \qquad (6.5)$$

$$\theta = \psi_{10}(x, 0) = x, \qquad\qquad -1 < x < 0, \qquad (6.6)$$

$$\theta = \psi_{11}(x, 0) = x, \qquad\qquad 0 < x < 1, \qquad (6.7)$$

$$\theta = \psi_{12}(x, 0) = x, \qquad\qquad 1 < x < \infty. \qquad (6.8)$$

The union of the domains of the above 12 smooth, invertable maps $\psi_1 \ldots \psi_{12}$, together with the maps themselves, form a differentiable, 1D manifold.

6.3   It is well known that for any symmetric matrix A (with real entries), there exists a matrix H for which the matrix $H^T A H$ is a diagonal matrix whose entries are the eigenvalues of A. [This exercise develops the transformation matrix from a coordinate system with arbitrary diagonal metric to a LIF.]

(a) Show that there is a matrix R such that $R^T H^T A H R$ is the same matrix as $H^T A H$ except with the eigenvalues rearranged in ascending order along the main diagonal from top to bottom.

Solution: Call the diagonal matrix $D = H^T A H$ and call the reordered matrix $\tilde{D}$,

$$\tilde{D} = R^T D R. \tag{6.9}$$

We will refer to the element in the ith row and jth column of a matrix A as a_{ij} or a^{ij} (no significance attached to the subscript versus superscript position other than it allows us to keep the summation convention). And likewise for other matrices; in particular $(r^T)^{ki}$ is the element of row k and colum i of the R^T matrix. Then

$$\tilde{d}^{kl} = (r^T)^{ki} d_{ij} r^{jl} \qquad \text{element version of eqn. (6.9) above}$$

$$= r^{ik} d_{ij} r^{jl} \qquad\qquad \text{by definition of transpose}$$

$$= r^{ik} d_{ii} r^{il}. \qquad\qquad \text{because } D \text{ is diagonal} \tag{6.10}$$

Suppose we want to move the diagonal element d_{II} into slot K for given (fixed) I and K, i.e. we want $\tilde{d}^{KK} = d_{II}$. We simply choose $r^{IK} = 1$ and $r^{iK} = 0 \ \forall i \neq I$. But does this prescription maintain the desired result that off-diagonal terms of $\tilde{D}$ are still zero? Yes, this is guaranteed because when $k \neq l$ in eqn. (6.10) above, we cannot have both $r^{ik} \neq 0$ and $r^{il} \neq 0$ since that would correspond to moving the diagonal element d_{ii} into two different slots $\tilde{d}^{kk}$ and $\tilde{d}^{ll}$. That is not allowed because each diagonal can only be moved into one slot.

6.3 (b) Show that there exists a third matrix N such that $N^T R^T H^T A H R N$ is a diagonal matrix whose entries on the diagonal are $-1, 0$, or $+1$.

Solution: That is, we are required to show that the diagonal elements can be scaled such that they are either $-1, 0$, or $+1$ using another matrix N as follows: $N^T \tilde{D} N$. As we found above the new elements will be

$$\tilde{\tilde{d}}^{kl} = (n^T)^{ki} \tilde{d}_{ij} n^{jl} = n^{ik} \tilde{d}_{ij} n^{jl} = n^{ik} \tilde{d}_{ii} n^{il}. \tag{6.11}$$

Suppose $\tilde{d}_{KK} \neq 0$ and we want the diagonal element $\tilde{d}_{KK}$ for given (fixed) K to be -1 or $+1$, i.e. we want

$$\tilde{\tilde{d}}^{KK} = 1 \ \mathrm{sgn}(\tilde{d}_{KK}) = \frac{\tilde{d}_{KK}}{|\tilde{d}_{KK}|}.$$

We simply choose $n^{KK} = 1/\sqrt{|\tilde{d}_{KK}|}$ and $n^{iK} = 0 \ \forall i \neq K$. But if $\tilde{d}_{KK} = 0$ we cannot do this, and must choose $n^{KK} = a$, where a is any number, and again set $n^{iK} = 0 \ \forall i \neq K$. We then end up with $\tilde{\tilde{d}}^{KK} = 0$.

6.3   (c) Show that if A has an inverse, none of the diagonal elements in (b) is zero.

Solution: The equation for the inverse is trivial for a diagonal matrix because one can find one equation with one unknown for each element of the inverse. This is immediately clear if we write the equation $DD^{-1} = I$, in element form and solve for the elements of D^{-1}:

$$d^{ij}\, d_{jk}^{-1} = d^{ii}\, d_{ik}^{-1} = \delta^{i}{}_{k}   \qquad\text{because } D \text{ is diagonal}$$
$$d_{ii}^{-1} = 1/d^{ii}, \quad d_{ik}^{-1} = 0 \quad i \neq k. \qquad\qquad (6.12)$$

So the inverse of a diagonal matrix is also diagonal but with diagonal elements equal to the inverse of the original elements. When the original matrix had zero for one or more diagonal elements, then the inverse does not exist because finding it would involve dividing by zero.

6.3   (d) Show from (a)–(c) that there exists a transformation matrix Λ that produces the Minkowski metric in pseudo-Cartesian coordinates,

$$\eta_{\alpha\beta} \equiv \begin{pmatrix} -1 & 0 & 0 & 0 \\ 0 & 1 & 0 & 0 \\ 0 & 0 & 1 & 0 \\ 0 & 0 & 0 & 1 \end{pmatrix}. \qquad\text{Schutz Eq. (6.2)} \qquad (6.13)$$

Solution: Recall the metric tensor $\mathbf{g}$ is symmetric by definition (for example if it is defined from the dot product of two vectors, the order of the vectors does not matter). So as stated in Exercise 6.3(a), this implies that the matrix of $\mathbf{g}$, i.e. $(g_{\alpha\beta})$, can be diagonalized. Furthermore $(g_{\alpha\beta})$ must have an inverse for the mapping from vectors to one-forms to be invertible. Then the results Exercise 6.3(a) through (c) show that we can reduce $(g_{\alpha\beta})$ to a matrix with either -1 or $+1$ on the diagonal. Minkowski spacetime has, by definition, a metric with *Lorentz signature*, i.e. a metric whose matrix representation has one negative eigenvalue and three positive eigenvalues (or one positive and three negative, depending upon convention). So we can choose coordinates, following the prescription in (a) through (c), such that the coordinate transform results in -1 as the first element of the diagonal, the remaining diagonal entries $+1$, and all off-diagonal entries zero. In particular the coordinate transformation matrix must be

$$\Lambda^{\alpha'}{}_{\alpha} = (HRN)^{\alpha'}{}_{\alpha} = h^{\alpha'}{}_{\sigma}\, r^{\sigma}{}_{\mu}\, n^{\mu}{}_{\alpha}. \qquad (6.14)$$

So then when we transform to the new coordinates the metric $g_{\alpha'\beta'}$ transforms to

$$g_{\alpha\beta} = \Lambda^{\alpha'}{}_{\alpha}\, g_{\alpha'\beta'} \Lambda^{\beta'}{}_{\beta} \qquad\qquad\text{general coordinate transformation}$$
$$= h^{\alpha'}{}_{\sigma}\, r^{\sigma}{}_{\mu}\, n^{\mu}{}_{\alpha}\, g_{\alpha'\beta'}\, h^{\beta'}{}_{\gamma}\, r^{\gamma}{}_{\nu}\, n^{\nu}{}_{\alpha} \qquad\text{substituted eqn. (6.14)}$$

$$= n^\mu{}_\alpha \, r^\sigma{}_\mu \, h^{\alpha'}{}_\sigma \, g_{\alpha'\beta'} \, h^{\beta'}{}_\gamma \, r^\gamma{}_\nu \, n^\nu{}_\alpha \qquad\qquad \text{rearranged}$$

$$(\eta_{\alpha\beta}) = N^T R^T H^T (g_{\alpha'\beta'}) H R N. \qquad\qquad \text{matrix notation} \quad (6.15)$$

As we have just shown, because $(g_{\alpha\beta})$ is symmetric, we can choose N, R, H such that $g_{\alpha\beta}$ has only ± 1 or zero on the diagonal, zero elsewhere. And for a spacetime with signature $+2$ this will be the Minkowski metric, $g_{\alpha\beta} = \eta_{\alpha\beta}$. But note that this applies only at a given event on the spacetime manifold; at another event we might need a different transformation. For that reason we say that GR always permits *local inertial frames* or *local Lorentz frames* and the stipulation *local* is important.

6.5 (a) Prove that

$$\Gamma^\mu{}_{\alpha\beta} = \Gamma^\mu{}_{\beta\alpha} \qquad\qquad (6.16)$$

in any coordinate system in a curved Riemannian space.

Hint: In principle a solution would be to use the expression for the Christoffel symbol in terms of the metric:

$$\Gamma^\alpha{}_{\mu\nu} = \frac{1}{2} g^{\alpha\beta} (g_{\beta\mu,\nu} + g_{\beta\nu,\mu} - g_{\mu\nu,\beta}). \qquad \text{Schutz Eq. (6.32)} \quad (6.17)$$

From the symmetry of the metric, $g_{\mu\nu} = g_{\nu\mu}$, and the invariance of $(g_{\beta\mu,\nu} + g_{\beta\nu,\mu})$ under an exchange $\mu \leftrightarrow \nu$, it immediately follows that $\Gamma^\alpha{}_{\mu\nu} = \Gamma^\alpha{}_{\nu\mu}$. But that misses the spirit of the exercise! Recall in §6.3 Schutz stated "It is left to Exer. 6.5 ... to demonstrate, by repeating the flat-space argument now in the locally inertial frame, that $\Gamma^\mu{}_{\beta\alpha}$ is indeed symmetric in any coordinate system, so that Eq. (6.32) is correct in any coordinates." This exercise is so important one really must do it. By the local flatness theorem on a general Riemann manifold, see Schutz §6.2, there is a local inertial (Lorentz) reference frame wherein the local physics is indistinguishable from that of SR. In a Lorentz frame spacetime is locally flat and one can construct a coordinate system with basis vectors that do not change with position, so the Christoffel symbols are zero. This is all one needs to reproduce the argument of §5.4 leading to Schutz Eq. (5.74), $\Gamma^\mu{}_{\alpha\beta} = \Gamma^\mu{}_{\beta\alpha}$.

6.5 (b) Use this to prove that eqn. (6.17) can be derived in the same manner as in flat space.

Solution: eqn. (6.17) is identical to eqn. (5.65), which was derived for flat space in §5.4. The argument leading to eqn. (5.65) can be repeated in curved Riemann space because it used three ingredients:

(i) the metric must have vanishing covariant derivative,

$$g_{\alpha\beta\,;\gamma} = 0. \qquad \text{Schutz Eq. (6.31), cf. Schutz Eq. (5.71)} \qquad (6.18)$$

(ii) the general expression for the covariant derivative of a $\binom{0}{2}$ tensor was applied to the metric,

$$g_{\alpha\beta\,;\mu} = g_{\alpha\beta,\mu} - \Gamma^{\nu}{}_{\alpha\mu}g_{\nu\beta} - \Gamma^{\nu}{}_{\beta\mu}g_{\alpha\nu}, \qquad \text{Schutz Eq. (5.72)} \qquad (6.19)$$

(iii) and $\Gamma^{\mu}{}_{\alpha\beta} = \Gamma^{\mu}{}_{\beta\alpha}$. All three ingredients apply at a point in a curved Riemannian space. Ingredient (i) is immediately true in flat space of course and it also holds locally in curved space because of the local flatness theorem. And ingredient (ii) is generally valid in curved spacetime. Finally we have (iii) from Exercise 6.5(a). Because all the ingredients carry over to curved Riemann space the metric can be used as in eqn. (6.17) to find the Christoffel symbols. This is a very useful equation in GR.

6.7   (a) Give the definition of the determinant of a matrix A in terms of cofactors of elements.

Solution: Say A is an arbitrary $n \times n$ matrix with a_{ij} the element at row i and column j. Then Laplace's formula for the determinant is:

$$\det(A) = \sum_{j=1}^{n} a_{ij}C_{ij} = \sum_{j=1}^{n} a_{ij}(-1)^{i+j}M_{ij} \qquad (6.20)$$

where C_{ij} is the cofactor, i is any row, and M_{ij} is the "minor," that is, the determinant of the matrix formed by removing the ith row and jth column from A.

6.7   (b) Differentiate the determinant of an arbitrary 2×2 matrix and show that it satisfies,

$$g_{,\mu} = g g^{\alpha\beta} g_{\beta\alpha,\mu}. \qquad \text{Schutz Eq. (6.39)} \qquad (6.21)$$

Solution: Here is a "brute force" solution. Let the arbitrary matrix be

$$A = \begin{pmatrix} a & b \\ c & d \end{pmatrix}.$$

Take the derivative with respect to x^{μ} of the determinant of A:

$$\frac{\partial}{\partial x^{\mu}} \det(A) = \frac{\partial}{\partial x^{\mu}}(ad - bc) = a_{,\mu}\,d + a\,d_{,\mu} - b_{,\mu}\,c - b\,c_{,\mu}. \qquad (6.22)$$

We need to show that eqn. (6.21) yields the result eqn. (6.22) when we replace g with the arbitrary 2×2 matrix A. First we write eqn. (6.21) in matrix form, with $A = (g_{\alpha\beta})$:

$$\det(A)_{,\mu} = \det(A) \operatorname{tr}\left(A^{-1} \frac{\partial}{\partial x^\mu} A\right) \qquad \text{tr( ) is "trace," sum of diagonal terms}$$

$$= \det(A) \operatorname{tr}\left(A^{-1} \begin{pmatrix} a_{,\mu} & b_{,\mu} \\ c_{,\mu} & d_{,\mu} \end{pmatrix}\right) \qquad \text{differentiated A component wise}$$

$$= \det(A) \operatorname{tr}\left(\frac{1}{\det(A)} \begin{pmatrix} d & -b \\ -c & a \end{pmatrix}\right.$$

$$\left. \times \begin{pmatrix} a_{,\mu} & b_{,\mu} \\ c_{,\mu} & d_{,\mu} \end{pmatrix}\right) \qquad \text{used eqn. (B.1)}$$

$$= a_{,\mu}\, d + a\, d_{,\mu} - b_{,\mu}\, c - b\, c_{,\mu}, \qquad (6.23)$$

as in eqn. (6.22) above.

6.7   (c) Generalize

$$g_{,\mu} = g\, g^{\alpha\beta} g_{\alpha\beta,\mu} \qquad\qquad \text{Schutz Eq. (6.39)} \qquad (6.24)$$

(by induction or otherwise) to arbitrary $n \times n$ matrices.

Solution: The most instructive derivation results from simply differentiating the expression for the determinant given in eqn. (6.20) above:

$$\frac{\partial \det(A)}{\partial x^\mu} = \frac{\partial}{\partial x^\mu} \sum_{j=1}^{n} a_{ij} C_{ij} \qquad \text{arbitrary fixed i}$$

$$= \sum_{k,l} \frac{\partial}{\partial a_{kl}} \left(\sum_{j=1}^{n} a_{ij} C_{ij}\right) \frac{\partial a_{kl}}{\partial x^\mu} \qquad \text{used chain rule}$$

$$= \sum_{k,l} \frac{\partial}{\partial a_{kl}} \left(\sum_{j=1}^{n} a_{kj} C_{kj}\right) \frac{\partial a_{kl}}{\partial x^\mu} \qquad \text{chose $i = k$}$$

$$= \sum_{k,l,j} \left(\frac{\partial a_{kj}}{\partial a_{kl}} C_{kj} + a_{kj} \frac{\partial C_{kj}}{\partial a_{kl}}\right) \frac{\partial a_{kl}}{\partial x^\mu}. \qquad \text{product rule} \qquad (6.25)$$

Recall that C_{kj} is proportional to the determinant of matrix formed from A with row k and column j removed, so it is independent of a_{kl}, giving $\partial C_{kj}/\partial a_{kl} = 0$. Hence the judicious choice of row $i = k$! So eqn. (6.25) simplies to

$$\frac{\partial \det(A)}{\partial x^\mu} = \sum_{k,l,j} \left(\frac{\partial a_{kj}}{\partial a_{kl}} C_{kj} \right) \frac{\partial a_{kl}}{\partial x^\mu} = \sum_{k,l,j} \delta_{jl} C_{kj} \frac{\partial a_{kl}}{\partial x^\mu}$$

$$= \sum_{k,j} C_{kj} \frac{\partial a_{kj}}{\partial x^\mu} \qquad \qquad \text{summed over } l$$

$$= \text{tr}\left(\text{adj}(A) \frac{\partial A}{\partial x^\mu} \right). \qquad \qquad \text{matrix notation} \quad (6.26)$$

The above result is known as Jacobi's formula. Here adj(A) is the adjugate of A, defined as the transpose of the matrix of cofactors of A. Now we can use,

$$A^{-1} = \frac{1}{\det(A)} \text{adj}(A), \qquad \text{Cramer's formula for the inverse}$$

to substitute for adj(A) in eqn. (6.26) to bring it closer to the form of eqn. (6.24):

$$\frac{\partial \det(A)}{\partial x^\mu} = \det(A)\text{tr}\left(A^{-1} \frac{\partial A}{\partial x^\mu} \right) \qquad \text{used Cramer's formula}$$

$$\frac{\partial g}{\partial x^\mu} = g g^{\alpha\beta} \frac{\partial g_{\beta\alpha}}{\partial x^\mu}. \qquad \qquad \text{tensor notation} \quad (6.27)$$

6.9   Show that

$$V^\alpha_{;\alpha} = \frac{1}{\sqrt{-g}} \left(\sqrt{-g} V^\alpha \right)_{,\alpha} \qquad \text{Schutz Eq. (6.42)} \qquad (6.28)$$

leads to,

$$V^\alpha_{;\alpha} = \frac{1}{r} \frac{\partial}{\partial r} \left(r V^\alpha \right) + \frac{\partial}{\partial \theta} V^\theta. \qquad \text{Schutz Eq. (5.56)} \qquad (6.29)$$

Derive the divergence formula for the metric [of Euclidean space in spherical coordinates]

$$(g_{ij}) = \begin{pmatrix} 1 & 0 & 0 \\ 0 & r^2 & 0 \\ 0 & 0 & r^2 \sin^2\theta \end{pmatrix}. \qquad \text{Schutz Eq. (6.19)} \qquad (6.30)$$

Solution: The first part amounts to showing that the general formula for the divergence of a velocity field is consistent with the special case derived in Schutz §5.3 for the Euclidean plane in polar coordinates. We start with Schutz Eq. (5.32) for the Euclidean metric in polar coordinates. The determinant is simply

$$g = \det(g_{\alpha\beta}) = \det \begin{pmatrix} 1 & 0 \\ 0 & r^2 \end{pmatrix} = r^2. \qquad (6.31)$$

Since this determinant is positive, while eqn. (6.28) applies to the case of a Lorentz metric where $g < 0$, we must rewrite eqn. (6.28) to apply to both positive or negative metric signature by replacing $\sqrt{-g} \to \sqrt{|g|}$. Substituting g into eqn. (6.28) we find

$$V^{\alpha}_{\ ;\alpha} = \frac{1}{\sqrt{|g|}}(\sqrt{|g|}\,V^{\alpha})_{,\alpha} = V^{\alpha}_{\ ,\alpha} + \frac{V^{\alpha}}{\sqrt{|g|}}(\sqrt{|g|})_{,\alpha} \qquad \text{used product rule}$$

$$= V^{r}_{\ ,r} + V^{\theta}_{\ ,\theta} + \frac{V^{r}}{\sqrt{r^2}}(\sqrt{r^2})_{,r} \qquad \text{substituted result eqn. (6.31)}$$

$$= V^{r}_{\ ,r} + V^{\theta}_{\ ,\theta} + \frac{V^{r}}{r} = \frac{1}{r}\frac{\partial}{\partial r}\left(rV^{\alpha}\right) + \frac{\partial}{\partial\theta}V^{\theta}, \qquad (6.32)$$

consistent with eqn. (6.29).

The second part amounts to repeating the exercise for 3D Euclidean space in spherical coordinates. Take the determinant of the matrix of the metric in eqn. (6.30),

$$g = \det\begin{pmatrix} 1 & 0 & 0 \\ 0 & r^2 & 0 \\ 0 & 0 & r^2\sin^2\theta \end{pmatrix} = r^4\sin^2\theta,$$

which has two non-zero gradient components,

$$g_{,r} = \frac{\partial\sqrt{r^4\sin^2\theta}}{\partial r} = 2r\sin\theta, \qquad g_{,\theta} = \frac{\partial\sqrt{r^4\sin^2\theta}}{\partial\theta} = r^2\cos\theta. \qquad (6.33)$$

Substituting eqn. (6.33) into eqn. (6.28), again replacing $\sqrt{-g}$ with $\sqrt{|g|}$, we find,

$$V^{i}_{\ ;i} = V^{r}_{\ ,r} + V^{\theta}_{\ ,\theta} + V^{\phi}_{\ ,\phi} + \frac{V^{r}}{r^2\sin\theta}2r\sin\theta + \frac{V^{\theta}}{r^2\sin\theta}r^2\cos\theta$$

$$= V^{r}_{\ ,r} + V^{\theta}_{\ ,\theta} + V^{\phi}_{\ ,\phi} + \frac{2V^{r}}{r} + \frac{V^{\theta}}{\tan\theta}. \qquad (6.34)$$

This is *not* the same as the formula for the divergence of a vector in spherical coordinates found elsewhere (e.g. Davis and Snider, 1979, Eq. (5.27)), see SP (6.14).

6.11 In this exercise we will determine the condition that a vector field $\vec{V}$ can be considered to be globally parallel on a manifold. More precisely, what guarantees that we can find a vector field $\vec{V}$ satisfying the equation

$$(\nabla\vec{V})^{\alpha}_{\ \beta} = V^{\alpha}_{\ ;\beta} = V^{\alpha}_{\ ,\beta} + \Gamma^{\alpha}_{\ \mu\beta}V^{\mu} = 0\ ? \qquad (6.35)$$

(a) A necessary condition, called the integrability condition for this equation, follows from the commuting of partial derivatives. Show that $V^{\alpha}_{\ ,\nu\beta} = V^{\alpha}_{\ ,\beta\nu}$ implies

$$(\Gamma^{\alpha}_{\ \mu\beta,\nu} - \Gamma^{\alpha}_{\ \mu\nu,\beta})V^{\mu} = (\Gamma^{\alpha}_{\ \mu\beta}\Gamma^{\mu}_{\ \sigma\nu} - \Gamma^{\alpha}_{\ \mu\nu}\Gamma^{\mu}_{\ \sigma\beta})V^{\sigma}. \qquad (6.36)$$

Solution: This problem is fairly straightforward once one thinks about "what would I do to eqn. (6.35) to obtain a term like ... in eqn. (6.36)." Obviously to obtain a term like $\Gamma^\alpha{}_{\mu\beta,\nu}$ one must take $\partial/\partial x^\nu$ of eqn. (6.35). Now rewrite this result with ν and β interchanged and subtract these two. This will eliminate $V^\alpha{}_{,\beta\nu}$ because of the commuting property of partial differentiation, as Schutz indicated. But then one must also deal with terms like $\Gamma^\alpha{}_{\mu\beta}V^\mu{}_{,\nu}$. For these one uses eqn. (6.35) again (without differentiating).

6.11 (b) By relabeling indices, work this into the form:[1]

$$(\Gamma^\alpha{}_{\mu\beta,\nu} - \Gamma^\alpha{}_{\mu\nu,\beta} + \Gamma^\alpha{}_{\sigma\nu}\Gamma^\sigma{}_{\mu\beta} - \Gamma^\alpha{}_{\sigma\beta}\Gamma^\sigma{}_{\mu\nu})\,V^\mu = 0. \qquad (6.37)$$

Solution: The first two terms in eqn. (6.37) and eqn. (6.36) are identical. Interchanging μ and σ on the RHS of eqn. (6.36), and bringing these terms to the LHS gives the remaining two terms in eqn. (6.37). Is that allowed? Yes because they are dummy indices. How can one spot these things?! Based on the sign of the third term in eqn. (6.37) it is clearly the final term in eqn. (6.36).

6.13 (a) Show that if $\vec{A}$ and $\vec{B}$ are parallel-transported along a curve, then $\mathbf{g}(\vec{A}, \vec{B}) = \vec{A} \cdot \vec{B}$ is constant on the curve.

Solution: A vector that is parallel-transported along a curve is moved in the direction of the tangent to the curve without rotating or changing its length. From this notion one should expect that the dot product of two vectors that were parallel-transported along a curve would not change. To demonstrate this mathematically, take the derivative along the curve (parameterized by λ) of the dot product:

$$\frac{d\,\mathbf{g}(\vec{A}, \vec{B})}{d\lambda} = \frac{d}{d\lambda}\left(g_{\alpha\beta}A^\alpha B^\beta\right) = A^\alpha B^\beta \frac{d}{d\lambda}\left(g_{\alpha\beta}\right) + g_{\alpha\beta}B^\beta\frac{d}{d\lambda}\left(A^\alpha\right) + g_{\alpha\beta}A^\alpha\frac{d}{d\lambda}\left(B^\beta\right).$$
$$(6.38)$$

All the derivatives are zero. The first term is the derivative of the metric along the curve:

$$\frac{d}{d\lambda}\left(g_{\alpha\beta}\right) = \frac{dx^\mu}{d\lambda}g_{\alpha\beta;\mu} = 0. \qquad \text{used eqn. (6.18)} \qquad (6.39)$$

The second and third terms are the derivatives of the vectors along the curve. These are zero because these vectors were assumed to be parallel-transported along the curve, $dA^\alpha/d\lambda = 0$ for A^α parallel-transported along the curve parameterized by λ, see Schutz Eq. (6.47).

[1] We have corrected a typo replacing $\Gamma^\sigma{}_{\sigma\nu}$ with $\Gamma^\alpha{}_{\sigma\nu}$.

6.13   (b) Conclude from the results of Exercise 6.13(a) that if a *geodesic* is spacelike (or timelike or null) at some point, it is necessarily spacelike (or timelike or null) at all points.

Solution: Vectors were defined as spacelike (or timelike or null) if their magnitude was $> 0(< 0, = 0)$, see Schutz §2.5. A geodesic is of course not a vector, but it does have a tangent vector at each point along the curve that gives the linear approximation to the displacement along the curve at the point, per unit of the parameter that parameterizes the curve. So it would be reasonable to call a geodesic spacelike at a point if its tangent vector $\vec{U}$ were of positive magnitude at that point, $\vec{U} \cdot \vec{U} = g_{\alpha\beta}U^\alpha U^\beta > 0$. Can this change as one moves along the curve? The geodesic is, by definition, the curve that parallel-transports its own tangent vector. But from (a) we have that *any* two vectors that are parallel-transported by an arbitrary (smooth) curve keep the same dot product. So the tangent vector, dotted with itself, does not change as it is parallel-transported along the geodesic.

6.14   Proper distance along a curve whose tangent is $\vec{V}$ is given by

$$l = \int_{\lambda_0}^{\lambda_1} |\vec{V} \cdot \vec{V}|^{1/2}\, d\lambda. \qquad \text{Schutz Eq. (6.8)} \qquad (6.40)$$

Show that if the curve is a geodesic, then proper length is an affine parameter. (Use results of Exercise 6.13.)

Solution: For a geodesic eqn. (6.40) for the proper length simplifies as follows:

$$l = \int_{\lambda_0}^{\lambda_1} |\vec{U} \cdot \vec{U}|^{1/2}\, d\lambda, \qquad \text{where } \vec{U} \text{ is tangent vector to geodesic}$$

$$= |\vec{U} \cdot \vec{U}|^{1/2} \int_{\lambda_0}^{\lambda_1} d\lambda. \qquad \text{used Exercise 6.13(a).} \qquad (6.41)$$

Thus the proper distance along the curve has the form,

$$l(\lambda) = |\vec{U} \cdot \vec{U}|^{1/2}(\lambda - \lambda_0)$$

$$= |\vec{U} \cdot \vec{U}|^{1/2}\lambda - \lambda_0|\vec{U} \cdot \vec{U}|^{1/2},$$

$$= a\lambda + b, \qquad (6.42)$$

where $a = |\vec{U} \cdot \vec{U}|^{1/2}$ is constant (see Exercise 6.13 (a)), and $b = -\lambda_0 a$. This has the same form as Schutz Eq. (6.52), which was shown to be an affine parameter in Exercise 6.12.

6.15 Use Exercises 6.13 and 6.14 to prove that the proper length of a geodesic between two points is unchanged to first order by small changes in the curve that do not change its endpoints.

Solution: Consider two spacetime events $\mathcal{A}$ and $\mathcal{B}$ connected by a geodesic $x_g^\alpha(\lambda)$ of proper length l, where λ is an affine parameter along the curve. Let $\delta x^\alpha(\lambda)$ be a small departure from the geodesic curve. This defines a new curve $x_p^\alpha(\lambda) = x_g^\alpha(\lambda) + \delta x^\alpha(\lambda)$ also parameterized with λ but whose tangent vector is

$$\vec{V} = \frac{\mathrm{d}x_p^\alpha}{\mathrm{d}\lambda} = \frac{\mathrm{d}}{\mathrm{d}\lambda}(x_g^\alpha(\lambda) + \delta x^\alpha(\lambda)). \tag{6.43}$$

The proper distance along $x_p^\alpha(\lambda)$ is given by

$$l + \delta l = \int_{\lambda_0}^{\lambda_1} \left| \vec{V} \cdot \vec{V}) \right|^{1/2} \mathrm{d}\lambda \qquad\qquad \text{used eqn. (6.40)}$$

$$= \int_{\lambda_0}^{\lambda_1} \left| \frac{\mathrm{d}x_g^\alpha}{\mathrm{d}\lambda} \cdot \frac{\mathrm{d}x_g^\alpha}{\mathrm{d}\lambda} + 2\frac{\mathrm{d}x_g^\alpha}{\mathrm{d}\lambda} \cdot \frac{\mathrm{d}\delta x^\alpha}{\mathrm{d}\lambda} + O(|\delta x^\alpha|^2) \right|^{1/2} \mathrm{d}\lambda \qquad \text{used eqn. (6.43)}$$

$$\simeq l + \int_{\lambda_0}^{\lambda_1} \left| \frac{\mathrm{d}x_g^\alpha}{\mathrm{d}\lambda} \cdot \frac{\mathrm{d}\delta x^\alpha}{\mathrm{d}\lambda} \right| \mathrm{d}\lambda \qquad\qquad \text{used binomial series}$$

$$= l + \left| \frac{\mathrm{d}x_g^\alpha}{\mathrm{d}\lambda} \cdot \left[\delta x^\alpha \right]_{\lambda_0}^{\lambda_1} \right| \qquad\qquad \text{tangent vector constant}$$

$$= l. \qquad\qquad \delta x^\alpha(\lambda_1) = \delta x^\alpha(\lambda_0) = 0$$

The final line used the given BCs. This proves that geodesics have extremal proper length.

6.17 (a) Prove that[2]

$$g_{\alpha\beta,\mu}(\mathcal{P}) \underset{\text{LIF}}{=} 0 \qquad\qquad \text{Schutz Eq. (6.5)} \tag{6.44}$$

at some event $\mathcal{P}$ implies that

$$g^{\alpha\beta}{}_{,\mu}(\mathcal{P}) \underset{\text{LIF}}{=} 0. \tag{6.45}$$

Solution: The metric tensor applied to its inverse gives the identity matrix, see SP (6.16). And the identity matrix is of course constant, so we have

$$g_{\alpha\mu}g^{\mu\beta} = g^\beta{}_\alpha = \delta^\beta{}_\alpha \qquad\qquad \text{used eqn. (6.148)}$$

$$(g_{\alpha\mu}g^{\mu\beta})_{,\gamma} = \delta^\beta{}_{\alpha,\gamma} = 0 \qquad \text{identity matrix is constant}$$

---

[2] We have altered the notation slightly by putting the LIF below the equality sign to remind the reader that we are working in a Local Inertial Frame.

$$g_{\alpha\mu,\gamma}\, g^{\mu\beta} + g_{\alpha\mu}\, g^{\mu\beta}{}_{,\gamma} = 0 \qquad\qquad\qquad\qquad \text{product rule}$$

$$g_{\alpha\mu}\, g^{\mu\beta}{}_{,\gamma} \underset{\text{LIF}}{=} 0. \qquad\qquad\qquad\qquad \text{used eqn. (6.44)} \qquad (6.46)$$

We are not finished yet because $g_{\alpha\beta}$ is a general tensor and there could in principle be several non-zero terms in each column that cancel to produce zero when multiplied by $g^{\mu\beta}{}_{,\gamma}$. To eliminate that possibility we simply multiply both sides of eqn. (6.46) by $g^{\nu\alpha}$ which gives

$$g^{\nu\alpha} g_{\alpha\mu}\, g^{\mu\beta}{}_{,\gamma} \underset{\text{LIF}}{=} 0$$

$$\delta^{\nu}{}_{\mu} g^{\mu\beta}{}_{,\gamma} \underset{\text{LIF}}{=} 0 \qquad\qquad\qquad\qquad \text{used eqn. (6.148)}$$

$$g^{\nu\beta}{}_{,\gamma} \underset{\text{LIF}}{=} 0, \qquad\qquad\qquad\qquad \text{summed over } \mu \qquad (6.47)$$

as we were required to prove.

6.17   (b) Use results of (a) to establish:[3]

$$\Gamma^{\alpha}{}_{\mu\nu,\sigma} \underset{\text{LIF}}{=} \frac{1}{2} g^{\alpha\beta} (g_{\beta\mu,\nu\sigma} + g_{\beta\nu,\mu\sigma} - g_{\mu\nu,\beta\sigma}). \qquad \text{Schutz Eq. (6.64)} \qquad (6.48)$$

Solution: Starting with eqn. (6.17)

$$\Gamma^{\alpha}{}_{\mu\nu} = \frac{1}{2} g^{\alpha\beta} (g_{\beta\mu,\nu} + g_{\beta\nu,\mu} - g_{\mu\nu,\beta}), \qquad\qquad\qquad\qquad (6.49)$$

we simply differentiate with respect to x^{σ},

$$\Gamma^{\alpha}{}_{\mu\nu,\sigma} = \frac{1}{2} \frac{\partial}{\partial x^{\sigma}} \left(g^{\alpha\beta} (g_{\beta\mu,\nu} + g_{\beta\nu,\mu} - g_{\mu\nu,\beta}) \right),$$

$$= \frac{1}{2} g^{\alpha\beta}{}_{,\sigma} (g_{\beta\mu,\nu} + g_{\beta\nu,\mu} - g_{\mu\nu,\beta}) + \frac{1}{2} g^{\alpha\beta} (g_{\beta\mu,\nu\sigma} + g_{\beta\nu,\mu\sigma} - g_{\mu\nu,\beta\sigma}),$$

$$\underset{\text{LIF}}{=} \frac{1}{2} g^{\alpha\beta} (g_{\beta\mu,\nu\sigma} + g_{\beta\nu,\mu\sigma} - g_{\mu\nu,\beta\sigma}). \qquad\qquad\qquad\qquad (6.50)$$

The final line is only true in a local inertial frame (LIF).

6.17   (c) Fill in the steps needed to establish:

$$R_{\alpha\beta\mu\nu} \underset{\text{LIF}}{=} \frac{1}{2} \left(g_{\alpha\nu,\beta\mu} - g_{\alpha\mu,\beta\nu} + g_{\beta\mu,\alpha\nu} - g_{\beta\nu,\alpha\mu} \right). \qquad \text{Schutz Eq. (6.68)} \quad (6.51)$$

---

[3] Again we have altered the notation slightly by putting the LIF below the equality sign to remind the reader that we are working in a Local Inertial Frame.

Solution: We start with the definition of the Riemann curvature tensor in terms of the Christoffel symbols,

$$R^{\alpha}{}_{\beta\mu\nu} := \Gamma^{\alpha}{}_{\beta\nu,\mu} - \Gamma^{\alpha}{}_{\beta\mu,\nu} + \Gamma^{\alpha}{}_{\sigma\mu}\Gamma^{\sigma}{}_{\beta\nu} - \Gamma^{\alpha}{}_{\sigma\nu}\Gamma^{\sigma}{}_{\beta\mu}. \quad \text{Schutz Eq. (6.63)} \quad (6.52)$$

Because we are in a LIF at point $\mathcal{P}$, the Christoffel symbols vanish at $\mathcal{P}$ leaving just the terms involving their derivatives:

$$R^{\alpha}{}_{\beta\mu\nu} \underset{\text{LIF}}{=} \Gamma^{\alpha}{}_{\beta\nu,\mu} - \Gamma^{\alpha}{}_{\beta\mu,\nu}. \tag{6.53}$$

For the derivative of the Christoffel symbol, we substitute from eqn. (6.48) above. To have the right indices for the first term of course we make the substitutions $\mu \rightarrow \beta$, $\sigma \rightarrow \mu$ and $\beta \rightarrow \sigma$. For the second term we simply interchange μ and ν in the first term and change the sign, giving

$$R^{\alpha}{}_{\beta\mu\nu} \underset{\text{LIF}}{=} \frac{1}{2}g^{\alpha\sigma}\left(g_{\sigma\beta,\nu\mu} + g_{\sigma\nu,\beta\mu} - g_{\beta\nu,\sigma\mu} \underline{-g_{\sigma\beta,\mu\nu}} - g_{\sigma\mu,\beta\nu} + g_{\beta\mu,\sigma\nu}\right).$$

$$\text{Schutz Eq. (6.65)}$$

Because partial derivatives commute the underlined terms above cancel giving

$$R^{\alpha}{}_{\beta\mu\nu} \underset{\text{LIF}}{=} \frac{1}{2}g^{\alpha\sigma}\left(g_{\sigma\nu,\beta\mu} - g_{\sigma\mu,\beta\nu} + g_{\beta\mu,\sigma\nu} - g_{\beta\nu,\sigma\mu}\right). \quad \text{changed order} \quad (6.54)$$

Finally we must lower the index. Change α to λ and multiply by $g_{\alpha\lambda}$ to give

$$R_{\alpha\beta\mu\nu} \underset{\text{LIF}}{=} g_{\alpha\lambda}R^{\lambda}{}_{\beta\mu\nu} = g_{\alpha\lambda}\frac{1}{2}g^{\lambda\sigma}\left(g_{\sigma\nu,\beta\mu} - g_{\sigma\mu,\beta\nu} + g_{\beta\mu,\sigma\nu} - g_{\beta\nu,\sigma\mu}\right),$$

$$\underset{\text{LIF}}{=} \delta_{\alpha}{}^{\sigma}\frac{1}{2}\left(g_{\sigma\nu,\beta\mu} - g_{\sigma\mu,\beta\nu} + g_{\beta\mu,\sigma\nu} - g_{\beta\nu,\sigma\mu}\right),$$

$$\underset{\text{LIF}}{=} \frac{1}{2}\left(g_{\alpha\nu,\beta\mu} - g_{\alpha\mu,\beta\nu} + g_{\beta\mu,\alpha\nu} - g_{\beta\nu,\alpha\mu}\right), \tag{6.55}$$

which is eqn. (6.51).

6.19 Prove that $R^{\alpha}{}_{\beta\mu\nu} = 0$ for polar coordinates in the Euclidean plane. Use the Christoffel symbols from eqn. (5.25) above or equivalent results.

Solution: First we find the number of independent components of $R^{\alpha}{}_{\beta\mu\nu}$ in two dimensions so we know when to stop calculating! [Refer to Exercise 6.18(b) for the case of four-dimensional space.] There is only one degree of freedom associated with the first pair of indices because $R_{r\theta\mu\nu} = -R_{\theta r\mu\nu}$ and $R_{\alpha\alpha\mu\nu} = 0$. And similarly only one degree of freedom associated with the last two indices since $R_{\alpha\beta r\theta} = -R_{\alpha\beta\theta r}$ and $R_{\alpha\beta\mu\mu} = 0$. Furthermore these two values are related by $R_{\alpha\beta\mu\nu} = R_{\mu\nu\alpha\beta}$. So there is only one independent value to compute, e.g. $R_{r\theta r\theta}$.

Starting with the definition of the Riemann tensor in terms of the Christoffel symbols, and using eqn. (5.25) for the Christoffel symbols of polar coordinates, we find (underlined terms are zero):

$$
\begin{aligned}
R_{r\theta r\theta} &= \Gamma^r{}_{\theta\theta,r} - \underline{\Gamma^r{}_{\theta r,\theta}} + \Gamma^\sigma{}_{\theta\theta}\Gamma^r{}_{\sigma r} - \Gamma^\sigma{}_{\theta r}\Gamma^r{}_{\sigma\theta} && \text{used eqn. (6.52)} \\
&= \Gamma^r{}_{\theta\theta,r} + \Gamma^r{}_{\theta\theta}\underline{\Gamma^r{}_{rr}} - \Gamma^\theta{}_{\theta r}\Gamma^r{}_{\theta\theta} && \text{used } \Gamma^r{}_{\theta r} = \Gamma^\theta{}_{\theta\theta} = 0 \\
&= \Gamma^r{}_{\theta\theta,r} - \Gamma^\theta{}_{\theta r}\Gamma^r{}_{\theta\theta} && \text{used } \Gamma^\mu{}_{rr} = 0 \\
&= \frac{\partial(-r)}{\partial r} - \frac{1}{r}(-r) = -1 + 1 = 0. && \text{used eqn. (5.25)}
\end{aligned}
$$

$$(6.56)$$

And this is of course what we expect since (despite the polar coordinates) we are in Euclidean space, which is flat. A necessary and sufficient condition for space to be flat is that the Riemann tensor vanishes, cf. Schutz Eq. (6.71).

6.21   Consider the sentences following

$$[\nabla_\alpha, \nabla_\beta]F^\mu{}_\nu = R^\mu{}_{\sigma\alpha\beta}F^\sigma{}_\nu + R_\nu{}^\sigma{}_{\alpha\beta}F^\mu{}_\sigma. \qquad \text{Schutz Eq. (6.78)} \qquad (6.57)$$

They were: "... *each* index gets a Riemann tensor on it, and each one comes in with a + sign. (They must all have the same sign because raising and lowering indices with **g** is unaffected by ∇_α, since $\nabla \mathbf{g} = 0$.)" Why does the argument in parentheses not apply to the signs in

$$V^\alpha{}_{;\beta} = V^\alpha{}_{,\beta} + \Gamma^\alpha{}_{\mu\beta}V^\mu \quad \text{and} \quad V_{\alpha;\beta} = V_{\alpha,\beta} - \Gamma^\mu{}_{\alpha\beta}V_\mu ?$$

Solution: To see why things work nicely for eqn. (6.57) above, raise the index ν (by changing it to dummy index γ and multiplying by $g^{\nu\gamma}$). Because this operation commutes with the covariant derivative we simply have

$$
\begin{aligned}
[\nabla_\alpha, \nabla_\beta]F^{\mu\nu} &= R^\mu{}_{\sigma\alpha\beta}F^{\sigma\nu} + R^{\nu\sigma}{}_{\alpha\beta}F^\mu{}_\sigma \\
&= R^\mu{}_{\sigma\alpha\beta}F^{\sigma\nu} + R^\nu{}_{\sigma\alpha\beta}F^{\mu\sigma}. \qquad \text{raise and lower } \sigma, \text{ cf. SP (3.2)}
\end{aligned}
$$

$$(6.58)$$

So we see that eqn. (6.57) above is compatible with an expression with two contravariant indices and is an obvious generalization of the case with one contravariant index, Schutz Eq. (6.77), following the prescription given. But the case of a covariant derivative of a one-form is quite different. When we attempt to raise the index we encounter a complication on the RHS; the metric tensor does *not* commute with the partial derivative resulting in terms involving its derivative:

$$g^{\gamma\alpha} V_{\gamma;\beta} = g^{\gamma\alpha}\left(V_{\gamma,\beta} - \Gamma^{\mu}{}_{\gamma\beta} V_{\mu}\right)$$

$$V^{\alpha}{}_{;\beta} = \left(V^{\alpha}{}_{,\beta} - V_{\alpha} g^{\gamma\alpha}{}_{,\beta} - \Gamma^{\mu\alpha}{}_{\beta} V_{\mu}\right). \tag{6.59}$$

So the parenthetical argument quoted above does not apply to the covariant derivative of vectors and one-forms.

6.23 Prove

$$R_{\alpha\beta\mu\nu,\lambda} \underset{\mathrm{LIF}}{=} \frac{1}{2}\left(g_{\alpha\nu,\beta\mu\lambda} - g_{\beta\nu,\alpha\mu\lambda} - g_{\alpha\mu,\beta\nu\lambda} + g_{\beta\mu,\alpha\nu\lambda}\right). \quad \text{Schutz Eq. (6.88)} \tag{6.60}$$

(Be careful: one cannot simply differentiate

$$R_{\alpha\beta\mu\nu} \underset{\mathrm{LIF}}{=} \frac{1}{2}\left(g_{\alpha\nu,\beta\mu} - g_{\alpha\mu,\beta\nu} + g_{\beta\mu,\alpha\nu} - g_{\beta\nu,\alpha\mu}\right) \quad \text{Schutz Eq. (6.68)} \tag{6.61}$$

since it is valid only at P, not in the neighborhood of P.)[4]

Solution: We seek the partial derivative of the Riemann curvature tensor. It is fine to work in local inertial coordinates, but one must be careful not to lose any derivatives. So we start with the most general expression, the definition of the Riemann curvature tensor given in eqn. (6.52). Only *after* differentiating we apply the simplifications of local inertial coordinates, namely $g_{\alpha\gamma,\lambda} \underset{\mathrm{LIF}}{=} 0$, and $\Gamma^{\alpha}{}_{\mu\nu} \underset{\mathrm{LIF}}{=} 0$:

$$R_{\alpha\beta\mu\nu,\lambda} = (g_{\alpha\gamma} R^{\gamma}{}_{\beta\mu\nu})_{,\lambda}$$

$$\underset{\mathrm{LIF}}{=} g_{\alpha\gamma} R^{\gamma}{}_{\beta\mu\nu,\lambda} \qquad\qquad \text{used eqn. (6.44)}$$

$$\underset{\mathrm{LIF}}{=} g_{\alpha\gamma}(\Gamma^{\gamma}{}_{\beta\nu,\mu\lambda} - \Gamma^{\gamma}{}_{\beta\mu,\nu\lambda} + \Gamma^{\gamma}{}_{\sigma\mu,\lambda}\Gamma^{\sigma}{}_{\beta\nu}$$
$$+ \Gamma^{\gamma}{}_{\sigma\mu}\Gamma^{\sigma}{}_{\beta\nu,\lambda} - \Gamma^{\gamma}{}_{\sigma\nu,\lambda}\Gamma^{\sigma}{}_{\beta\mu} - \Gamma^{\gamma}{}_{\sigma\nu}\Gamma^{\sigma}{}_{\beta\mu,\lambda})$$

$$\underset{\mathrm{LIF}}{=} g_{\alpha\gamma}(\Gamma^{\gamma}{}_{\beta\nu,\mu\lambda} - \Gamma^{\gamma}{}_{\beta\mu,\nu\lambda}). \tag{6.62}$$

We now use eqn. (6.17), which is applicable in any coordinate system, to write Christoffel symbols in terms of the metric tensor and differentiate with respect to x^{σ}:

$$\Gamma^{\gamma}{}_{\mu\nu,\sigma} = \frac{1}{2}g^{\gamma\beta}(g_{\beta\mu,\nu\sigma} + g_{\beta\nu,\mu\sigma} - g_{\mu\nu,\beta\sigma}) + \frac{1}{2}g^{\gamma\beta}{}_{,\sigma}(g_{\beta\mu,\nu} + g_{\beta\nu,\mu} - g_{\mu\nu,\beta}). \tag{6.63}$$

---

[4] We have fixed a typo, changing "Eq. (6.67)" to "Eq. (6.68)".

(As an aside, we note that could eliminate the terms that are zero in the LIF by eqns. (6.44) and 6.45), leading to:

$$\Gamma^{\gamma}{}_{\mu\nu,\sigma} \underset{\text{LIF}}{=} \frac{1}{2} g^{\gamma\beta} \left(g_{\beta\mu,\nu\sigma} + g_{\beta\nu,\mu\sigma} - g_{\mu\nu,\beta\sigma} \right), \tag{6.64}$$

in agreement with Schutz Eq. (6.64).) To be on the safe side, return to eqn. (6.63) and differentiate again:

$$\Gamma^{\gamma}{}_{\mu\nu,\sigma\lambda} = \frac{1}{2} g^{\gamma\beta} \left(g_{\beta\mu,\nu\sigma\lambda} + g_{\beta\nu,\mu\sigma\lambda} - g_{\mu\nu,\beta\sigma\lambda} \right)$$

$$+ \frac{1}{2} g^{\gamma\beta}{}_{,\sigma\lambda} \left(g_{\beta\mu,\nu} + g_{\beta\nu,\mu} - g_{\mu\nu,\beta} \right) + \frac{1}{2} \underline{g^{\gamma\beta}{}_{,\sigma} \left(g_{\beta\mu,\nu\lambda} + g_{\beta\nu,\mu\lambda} - g_{\mu\nu,\beta\lambda} \right)}$$

$$\underset{\text{LIF}}{=} \frac{1}{2} g^{\gamma\beta} \left(g_{\beta\mu,\nu\sigma\lambda} + g_{\beta\nu,\mu\sigma\lambda} - g_{\mu\nu,\beta\sigma\lambda} \right). \tag{6.65}$$

The underlined terms above did not contribute because there was always a common factor with at most one derivative of the metric, which vanishes in the LIF. So the safe side led to a bit of extraneous work – we could have differentiated eqn. (6.64) straight away. Armed with this second derivative of the Christoffel symbols, we can substitute eqn. (6.65) into eqn. (6.62), giving:

$$R_{\alpha\beta\mu\nu,\lambda} \underset{\text{LIF}}{=} g_{\alpha\gamma} \frac{1}{2} g^{\gamma\sigma} \left[\underline{g_{\sigma\beta,\nu\mu\lambda}} + g_{\sigma\nu,\beta\mu\lambda} - g_{\beta\nu,\sigma\mu\lambda} \right.$$

$$\left. - (\underline{g_{\sigma\beta,\mu\nu\lambda}} + g_{\sigma\mu,\beta\nu\lambda} - g_{\beta\mu,\sigma\nu\lambda}) \right] \qquad \text{underlined terms cancel}$$

$$\underset{\text{LIF}}{=} \frac{1}{2} \delta^{\sigma}{}_{\alpha} [g_{\sigma\nu,\beta\mu\lambda} - g_{\beta\nu,\sigma\mu\lambda} - (g_{\sigma\mu,\beta\nu\lambda} - g_{\beta\mu,\sigma\nu\lambda})] \qquad \text{used eqn. (6.148)}$$

$$\underset{\text{LIF}}{=} \frac{1}{2} \left(g_{\alpha\nu,\beta\mu\lambda} - g_{\beta\nu,\alpha\mu\lambda} - g_{\alpha\mu,\beta\nu\lambda} + g_{\beta\mu,\alpha\nu\lambda} \right), \qquad \text{summed over } \sigma$$

which is eqn. (6.60).

6.25 (a) Prove that the Ricci tensor $R^{\mu}{}_{\alpha\mu\beta}$ is the only independent contraction of $R^{\alpha}{}_{\beta\mu\nu}$ since all others are multiples of it (or they are zero as pointed out in the text).

Solution: We simply step through the possibilities and determine their values based upon the symmetry relations

$$R_{\alpha\beta\mu\nu} = -R_{\beta\alpha\mu\nu} = -R_{\alpha\beta\nu\mu} = R_{\mu\nu\alpha\beta}. \qquad \text{Schutz Eq. (6.69)} \tag{6.66}$$

An important principle here is that we can only use the Riemann tensor symmetry relations when the indices are all in the same position (either all lower or all upper); if you're not sure why, see Exercise 3.24(b)!

The contraction of the first and second indices gives

$$R^\alpha{}_{\alpha\mu\nu} = g^{\alpha\beta} R_{\alpha\beta\mu\nu} = -g^{\alpha\beta} R_{\beta\alpha\mu\nu} \qquad \text{used eqn. (6.66)}$$

$$= -R^\alpha{}_{\alpha\mu\nu} = 0, \quad \forall \mu, \nu, \tag{6.67}$$

since zero is the only number equal to its own negative. Furthermore, this also implies $R_\alpha{}^\alpha{}_{\mu\nu} = 0$, see SP3.2; or you can see this quickly via $R^\alpha{}_{\alpha\mu\nu} = g^{\alpha\beta} R_{\alpha\beta\mu\nu} = R_\alpha{}^\alpha{}_{\mu\nu} = 0$. By similar reasoning contracting the last two indices gives

$$R_{\alpha\beta}{}^\mu{}_\mu = 0, \quad \forall \alpha, \beta.$$

It remains to consider $R^\alpha{}_{\mu\nu\alpha}, R_\alpha{}^\mu{}_{\mu\beta}, R_\alpha{}^\mu{}_{\beta\mu}$. These candidates were identified by stepping through the possibilities systematically: first and second, first and third, first and fourth, (that is all for those involving the first index), second and third (first and second already considered), second and fourth, third and fourth. That is all.

Two of the remaining candidates give -1 times the Ricci tensor. Contracting the first and last indices we have

$$-R^\mu{}_{\alpha\beta\mu} = -g^{\sigma\mu} R_{\sigma\alpha\beta\mu} = g^{\sigma\mu} R_{\sigma\alpha\mu\beta} \qquad \text{by symmetry in eqn. (6.66)}$$

$$= R^\mu{}_{\alpha\mu\beta} = R_{\alpha\beta}. \qquad \text{definition of Ricci tensor} \tag{6.68}$$

And contracting the second and third indices gives

$$-R_\alpha{}^\mu{}_{\mu\beta} = -g^{\sigma\mu} R_{\alpha\sigma\mu\beta} = g^{\sigma\mu} R_{\sigma\alpha\mu\beta}, \qquad \text{by symmetry in eqn. (6.66)}$$

$$= R^\mu{}_{\alpha\mu\beta} = R_{\alpha\beta}. \qquad \text{definition of Ricci tensor} \tag{6.69}$$

Finally contracting the second and fourth indices gives the same result as the standard first and third:

$$R_\alpha{}^\mu{}_{\beta\mu} = -R_\alpha{}^\mu{}_{\mu\beta} \qquad \text{by symmetry in eqn. (6.66)}$$

$$= R_{\alpha\beta}. \qquad \text{used eqn. (6.69)} \tag{6.70}$$

6.25 (b) Show that the Ricci tensor is symmetric.

Solution: Starting from the definition of the Ricci tensor we have have

$$R_{\alpha\beta} = g^{\sigma\mu} R_{\sigma\alpha\mu\beta} = g^{\sigma\mu} R_{\mu\beta\sigma\alpha} \qquad \text{by symmetry in eqn. (6.66)}$$

$$= R^\sigma{}_{\beta\sigma\alpha} = R_{\beta\alpha}, \tag{6.71}$$

which proves it is symmetric.

6.27 Fill in the algebra necessary to establish:

$$g^{\alpha\mu} R_{\alpha\beta\lambda\mu;\nu} = -g^{\alpha\mu} R_{\alpha\beta\mu\lambda;\nu} = -R_{\beta\lambda;\nu}, \qquad \text{Schutz Eq. (6.95)} \qquad (6.72)$$

$$(2 R^{\mu}_{\ \lambda} - \delta^{\mu}_{\ \lambda} R)_{;\mu} = 0, \qquad\qquad\qquad \text{Schutz Eq. (6.97)} \qquad (6.73)$$

$$G^{\alpha\beta}_{\ \ ;\beta} = 0. \qquad\qquad\qquad\qquad \text{Schutz Eq. (6.99)} \qquad (6.74)$$

Solution: Eqn. (6.72) above is the middle term in the *contracted Bianchi identities*

$$R_{\beta\nu;\lambda} + (-R_{\beta\lambda;\nu}) + R^{\mu}_{\ \beta\nu\lambda;\mu} = 0, \qquad \text{Schutz Eq. (6.93)} \qquad (6.75)$$

which are obtained by applying the Ricci contraction (contract on the first and third indices) to the Bianchi identities, Schutz Eq. (6.90). The detailed development for this middle term is:

$$g^{\alpha\mu} R_{\alpha\beta\lambda\mu;\nu} = (g^{\alpha\mu} R_{\alpha\beta\lambda\mu})_{;\nu} \qquad\qquad \text{used Schutz Eq. (6.94)}$$

$$= (-g^{\alpha\mu} R_{\alpha\beta\mu\lambda})_{;\nu} \qquad\qquad \text{used eqn. (6.66)}$$

$$= (-R^{\mu}_{\ \beta\mu\lambda})_{;\nu} = -R_{\beta\lambda;\nu}, \qquad \text{used eqn. (6.69)} \qquad (6.76)$$

which establishes eqn. (6.72). Although not asked for this, note that the first and third terms in eqn. (6.75) follow immediately from multiplication of the Bianchi identities by the inverse metric tensor, $g^{\alpha\mu}$.

Eqn. (6.73) above is the *twice contracted Bianchi identities*. To establish eqn. (6.73) we contract eqn. (6.75) using the inverse metric as follows:

$$0 = g^{\beta\nu} [R_{\beta\nu;\lambda} - R_{\beta\lambda;\nu} + R^{\mu}_{\ \beta\nu\lambda;\mu}] \qquad \text{line before Schutz Eq. (6.96)}$$

$$= R_{;\lambda} + g^{\beta\nu} [-R_{\beta\lambda;\nu} + R^{\mu}_{\ \beta\nu\lambda;\mu}] \qquad \text{used Schutz Eqs. (6.92), (6.94)}$$

$$= R_{;\lambda} - R^{\mu}_{\ \lambda;\mu} + g^{\beta\nu} R^{\mu}_{\ \beta\nu\lambda;\mu}. \qquad \text{used Schutz Eq. (6.94), relabled } \nu \to \mu$$

$$(6.77)$$

The third term is more involved. The guiding idea here is that we want to obtain a Ricci contraction so we need to perform index calisthenics to get the contraction on the first and third indices, remembering that we can only use the Riemann tensor symmetry relations in eqn. (6.66) when the indices are the same (e.g. lower) position.

$$g^{\beta\nu} R^{\mu}_{\ \beta\nu\lambda;\mu} = g^{\beta\nu} R_{\mu\beta\nu\lambda}^{\quad\ ;\mu} \qquad\qquad \text{cf. SP (3.2)}$$

$$= -g^{\beta\nu} R_{\beta\mu\nu\lambda}^{\quad\ ;\mu} \qquad\qquad \text{used eqn. (6.66)}$$

$$= -R^{\nu}_{\ \mu\nu\lambda}^{\quad ;\mu} = -R_{\mu\lambda}^{\quad ;\mu} \qquad \text{used eqn. (6.69)}$$

$$= -R^{\mu}_{\ \lambda;\mu}. \qquad\qquad\qquad \text{cf. SP (3.2)} \qquad (6.78)$$

Substituting eqn. (6.78) for the third term in eqn. (6.77) gives:

$$0 = R_{;\lambda} - R^{\mu}_{\ \lambda;\mu} - R^{\mu}_{\ \lambda;\mu}, \qquad \text{Schutz Eq. (6.96)} \qquad (6.79)$$

from which we immediately obtain

$$2 R^{\mu}{}_{\lambda;\mu} - R_{;\lambda} = 0. \tag{6.80}$$

The Kronecker delta commutes with the covariant derivative so we can write,

$$R_{;\lambda} = \delta^{\mu}{}_{\lambda} R_{;\mu} = (\delta^{\mu}{}_{\lambda} R)_{;\mu}. \tag{6.81}$$

Substituting eqn. (6.81) into eqn. (6.80) gives the twice contracted Bianchi identites, eqn. (6.73).

Eqn. (6.74) expresses that the Einstein tensor $G^{\alpha\beta}$ is divergence free. This can be established from the definition

$$G^{\alpha\beta} \equiv R^{\alpha\beta} - \frac{1}{2}g^{\alpha\beta} R, \qquad \text{Schutz Eq. (6.98)} \tag{6.82}$$

and the twice contracted bianchi identities eqn. (6.73) as follows. Multiply eqn. (6.73) by $g^{\lambda\nu}/2$:

$$g^{\lambda\nu}\left(R^{\mu}{}_{\lambda} - \frac{1}{2}\delta^{\mu}{}_{\lambda}R\right)_{;\mu} = \left[g^{\lambda\nu}\left(R^{\mu}{}_{\lambda} - \frac{1}{2}\delta^{\mu}{}_{\lambda}R\right)\right]_{;\mu} \qquad \text{used Schutz Eq. (6.94)}$$

$$= \left[R^{\mu\nu} - \frac{1}{2}g^{\mu\nu}R\right]_{;\mu} = G^{\mu\nu}{}_{;\mu} = G^{\nu\mu}{}_{;\mu}, \qquad \text{used eqn. (6.82)}$$

$$\tag{6.83}$$

which establishes eqn. (6.74). The final step used the symmetry $G^{\alpha\beta} = G^{\beta\alpha}$.

6.29 In [spherical] polar coordinates, calculate the Riemann curvature tensor of the sphere of unit radius, whose metric is given in Exercise 6.28. (Note that in two dimensions there is only one independent component, by the same arguments as in Exercise 6.18(b). So calculate $R_{\theta\phi\theta\phi}$ and obtain all other components in terms of it.)

Solution: This is a great exercise. Working through this covers several key ideas we need for GR in 4D spacetime but in the much less computationally demanding and easily visualized setting of 2D.

To calculate the Riemann tensor we need the Christoffel symbols. Now we could calculate these using their definition involving the partial derivatives of the basis vectors eqn. (5.48), or from the metric using eqn. (6.17) above. In this case it's much easier to use eqn. (6.17) since, as pointed out in the question, we already have the metric for the surface of a sphere in spherical coordinates. Our 2D manifold is the surface of the unit sphere but we keep r as a variable since it's no extra effort and it gains us a more general result. From Exercise 6.28, $(g_{\alpha\beta}) = \mathrm{diag}(r^2, r^2 \sin^2\theta)$ in the coordinates $x^\alpha = (\theta, \phi)$. As a word of warning, you might get the false impression that in general we can infer the metric on a lower-dimensional submanifold by simply ignoring the unused dimensions; that works here, ignoring the r dimension, but in general one must be cautious; see SP7.7. We also need the inverse metric; fortunately this is easy for a diagonal metric: $(g^{\alpha\beta}) = \text{diag}(r^{-2}, r^{-2}\sin^{-2}\theta)$, (see Exercise 6.3 if that's not obvious.)

It is easiest to use eqn. (6.17) above to calculate the Christoffel symbols for this metric. Only three are non-zero. The first is

$$\Gamma^{\theta}{}_{\phi\phi} = \frac{1}{2}g^{\theta\sigma}\left(2g_{\sigma\phi,\phi} - g_{\phi\phi,\sigma}\right) = \frac{1}{2}g^{\theta\theta}\left(2g_{\theta\phi,\phi}^{\;\;0} - g_{\phi\phi,\theta}\right) \quad \text{used diagonal metric}$$

$$= -\frac{1}{2}r^{-2}\frac{\partial r^2\sin^2\theta}{\partial\theta} = -\sin\theta\,\cos\theta. \tag{6.84}$$

Consider next

$$\Gamma^{\phi}{}_{\phi\theta} = \frac{1}{2}g^{\phi\sigma}\left(g_{\sigma\phi,\theta} + g_{\sigma\theta,\phi} - g_{\phi\theta,\sigma}\right) = \frac{1}{2}g^{\phi\phi}\left(g_{\phi\phi,\theta}\right) \quad \text{used diagonal metric}$$

$$= \frac{1}{2}r^{-2}\sin^{-2}\theta\frac{\partial r^2\sin^2\theta}{\partial\theta} = \cot\theta = \Gamma^{\phi}{}_{\theta\phi}. \quad \text{used eqn. (6.16)}$$

$$\tag{6.85}$$

Substitute these into the general expression for the Riemann curvature tensor, eqn. (6.52), to find the only independent, non-zero component

$$R^{\theta}{}_{\phi\theta\phi} = \Gamma^{\theta}{}_{\phi\phi,\theta} - \Gamma^{\theta}{}_{\phi\theta,\phi} + \Gamma^{\theta}{}_{\sigma\theta}\Gamma^{\sigma}{}_{\phi\phi} - \Gamma^{\theta}{}_{\sigma\phi}\Gamma^{\sigma}{}_{\phi\theta}$$

$$= \frac{\partial(-\sin\theta\,\cos\theta)}{\partial\theta} + 0 + 0 - (-\sin\theta\,\cos\theta)\frac{\cos\theta}{\sin\theta} = \sin^2\theta$$

$$R_{\theta\phi\theta\phi} = g_{\alpha\theta}R^{\alpha}{}_{\phi\theta\phi} = g_{\theta\theta}R^{\theta}{}_{\phi\theta\phi} = r^2\sin^2\theta. \tag{6.86}$$

From this and the symmetry relations, eqn. (6.66), we can find the other components:

$$R_{\phi\theta\theta\phi} = -r^2\sin^2\theta \qquad R_{\theta\phi\phi\theta} = -r^2\sin^2\theta \qquad R_{\phi\theta\phi\theta} = r^2\sin^2\theta$$

$$R_{\alpha\alpha\mu\nu} = 0 \qquad\qquad R_{\alpha\beta\mu\mu} = 0, \tag{6.87}$$

which agrees with that found with Maple™, see accompanying worksheet.

6.31   Show that covariant differentiation obeys the usual product rule, e.g.

$$(V^{\alpha\beta}\,W_{\beta\gamma})_{;\mu} = V^{\alpha\beta}{}_{;\mu}\,W_{\beta\gamma} + V^{\alpha\beta}\,W_{\beta\gamma;\mu}.$$

Hint: Use a locally inertial frame.

Solution: In a locally inertial frame, the Christoffel symbols vanish and covariant derivatives equal partial derivatives, so

$$(V^{\alpha\beta} W_{\beta\gamma})_{;\mu} \underset{\text{LIF}}{=} (V^{\alpha\beta} W_{\beta\gamma})_{,\mu} \qquad\qquad \text{in a locally inertial frame}$$

$$= \frac{\partial}{\partial x^\mu} \left(\sum_\beta V^{\alpha\beta} W_{\beta\gamma} \right) \qquad\qquad \text{suspend summation convention}$$

$$= \sum_\beta \frac{\partial}{\partial x^\mu} \left(V^{\alpha\beta} W_{\beta\gamma} \right) \qquad\qquad \text{partial derivative commutes with sum}$$

$$= \sum_\beta \left(W_{\beta\gamma} \frac{\partial}{\partial x^\mu} V^{\alpha\beta} + V^{\alpha\beta} \frac{\partial}{\partial x^\mu} W_{\beta\gamma} \right) \qquad\qquad \text{regular product rule}$$

$$= \sum_\beta \left(W_{\beta\gamma} V^{\alpha\beta}{}_{,\mu} + V^{\alpha\beta} W_{\beta\gamma,\mu} \right) \qquad\qquad \text{notation change only}$$

$$= W_{\beta\gamma} V^{\alpha\beta}{}_{,\mu} + V^{\alpha\beta} W_{\beta\gamma,\mu} \qquad\qquad \text{reinvoke summation convention}$$

$$= W_{\beta\gamma} V^{\alpha\beta}{}_{;\mu} + V^{\alpha\beta} W_{\beta\gamma;\mu}. \qquad\qquad \text{in a locally inertial frame}$$

The last equality is a valid tensor equation, valid in all reference frames.

6.33 A "three-sphere" is the three-dimensional surface in four-dimensional Euclidean space (coordinates x, y, z, w), given by the equation $x^2 + y^2 + z^2 + w^2 = r^2$, where r is the radius of the three-sphere.

6.33 (a) Define new coordinates (r, θ, ϕ, χ) by the equations

$$w = r\,\cos(\chi), \qquad\qquad\qquad z = r\,\sin(\chi)\,\cos(\theta),$$
$$y = r\,\sin(\chi)\,\sin(\theta)\,\sin(\phi), \qquad x = r\,\sin(\chi)\,\sin(\theta)\,\cos(\phi). \qquad (6.88)$$

Show that (θ, ϕ, χ) are coordinates for the sphere. These generalize the familiar polar coordinates.

Solution: If we simply substitute eqn. (6.88) into the equation for the three-sphere, we find that the equation is satisfied for fixed r for all values of (θ, ϕ, χ). So these coordinates can vary and we stay on the three-sphere. To show that these are truly coordinates, we must also show that the transformation defined by eqn. (6.88) is not singular, cf. eqn. (5.6). After a considerable amount of algebra, one finds the determinant,

$$\det \begin{pmatrix} \frac{\partial w}{\partial r} & \frac{\partial w}{\partial \theta} & \frac{\partial w}{\partial \phi} & \frac{\partial w}{\partial \chi} \\ \frac{\partial z}{\partial r} & \frac{\partial z}{\partial \theta} & \cdots & \\ \frac{\partial y}{\partial r} & \cdots & & \\ \frac{\partial x}{\partial r} & \cdots & & \frac{\partial x}{\partial \chi} \end{pmatrix} = -r^3 \sin^2(\chi)\,\sin(\theta).$$

So just as in spherical-polar coordinates there are singular points at the poles $\theta = 0$ and $\theta = \pi$, and additionally at $\chi = 0$ and $\chi = \pi$. But the transformation is generally non-singular.

6.33 (b) Show that the metric of the three-sphere of radius r has components in these coordinates $g_{\chi\chi} = r^2$, $g_{\theta\theta} = r^2 \sin^2\chi$, $g_{\phi\phi} = r^2 \sin^2\chi \sin^2\theta$, all other components vanishing. (Use the same method as in Exercise 6.28.)

Solution: There are only six independent terms (because of symmetry, $g_{\bar\alpha\bar\beta} = g_{\bar\beta\bar\alpha}$). We will use an overbar to indicate indices on the basis in (θ, ϕ, χ), with $x^{\bar 1} = \theta$, $x^{\bar 2} = \phi$, $x^{\bar 3} = \chi$. And indices without overbar indicate the original coordinates in (x, y, z, w). Then in general

$$g_{\bar\alpha\bar\beta} = g_{\alpha\beta} \Lambda^\alpha{}_{\bar\alpha} \Lambda^\beta{}_{\bar\beta}, \tag{6.89}$$

where $\Lambda^\alpha{}_{\bar\alpha} = \partial x^\alpha / \partial x^{\bar\alpha}$. The metric tensor in the 4D Euclidean space in the Cartesian coordinates (x, y, z, w) is

$$g_{\alpha\beta} = \begin{cases} +1 & \text{if } \alpha = \beta \\ 0 & \text{if } \alpha \neq \beta. \end{cases} \tag{6.90}$$

The calculus is tedious but straightforward. For instance,

$$\begin{aligned} g_{\bar 1 \bar 1} \equiv g_{\theta\theta} &= g_{xx}\left(\frac{\partial x}{\partial\theta}\right)^2 + g_{yy}\left(\frac{\partial y}{\partial\theta}\right)^2 + g_{zz}\left(\frac{\partial z}{\partial\theta}\right)^2 + g_{ww}\left(\frac{\partial w}{\partial\theta}\right)^2 \\ &= r^2 \sin^2\chi \cos^2\theta \sin^2\phi + r^2 \sin^2\chi \cos^2\theta \cos^2\phi + r^2 \sin^2\chi \sin^2\theta \\ &= r^2 \sin^2\chi. \end{aligned} \tag{6.91}$$

In a similar manner one can easily show the off-diagonal terms are zero.

6.35 Compute 20 independent components of $R_{\alpha\beta\mu\nu}$ for a manifold with line element,

$$ds^2 = -e^{2\Phi(r)} dt^2 + e^{2\Lambda(r)} dr^2 + r^2 d\theta^2 + r^2 \sin^2\theta d\phi^2, \tag{6.92}$$

where $\Phi(r)$ and $\Lambda(r)$ are arbitrary functions of the coordinate r alone. (First, identify the coordinates and the components $g_{\alpha\beta}$; then compute $g^{\alpha\beta}$ and the Christoffel symbols. Then decide on the indices of the 20 components of $R_{\alpha\beta\mu\nu}$ you wish to calculate, and compute them. Remember that you can deduce the remaining 236 components from those 20.)

Solution: This problem is instructive on several levels. It should be done after (or concurrently with) Exercise 6.18, for it helps clarify the symmetry relations and the implied reduction in degrees of freedom of the Riemann curvature tensor. It also helps to reveal how much information is packed in the line element equation! Later we will learn that the line element eqn. (6.92) is the general form for a static spherically symmetric spacetime and leads to the Schwarzschild metric, which represents the simplest solution of the full Einstein equations. It is therefore extremely useful to know.

(i) The coordinates are $\{t, r, \theta, \phi\}$. This is clear because these form the differential variables of the line element eqn. (6.92). Inspection reveals the metric tensor is

$$(g_{\alpha\beta}) = \begin{pmatrix} -e^{2\Phi} & 0 & 0 & 0 \\ 0 & e^{2\Lambda} & 0 & 0 \\ 0 & 0 & r^2 & 0 \\ 0 & 0 & 0 & r^2\sin^2\theta \end{pmatrix}.$$

A fair question is "why is the metric tensor diagonal." The answer is that there are no cross-terms like say $(dr\ d\theta)$ or $(dt\ d\phi)$, etc. in the line element eqn. (6.92).

(ii) The inverse of the metric tensor:

$$(g^{\alpha\beta}) = \begin{pmatrix} -e^{-2\Phi} & 0 & 0 & 0 \\ 0 & e^{-2\Lambda} & 0 & 0 \\ 0 & 0 & r^{-2} & 0 \\ 0 & 0 & 0 & r^{-2}\sin^{-2}\theta \end{pmatrix}.$$

(After working Exercise 6.3, computing the inverse of a diagonal matrix should be automatic.)

The Christoffel symbols can be computed from the metric tensor using eqn. (6.17). One needs the first derivatives of the metric tensor:

$$g_{tt,r} = \frac{\partial}{\partial r}[-e^{2\Phi}] = -2e^{2\Phi}\,\Phi', \qquad g_{rr,r} = \frac{\partial}{\partial r}[e^{2\Lambda}] = 2e^{2\Lambda}\,\Lambda',$$

$$g_{\theta\theta,r} = \frac{\partial}{\partial r}[r^2] = 2r, \qquad g_{\phi\phi,r} = \frac{\partial}{\partial r}[r^2\sin^2\theta] = 2r\sin^2\theta,$$

$$g_{\phi\phi,\theta} = \frac{\partial}{\partial\theta}[r^2\sin^2\theta]$$
$$= 2r^2\sin(\theta)\cos(\theta) = r^2\sin(2\theta). \tag{6.93}$$

All other first derivatives of the metric tensor are zero. The resulting non-zero Christoffel symbols are:

$$\Gamma^0{}_{01} = \Phi' \qquad\qquad \Gamma^1{}_{00} = e^{-2\Lambda}\,e^{2\Phi}\,\Phi' \qquad \Gamma^1{}_{11} = \Lambda'$$

$$\Gamma^1{}_{22} = -e^{-2\Lambda}\,r \qquad\qquad \Gamma^1{}_{33} = -e^{-2\Lambda}\,r\sin^2\theta \qquad \Gamma^2{}_{12} = \frac{1}{r}$$

$$\Gamma^2{}_{33} = -\sin(\theta)\cos(\theta) \qquad \Gamma^3{}_{13} = \frac{1}{r} \qquad\qquad\qquad \Gamma^3{}_{23} = \frac{\cos(\theta)}{\sin(\theta)}. \tag{6.94}$$

(iii) Deciding the 20 terms of $R_{\alpha\beta\mu\nu}$ to calculate. This is not as simple as it might sound. Here it helps tremendously if one has solved Exercise 6.18.

Recall $R_{\alpha\alpha\mu\nu} = 0 = R_{\alpha\beta\nu\nu}$ because of the symmetry relations expressed in eqn. (6.66) (see Exercise 6.18 or 6.25(a)).

We organize the terms as recommended in the hint for Exercise 6.18: we choose pairs of $\alpha \neq \beta$ (there are 6 of them accounting for the fact that order does not matter), and similarly there are 6 pairs of $\mu \neq \nu$. These would give $6 \times 6 = 36$ elements, but, because of the symmetry $R_{\alpha\beta\mu\nu} = R_{\mu\nu\alpha\beta}$, we must divide the number of off-diagonal elements by two to get $5 \times 6/2 + 6 = 21$. (We will deal with the reduction to 20 by the cyclic identity in Schutz Eq. (6.70) below.) It is easiest to write down all $6 \times 6 = 36$ terms and then eliminate the lower diagonal:

$$
\begin{array}{cccccc}
R_{trtr} & R_{trt\theta} & R_{trt\phi} & R_{trr\theta} & R_{trr\phi} & \underline{R_{tr\theta\phi}} \\
R_{t\theta tr} & R_{t\theta t\theta} & R_{t\theta t\phi} & R_{t\theta r\theta} & \underline{R_{t\theta r\phi}} & R_{t\theta\theta\phi} \\
R_{t\phi tr} & R_{t\phi t\theta} & R_{t\phi t\phi} & \underline{R_{t\phi r\theta}} & R_{t\phi r\phi} & R_{t\phi\theta\phi} \\
R_{r\theta tr} & R_{r\theta t\theta} & R_{r\theta t\phi} & R_{r\theta r\theta} & R_{r\theta r\phi} & R_{r\theta\theta\phi} \\
R_{r\phi tr} & R_{r\phi t\theta} & R_{r\phi t\phi} & R_{r\phi r\theta} & R_{r\phi r\phi} & R_{r\phi\theta\phi} \\
R_{\theta\phi tr} & R_{\theta\phi t\theta} & R_{\theta\phi t\phi} & R_{\theta\phi r\theta} & R_{\theta\phi r\phi} & R_{\theta\phi\theta\phi}.
\end{array}
\tag{6.95}
$$

Note that we are not writing them down randomly. Instead, we step the second pair of indices, i.e. $\mu\nu$, systematically by increasing the ν most rapidly with increasing column, μ more slowly with increasing column. Similarly we increase the first pair of indices, i.e. $\alpha\beta$, with row, and β more rapidly than α. These were arbitrary choices of course, but having a system and sticking to it makes it easy.

Recall only the upper diagonal is necessary to determine the tensor because of the symmetry $R_{\alpha\beta\mu\nu} = R_{\mu\nu\alpha\beta}$:

$$
\begin{array}{cccccc}
R_{trtr} & R_{trt\theta} & R_{trt\phi} & R_{trr\theta} & R_{trr\phi} & \underline{R_{tr\theta\phi}} \\
& R_{t\theta t\theta} & R_{t\theta t\phi} & R_{t\theta r\theta} & \underline{R_{t\theta r\phi}} & R_{t\theta\theta\phi} \\
& & R_{t\phi t\phi} & \underline{R_{t\phi r\theta}} & R_{t\phi r\phi} & R_{t\phi\theta\phi} \\
& & & R_{r\theta r\theta} & R_{r\theta r\phi} & R_{r\theta\theta\phi} \\
& & & & R_{r\phi r\phi} & R_{r\phi\theta\phi} \\
& & & & & R_{\theta\phi\theta\phi}.
\end{array}
\tag{6.96}
$$

Now we must also impose the cyclic indentity, Schutz Eq. (6.70), which you will have found in Exercise 6.18 only applies to the case when none of the indices are equal. There are three such terms, underlined above. One of these can be determined from the other two.

We will evaluate a few of these in full detail. It is important to use eqn. (6.52), which is true in all coordinate bases and *not* eqn. (6.51) which is only true in a local inertial frame. From eqn. (6.52) and the Christoffel symbols eqn. (6.94) above we find for

$$
\begin{aligned}
R_{trtr} &= g_{tt} R^t{}_{rtr} \\
&= -e^{2\Phi}\left[-\Gamma^0{}_{01,r} + \Gamma^0{}_{10}\Gamma^1{}_{11} - \Gamma^0{}_{01}\Gamma^0{}_{01} \right] \\
&= e^{2\Phi}\left[(\Phi')^2 + \Phi'' - \Phi'\Lambda' \right].
\end{aligned}
\tag{6.97}
$$

It is important to note that if one were to use:

$$R_{\alpha\beta\mu\nu} \underset{\text{LIF}}{=} \frac{1}{2}\left(g_{\alpha\nu,\beta\mu} - g_{\alpha\mu,\beta\nu} + g_{\beta\mu,\alpha\nu} - g_{\beta\nu,\alpha\mu}\right), \qquad \text{eqn. (6.51)}$$

one would miss the cross term $(-e^{2\Phi}\Phi'\Lambda')$ because

$$R_{trtr} \underset{\text{LIF}}{=} \frac{1}{2}\left(g_{tr,rt} - g_{tt,rr} + g_{rt,tr} - g_{rr,tt}\right)$$

$$R_{trtr} \neq \frac{1}{2}\left[0 - (-4e^{2\Phi}(\Phi')^2 - 2e^{2\Phi}\Phi'') + 0 - 0\right]$$

$$= 2e^{2\Phi}(\Phi')^2 + e^{2\Phi}\Phi''. \tag{6.98}$$

For the next one $R_{t\theta t\theta}$,

$$R_{t\theta t\theta} = g_{tt}\left[\Gamma^0{}_{22,t} - \Gamma^0{}_{20,\theta} + \Gamma^0{}_{\sigma 0}\Gamma^\sigma{}_{22} - \Gamma^0{}_{\sigma 2}\Gamma^\sigma{}_{20}\right] = g_{tt}\left[\Gamma^0{}_{\sigma 0}\Gamma^\sigma{}_{22}\right]$$

$$= g_{tt}\left[\Gamma^0{}_{10}\Gamma^1{}_{22}\right] = -e^{2\Phi}\left[\Phi'(-re^{-2\Lambda})\right]$$

$$= +r\Phi'e^{2\Phi-2\Lambda}. \tag{6.99}$$

After a lot of simple calculus and algebra one finds that only the diagonal elements of eqn. (6.96) are non-zero,

$$(R_{\alpha\beta\mu\nu}) = \begin{pmatrix} R_{trtr} & 0 & 0 & 0 & 0 & 0 \\ 0 & R_{t\theta t\theta} & 0 & 0 & 0 & 0 \\ 0 & 0 & R_{t\phi t\phi} & 0 & 0 & 0 \\ 0 & 0 & 0 & R_{r\theta r\theta} & 0 & 0 \\ 0 & 0 & 0 & 0 & R_{r\phi r\phi} & 0 \\ 0 & 0 & 0 & 0 & 0 & R_{\theta\phi\theta\phi} \end{pmatrix}. \tag{6.100}$$

The other $256 - 36 = 220$ terms are determined by symmetry relations in eqn. (6.66). These six non-zero terms are

$$R_{trtr} = e^{2\Phi}\left[(\Phi')^2 + \Phi'' - \Phi'\Lambda'\right] \qquad R_{t\theta t\theta} = r\Phi'e^{2\Phi}e^{-2\Lambda}$$

$$R_{t\phi t\phi} = \sin^2\theta\, r\Phi'e^{2\Phi}e^{-2\Lambda} \qquad R_{r\theta r\theta} = r\Lambda'$$

$$R_{r\phi r\phi} = r\Lambda'\sin^2\theta \qquad R_{\theta\phi\theta\phi} = -r^2\Big(\cos^2(\theta) - 1 + e^{-2\Lambda}$$

$$- \cos^2(\theta)e^{-2\Lambda}\Big)$$

$$= r^2\sin^2\theta\left(1 - e^{-2\Lambda}\right). \tag{6.101}$$

The answers above for both the Christoffel symbols and Riemann curvature tensor disagree with those provided by (Schutz, 1985, Appendix B) but were verified with Maple™, see accompanying worksheet.

6.37 (a) "Proper volume" of a two-dimensional manifold is usually called "proper area." Using the metric in Exercise 6.28, integrate

$$dx^0\, dx^1\, dx^2\, dx^3 = \left(- \det(g_{\alpha'\beta'})\right)^{1/2} dx^{0'}\, dx^{1'}\, dx^{2'}\, dx^{3'}$$
$$= \left(-g'\right)^{1/2} dx^{0'}\, dx^{1'}\, dx^{2'}\, dx^{3'} \qquad \text{Schutz Eq. (6.18)}$$
$$(6.102)$$

to find the proper area of a sphere of radius r.[5]

Solution: First note that here $\det(g') > 0$, cf. eqn. (6.30), so we must change the sign under the radical:

$$\int\int dx^1\, dx^2 = \int_0^\pi \int_0^{2\pi} g'^{1/2}\, d\phi\, d\theta = \int_0^\pi \int_0^{2\pi} (r^2\, r^2 \sin^2\theta)^{1/2}\, d\phi\, d\theta$$
$$= \int_0^\pi \int_0^{2\pi} (r^2 \sin\theta)\, d\phi\, d\theta = r^2 2\pi \int_0^\pi \sin\theta\, d\theta$$
$$= -r^2 2\pi [\cos\theta]_0^\pi = 4\pi r^2. \qquad (6.103)$$

6.37 (b) Do the analogous calculation for the three-sphere of Exercise 6.33.

Solution: Again the determinant is positive,

$$\det(g') = g_{\chi\chi}\, g_{\theta\theta}\, g_{\phi\phi} = r^2 \cdot r^2 \sin^2\chi \cdot r^2 \sin^2\chi \sin^2\theta = r^6\, (\sin^4\chi)(\sin^2\theta),$$

so again we take the positive sign under the radical:

$$\int\int dx^1\, dx^2\, dx^3 = \int_0^\pi \int_0^\pi \int_0^{2\pi} g'^{1/2}\, d\chi\, d\theta\, d\phi$$
$$= r^3 \int_0^\pi \int_0^\pi \int_0^{2\pi} (\sin^2\chi)(\sin\theta)\, d\chi\, d\theta\, d\phi$$
$$= 4\pi r^3 \int_0^\pi (\sin^2\chi)d\chi = 4\pi r^3 \frac{1}{2} [\chi - \sin\chi\, \cos\chi]_0^\pi = 2\pi^2 r^3.$$
$$(6.104)$$

One question that might arise is "how do we know the range of the angle χ?" Inspection of eqn. (6.88) reveals that in order for w to extend from $w = -r$ to $w = +r$, when $z = y = x = 0$, we require $\cos\chi$ to pass from -1 to $+1$, so in analogy with the polar angle θ of spherical coordinates, we take $\chi \in [0, \pi]$.

[5] We have altered the notation slightly by replacing $g \to g'$ on the RHS of eqn. (6.102) just to emphasize that it's the determinant of the metric in the $x^{\alpha'}$ coordinate system.

6.39 (a) For any two vector fields $\vec{U}$ and $\vec{V}$, their *Lie bracket* is defined to be the vector field $[\vec{U}, \vec{V}]$ with components,

$$[\vec{U}, \vec{V}]^\alpha = U^\beta \nabla_\beta V^\alpha - V^\beta \nabla_\beta U^\alpha. \qquad \text{Schutz Eq. (6.100)} \qquad (6.105)$$

Show that,

$$[\vec{U}, \vec{V}] = -[\vec{V}, \vec{U}] \qquad (6.106)$$

$$[\vec{U}, \vec{V}]^\alpha = U^\beta \partial V^\alpha / \partial x^\beta - V^\beta \partial U^\alpha / \partial x^\beta. \qquad (6.107)$$

This is one tensor field in which partial derivatives need not be accompanied by Christoffel symbols!

Solution: The Lie bracket is clearly antisymmetric in the order of the arguments; exchanging $\vec{U}$ for $\vec{V}$ changes the sign in eqn. (6.105) so eqn. (6.106) follows immediately from the definition eqn. (6.105).

Regarding eqn. (6.107) we start with the definition eqn. (6.105), adding ( ) on the LHS just to be consistent with the notation introduced in eqn. (2.3):

$$\left([\vec{U}, \vec{V}]\right)^\alpha = U^\beta \, \nabla_\beta V^\alpha - V^\beta \, \nabla_\beta U^\alpha$$

$$= U^\beta \, V^\alpha{}_{;\beta} - V^\beta \, U^\alpha{}_{;\beta} \qquad\qquad\qquad \text{notation change only}$$

$$= U^\beta \, (V^\alpha{}_{,\beta} + \Gamma^\alpha{}_{\mu\beta} V^\mu) - V^\beta \, (U^\alpha{}_{,\beta} + \Gamma^\alpha{}_{\mu\beta} U^\mu) \qquad \text{used Schutz Eq. 6.33}$$

$$= U^\beta \, V^\alpha{}_{,\beta} - V^\beta \, U^\alpha{}_{,\beta} + \underline{U^\beta \, \Gamma^\alpha{}_{\mu\beta} V^\mu - V^\beta \, \Gamma^\alpha{}_{\mu\beta} U^\mu}. \qquad \text{rearranging only}$$

$$(6.108)$$

So we only need to show that the underlined terms cancel. And this is so because

$$\underline{U^\beta \, \Gamma^\alpha{}_{\mu\beta} V^\mu - V^\beta \, \Gamma^\alpha{}_{\mu\beta} U^\mu} = \left(\Gamma^\alpha{}_{\beta\mu} - \Gamma^\alpha{}_{\mu\beta}\right) U^\mu V^\beta \quad \text{relabeled dummy indices on first term}$$

$$= 0. \qquad\qquad\qquad \text{use eqn. (6.16)}$$

6.39 (b) Show that $[\vec{U}, \vec{V}]$ is a derivative operator on $\vec{V}$ along $\vec{U}$, i.e. show that for any scalar f,

$$[\vec{U}, f\vec{V}] = f[\vec{U}, \vec{V}] + \vec{V}(\vec{U} \cdot \nabla f). \qquad \text{Schutz Eq. (6.101)} \qquad (6.109)$$

This is sometimes called the *Lie derivative* with respect to $\vec{U}$ and is denoted by,

$$[\vec{U}, \vec{V}] := \pounds_{\vec{U}} \vec{V}, \qquad \vec{U} \cdot \nabla f := \pounds_{\vec{U}} f. \qquad \text{Schutz Eq. (6.102)} \qquad (6.110)$$

Then eqn. (6.109) would be written in the more conventional form of the Leibniz rule for the derivative operator $\pounds_{\vec{U}}$:

$$\pounds_{\vec{U}}(f\vec{V}) = f\pounds_{\vec{U}} \vec{V} + \vec{V}\pounds_{\vec{U}} f. \qquad \text{Schutz Eq. (6.103)} \qquad (6.111)$$

The result of Exercise 6.39(a) shows that this derivative operator may be defined without a connection or metric, and is therefore very fundamental. See Schutz (1980) for an introduction.

Solution: Simply write out the LHS of eqn. (6.109) in component form, treating $(f\vec{V})$ as the second vector:

$$
\left([\vec{U}, f\vec{V}]\right)^{\alpha} = U^{\beta}\frac{\partial(fV^{\alpha})}{\partial x^{\beta}} - (fV^{\beta})\frac{\partial(U^{\alpha})}{\partial x^{\beta}} \qquad \text{used eqn. (6.107)}
$$

$$
= fU^{\beta}\frac{\partial(V^{\alpha})}{\partial x^{\beta}} + V^{\alpha}U^{\beta}\frac{\partial(f)}{\partial x^{\beta}} - (fV^{\beta})\frac{\partial(U^{\alpha})}{\partial x^{\beta}} \qquad \text{product rule}
$$

$$
= f\left([\vec{U}, \vec{V}]\right)^{\alpha} + V^{\alpha}U^{\beta}\frac{\partial(f)}{\partial x^{\beta}} \qquad \text{used eqn. (6.107)}
$$

$$
= f\left([\vec{U}, \vec{V}]\right)^{\alpha} + V^{\alpha}\vec{U}\cdot\nabla(f), \qquad \text{because scalar, } f_{;\beta} = f_{,\beta}
$$

$$\tag{6.112}$$

which is the α component of the RHS of eqn. (6.109).

6.39 (c) Calculate the components of the Lie derivative of a one-form field $\tilde{\omega}$ from the knowledge that, for any vector field $\vec{V}$, $\tilde{\omega}(\vec{V})$ is a scalar like f above, and from the definition that $\pounds_{\vec{U}}\tilde{\omega}$ is a one-form field:

$$
\pounds_{\vec{U}}[\tilde{\omega}(\vec{V})] = \left(\pounds_{\vec{U}}\tilde{\omega}\right)(\vec{V}) + \tilde{\omega}\left(\pounds_{\vec{U}}\vec{V}\right).
$$

Solution: First we isolate the term of interest, $\left(\pounds_{\vec{U}}\tilde{\omega}\right)(\vec{V})$:

$$
\left(\pounds_{\vec{U}}\tilde{\omega}\right)(\vec{V}) = \pounds_{\vec{U}}[\tilde{\omega}(\vec{V})] - \tilde{\omega}\left(\pounds_{\vec{U}}\vec{V}\right). \tag{6.113}
$$

The first term in the RHS of eqn. (6.113) is the Lie derivative of a scalar, so takes the form:

$$
\pounds_{\vec{U}}[\tilde{\omega}(\vec{V})] = \vec{U}\cdot\nabla[\tilde{\omega}(\vec{V})] = U^{\alpha}\left(\omega_{\beta}V^{\beta}\right)_{,\alpha}
$$

$$
= U^{\alpha}\left(\omega_{\beta}V^{\beta}{}_{,\alpha} + V^{\beta}\omega_{\beta,\alpha}\right) \qquad \text{product rule}
$$

$$
= \underline{U^{\alpha}\omega_{\beta}V^{\beta}{}_{,\alpha}} + U^{\alpha}V^{\beta}\omega_{\beta,\alpha}. \qquad \text{expand} \qquad (6.114)
$$

The second term on the RHS of eqn. (6.113) is

$$
-\tilde{\omega}\left(\pounds_{\vec{U}}\vec{V}\right) = -\omega_{\beta}\left(\pounds_{\vec{U}}\vec{V}\right)^{\beta}
$$

$$
= -\omega_{\beta}\left(U^{\alpha}V^{\beta}{}_{,\alpha} - V^{\alpha}U^{\beta}{}_{,\alpha}\right)
$$

$$
= -\underline{U^{\alpha}\omega_{\beta}V^{\beta}{}_{,\alpha}} + \omega_{\beta}V^{\alpha}U^{\beta}{}_{,\alpha}. \qquad \text{expand} \qquad (6.115)
$$

Note the underlined terms cancel. This leaves:

$$\left(\pounds_{\vec{U}}\tilde{\omega}\right)\left(\vec{V}\right) = \left(\pounds_{\vec{U}}\tilde{\omega}\right)_{\beta}\left(\vec{V}\right)^{\beta} = \left(\pounds_{\vec{U}}\tilde{\omega}\right)_{\beta}V^{\beta}$$

$$= U^{\alpha}V^{\beta}\omega_{\beta,\alpha} + \omega_{\beta}V^{\alpha}U^{\beta}{}_{,\alpha}$$

$$= \left(U^{\alpha}\omega_{\beta,\alpha} + \omega_{\alpha}U^{\alpha}{}_{,\beta}\right)V^{\beta}. \qquad \text{relabeled dummy index} \qquad (6.116)$$

So the β component of the Lie derivative of a one-form must be

$$\left(\pounds_{\vec{U}}\tilde{\omega}\right)_{\beta} = \left(U^{\alpha}\omega_{\beta,\alpha} + \omega_{\alpha}U^{\alpha}{}_{,\beta}\right),$$

in agreement with (Baumgarte and Shapiro, 2010, Eq. (A.8)). In fact Baumgarte and Shapiro (2010, Appendix A.1) show that the Lie derivative of a $\binom{1}{1}$ rank tensor $T^{\alpha}{}_{\beta}$ is

$$\pounds_{\vec{U}}T^{\alpha}{}_{\beta} = U^{\sigma}T^{\alpha}{}_{\beta,\sigma} - T^{\sigma}{}_{\beta}U^{\alpha}{}_{,\sigma} + T^{\alpha}{}_{\sigma}U^{\sigma}{}_{,\beta}, \qquad (6.117)$$

which can be extended to a tensor of arbitrary rank by including terms involving derivatives of U^{α} as in eqn. (6.117) for each index, with a positive sign for the lower index terms and negative for the upper index terms.

6.2 Supplementary problems

SP 6.1 What are the numerical values of the elements of the Riemann curvature tensor, $R_{\alpha\alpha\mu\nu}$ and $R_{\alpha\beta\mu\mu}$. Hint: Think about the implications of the symmetry relations contained in eqn. (6.66). I recommend doing this problem before attempting Exercise 6.18(b).

Solution

$R_{\alpha\alpha\mu\nu} = 0$ and $R_{\alpha\beta\mu\mu} = 0$ because we must have

$$R_{\alpha\beta\mu\nu} = -R_{\beta\alpha\mu\nu}. \qquad \text{see eqn. (6.66)}$$

For the case where $\alpha = \beta$ we must have that this element has a numerical value equal to -1 times itself. The only number satisfying this equation is zero.

SP 6.2 Use calculus of variations to prove that geodesic curves have extremal "length."

Hint

The length of the geodesic, a measure of either proper time if it's timelike or proper length if it's spacelike, is given by

$$l = \int_{\lambda_0}^{\lambda_1}\left|g_{\alpha\beta}\frac{dx^{\alpha}}{d\lambda}\frac{dx^{\beta}}{d\lambda}\right|^{1/2}d\lambda. \qquad (6.118)$$

Use the Euler–Lagrange equations to find the conditions for l to be extremal for fixed end points λ_0 and λ_1. See Boas (1983, Chapter 9) for a gentle introduction to calculus of variations and the Euler–Lagrange equations, or Hobson et al. (2006, Appendix C of Chapter 3) for a quick refresher.

SP 6.3 Generalize

$$\nabla_\alpha \nabla_\beta V^\mu \underset{\text{LIF}}{=} V^\mu{}_{,\beta\mu} + \Gamma^\mu{}_{\nu\beta,\alpha} V^\nu \qquad \text{Schutz Eq. (6.73)} \qquad (6.119)$$

to the case of $\binom{1}{1}$ tensor, $F^\mu{}_\nu$.

Solution

This is a straightforward generalization of the argument leading to Schutz Eq. (6.73), the details of which were requested in Exercise 6.20. The covariant derivative of a $\binom{1}{1}$ tensor can be inferred from Schutz Eqs. (6.34) and (6.35):

$$\nabla_\beta F^\mu{}_\nu = F^\mu{}_{\nu;\beta} = F^\mu{}_{\nu,\beta} + \Gamma^\mu{}_{\sigma\beta} F^\sigma{}_\nu - \Gamma^\sigma{}_{\nu\beta} F^\mu{}_\sigma. \qquad (6.120)$$

Now we simply apply the gradient operator another time, initially for compactness without expanding $F^\mu{}_{\nu;\beta}$, giving

$$\nabla_\alpha(\nabla_\beta F^\mu{}_\nu) = (F^\mu{}_{\nu;\beta})_{;\alpha} = (F^\mu{}_{\nu;\beta})_{,\alpha} + \Gamma^\mu{}_{\sigma\alpha} F^\sigma{}_{\nu;\beta} - \Gamma^\sigma{}_{\nu\alpha} F^\mu{}_{\sigma;\beta} - \Gamma^\sigma{}_{\beta\alpha} F^\mu{}_{\nu;\sigma}$$

$$\underset{\text{LIF}}{=} (F^\mu{}_{\nu;\beta})_{,\alpha}, \qquad (6.121)$$

where the last step used the fact that Christoffel symbols are zero in local inertial coordinates. But their gradients are not necessarily zero, so

$$\nabla_\alpha(\nabla_\beta F^\mu{}_\nu) \underset{\text{LIF}}{=} (F^\mu{}_{\nu;\beta})_{,\alpha}$$

$$= F^\mu{}_{\nu,\beta\alpha} + (\Gamma^\mu{}_{\sigma\beta} F^\sigma{}_\nu - \Gamma^\sigma{}_{\nu\beta} F^\mu{}_\sigma)_{,\alpha} \qquad \text{used eqn. (6.120)) above}$$

$$= F^\mu{}_{\nu,\beta\alpha} + \Gamma^\mu{}_{\sigma\beta,\alpha} F^\sigma{}_\nu - \Gamma^\sigma{}_{\nu\beta,\alpha} F^\mu{}_\sigma, \qquad \text{Christoffel symbols are zero}$$

$$(6.122)$$

which generalizes eqn. (6.119).

SP 6.4 Schutz Eq. (6.62), when written using the Riemann curvature tensor given by eqn. (6.52) above, gives the change in the components of the vector δV^α when it is parallel transported around the loop with sides $\delta a \vec{e}_\mu$ and $\delta b \vec{e}_\nu$:

$$\delta V^\alpha = \delta a\, \delta b\, R^\alpha{}_{\mu\lambda\sigma} V^\mu. \qquad (6.123)$$

At first this appears strange because the indices do not match on both sides of the equal sign. Recall that expressions written for just the components, like V^β are really shorthand notation for the full vector $\vec{V} = V^\beta \vec{e}_\beta$. To help clarify eqn. (6.123), write it first in terms of vectors like $\vec{\delta a}$, and tensors $\mathbf{R}$, and then introduce the appropriate basis vectors and one-forms, by replacing $\vec{\delta a}$ with $\delta a^\sigma \vec{e}_\sigma$ etc.

SP 6.5   Derive eqn. (6.57) above

$$[\nabla_\alpha, \nabla_\beta]F^\mu{}_\nu = R^\mu{}_{\sigma\alpha\beta}F^\sigma{}_\nu + R_\nu{}^\sigma{}_{\alpha\beta}F^\mu{}_\sigma \qquad \text{Schutz Eq. (6.78)} \qquad (6.124)$$

in a manner analogous to the derivation of Schutz Eq. (6.77) i.e. simply expanding and calculating all the terms. Use the results of SP6.3 above. This problem provides an alternative to Exercise 6.21 wherein one is to explain the positive signs in eqn. (6.124).

Solution

Using result eqn. (6.122) from SP6.3 above, we start by changing the order of the derivatives by changing the order of the indices α and β. Taking the difference between the two derivatives we obtain (underlined terms cancel):

$$[\nabla_\alpha, \nabla_\beta]F^\mu{}_\nu \equiv \nabla_\alpha(\nabla_\beta F^\mu{}_\nu) - \nabla_\beta(\nabla_\alpha F^\mu{}_\nu)$$

$$\underset{\text{LIF}}{=} \underline{F^\mu{}_{,\beta\alpha}} + \Gamma^\mu{}_{\sigma\beta,\alpha}F^\sigma{}_\nu - \Gamma^\sigma{}_{\nu\beta,\alpha}F^\mu{}_\sigma - (\underline{F^\mu{}_{,\alpha\beta}} + \Gamma^\mu{}_{\sigma\alpha,\beta}F^\sigma{}_\nu - \Gamma^\sigma{}_{\nu\alpha,\beta}F^\mu{}_\sigma)$$

$$= (\Gamma^\mu{}_{\sigma\beta,\alpha} - \Gamma^\mu{}_{\sigma\alpha,\beta})F^\sigma{}_\nu - (\Gamma^\sigma{}_{\nu\beta,\alpha} - \Gamma^\sigma{}_{\nu\alpha,\beta})F^\mu{}_\sigma.$$

collected common factors

Use the definition of the Riemann tensor in terms of Christoffel symbols eqn. (6.52) above but in a reference frame where the Christoffel symbols all vanish,

$$[\nabla_\alpha, \nabla_\beta]F^\mu{}_\nu = R^\mu{}_{\sigma\alpha\beta}F^\sigma{}_\nu - R^\sigma{}_{\nu\alpha\beta}F^\mu{}_\sigma. \qquad (6.125)$$

Using the symmetry properties of the Riemann tensor eqn. (6.66) we see this is equivalent to eqn. (6.124). If the final step concerns you, see SP6.7.

SP 6.6   For an n-dimensional spacetime manifold to possess a global basis we require a set of n linearly independent vector fields. In some cases this is not possible. For instance, show that the surface of a sphere in three-dimensional space (the two-dimensional surface called a two-sphere denoted S^2) does not possess a global basis.

SP 6.7   Do the symmetry relations in eqn. (6.66) apply also when an index is raised? Prove that $R^\alpha{}_{\beta\mu\nu} = -R_\alpha{}^\beta{}_{\mu\nu}$. This result will be useful for SP6.5.

Solution

Yes these symmetry relations also apply when an index is raised. For example,

$$
\begin{aligned}
-R_\alpha{}^\beta{}_{\mu\nu} &= -g^{\sigma\beta} R_{\alpha\sigma\mu\nu} \\
&= g^{\sigma\beta} R_{\sigma\alpha\mu\nu}, \quad \text{by Schutz Eq. (6.69)} \\
&= R^\beta{}_{\alpha\mu\nu}.
\end{aligned} \tag{6.126}
$$

SP 6.8 Show that the Ricci tensor and Ricci scalar of the metric of Exercise 6.35 (which turns out to be that of a static spherically symmetric spacetime, and leads to the Schwarzschild metric) are

$$
R_{tt} = -e^{(2\phi-2\Lambda)}\left(\Lambda'\Phi' - \Phi'^2 - \Phi'' - \frac{2\Phi'}{r}\right) \quad R_{rr} = -(-\Lambda'\Phi' + \Phi'^2 + \Phi'' - \frac{2\Lambda'}{r})
$$

$$
R_{\theta\theta} = -(-1 + e^{-2\Lambda}[1 - r(\Lambda' - \Phi')]) \qquad R_{\phi\phi} = -\sin^2\theta\left(e^{-2\Lambda}[1 + r(\Phi' - \Lambda')] - 1\right) \tag{6.127}
$$

and

$$
R = 2e^{-2\Lambda}\left(-\Lambda'\Phi' + \Phi'^2 + \Phi'' + \frac{2(\Phi' - \Lambda')}{r} + \frac{1 - e^{2\Lambda}}{r^2}\right). \tag{6.128}
$$

SP 6.9 Redo the derivation of the expression for geodesic deviation in terms of the Riemann tensor. Start with two neighboring geodesic curves with coordinates $x^\alpha(\lambda)$ and $\bar{x}^\alpha(\lambda)$ such that

$$
\bar{x}^\alpha(\lambda) = x^\alpha(\lambda) + \xi^\lambda(\lambda), \qquad \left|\frac{d\xi^\alpha}{d\lambda}\right| \ll \left|\frac{dx^\alpha}{d\lambda}\right|. \tag{6.129}
$$

Use a LIF centered on a point $\mathcal{P}$ on $x^\alpha(\lambda)$ where the tangent vectors are parallel. Use a Taylor series about $\mathcal{P}$ for the Christoffel symbols to show that the difference between the two geodesic equations can be written:

$$
\frac{d^2\xi^\alpha}{d\lambda^2} + \Gamma^\alpha{}_{\mu\nu,\beta}\left[\frac{dx^\mu}{d\lambda}\frac{dx^\nu}{d\lambda}\frac{d\xi^\beta}{d\lambda} + O\left(\left|\frac{d\xi^\alpha}{d\lambda}\right|^2\right)\right]_{\text{LIF}} = 0. \tag{6.130}
$$

Combine this with the covariant derivative of ξ^α at $\mathcal{P}$ to obtain the geodesic deviation equation

$$
\nabla_{\vec{V}}\nabla_{\vec{V}}\xi^\alpha = R^\alpha{}_{\mu\nu\beta}V^\mu V^\nu\xi^\beta. \qquad \text{Schutz Eq. (6.87)} \tag{6.131}
$$

Solution

Evaluate the general geodesic equation

$$
\frac{d}{d\lambda}\left(\frac{dx^\alpha}{d\lambda}\right) + \Gamma^\alpha{}_{\mu\nu}\frac{dx^\mu}{d\lambda}\frac{dx^\nu}{d\lambda} = 0, \qquad \text{Schutz Eq. (6.51)} \tag{6.132}
$$

at the point $\mathcal{P}$ in the LIF:

$$\frac{\mathrm{d}}{\mathrm{d}\lambda}\left(\frac{\mathrm{d}x^\alpha}{\mathrm{d}\lambda}\right)\underset{\mathrm{LIF}}{=} 0. \tag{6.133}$$

Now apply eqn. (6.132) to the second geodesic at a neighboring point still using the LIF at point $\mathcal{P}$ (and use a Taylor series approximation for $\Gamma^\alpha{}_{\mu\nu}$):

$$\frac{\mathrm{d}}{\mathrm{d}\lambda}\left(\frac{\mathrm{d}\bar{x}^\alpha}{\mathrm{d}\lambda}\right)+\Gamma^\alpha{}_{\mu\nu}\frac{\mathrm{d}\bar{x}^\mu}{\mathrm{d}\lambda}\frac{\mathrm{d}\bar{x}^\nu}{\mathrm{d}\lambda}=0,$$

$$\frac{\mathrm{d}}{\mathrm{d}\lambda}\left(\frac{\mathrm{d}x^\alpha+\xi^\alpha}{\mathrm{d}\lambda}\right)+\Gamma^\alpha{}_{\mu\nu,\beta}\,\xi^\beta\frac{\mathrm{d}(x^\mu+\xi^\mu)}{\mathrm{d}\lambda}\frac{\mathrm{d}(x^\nu+\xi^\nu)}{\mathrm{d}\lambda}\underset{\mathrm{LIF\,at\,}\mathcal{P}}{=}0.\quad\text{used eqn. (6.129)}$$
$$\tag{6.134}$$

Substract eqn. (6.133) from eqn. (6.134) keeping only terms of first order in the separation vector:

$$\frac{\mathrm{d}^2\xi^\alpha}{\mathrm{d}\lambda^2}+\Gamma^\alpha{}_{\mu\nu,\beta}\left[\frac{\mathrm{d}x^\mu}{\mathrm{d}\lambda}\frac{\mathrm{d}x^\nu}{\mathrm{d}\lambda}\xi^\beta+O\left(\left|\frac{\mathrm{d}\xi^\alpha}{\mathrm{d}\lambda}\right|^2\right)\right]\underset{\mathrm{LIF}}{=}0. \tag{6.135}$$

Now think of ξ^α as any old vector defined along the curve parameterized by λ. Its first derivative in the direction of the tangent vector $V^\alpha=\mathrm{d}x^\alpha/\mathrm{d}\lambda$, is

$$\nabla_{\vec{V}}\xi^\alpha=\frac{\mathrm{d}\xi^\alpha}{\mathrm{d}\lambda}+\Gamma^\alpha{}_{\mu\nu}\,\xi^\mu V^\nu\equiv\frac{D\xi^\alpha}{D\lambda}, \tag{6.136}$$

where the second equality introduces the notation of Hobson et al. (2006, §3.14), who call this the *intrinsic derivative* of the components of ξ^α along the curve x^α. You'll find this notation in other texts, (e.g. Weinberg, 1972, Eq. (4.9.3)). This is a good idea because some confusion can arise here, as discussed further in SP3.15. The second derivative in the direction of the tangent vector $V^\alpha=\mathrm{d}x^\alpha/\mathrm{d}\lambda$, is

$$\nabla_{\vec{V}}\nabla_{\vec{V}}\xi^\alpha=\frac{D^2\xi^\alpha}{D\lambda^2}\underset{\mathrm{LIF}}{=}\frac{\mathrm{d}^2\xi^\alpha}{\mathrm{d}^2\lambda}+\frac{\mathrm{d}}{\mathrm{d}\lambda}\left(\Gamma^\alpha{}_{\mu\nu}\,\xi^\mu V^\nu\right)$$

$$=\frac{\mathrm{d}^2\xi^\alpha}{\mathrm{d}^2\lambda}+\Gamma^\alpha{}_{\mu\nu,\gamma}\,\xi^\mu V^\nu V^\gamma. \tag{6.137}$$

We immediately evaluated this second intrinsic derivative in the LIF to avoid the Christoffel symbols and then used $\mathrm{d}\Gamma^\alpha{}_{\mu\nu}/\mathrm{d}\lambda=\Gamma^\alpha{}_{\mu\nu,\gamma}V^\gamma$. Now combine eqn. (6.137) and eqn. (6.135) with eqn. (6.53) to give, after a bit of algebra, the geodesic deviation equation eqn. (6.131). [This solution followed closely Hobson et al. (2006, §7.13).]

SP 6.10   Why is it generally important to use eqn. (6.52) above, i.e.

$$R^\alpha{}_{\beta\mu\nu}:=\Gamma^\alpha{}_{\beta\nu,\mu}-\Gamma^\alpha{}_{\beta\mu,\nu}+\Gamma^\alpha{}_{\sigma\mu}\Gamma^\sigma{}_{\beta\nu}-\Gamma^\alpha{}_{\sigma\nu}\Gamma^\sigma{}_{\beta\mu}, \tag{6.138}$$

for the computation of the Riemann tensor components and not the one in terms of the metric, given by,

$$R_{\alpha\beta\mu\nu} = \frac{1}{2}\left(g_{\alpha\nu,\beta\mu} - g_{\alpha\mu,\beta\nu} + g_{\beta\mu,\alpha\nu} - g_{\beta\nu,\alpha\mu}\right)? \qquad\qquad \text{eqn. (6.51)}$$

SP 6.11 Consider the geodesic $x_g^{\alpha}(\lambda)$, where λ is an affine parameter, with tangent vector $\vec{U}$. Another curve with the same path, $x_n^{\alpha}(\phi) = x_g^{\alpha}(\lambda)$, but parameterized by a non-affine parameter, say $\phi = \lambda^2$, is not a geodesic (as stated in Schutz §6.4). That is, while the tangent vector $\vec{U}$ does not change along the path, the tangent vector to $x_n^{\alpha}(\phi)$, say $\vec{V}$, apparently does. Make sense of this by first finding $\vec{V}$ in terms of $\vec{U}$ for general $\phi(\lambda)$. Then find the rate of change of $\vec{V}$ with respect to ϕ, expressing your answer as a function of $\vec{U}$. What happens to the magnitude and direction of $\vec{V}$ as one moves along the path of the geodesic?

Solution

Recall that a curve was defined in Schutz §5.2 as a path through spacetime that is parameterized by a real parameter. And in Schutz §6.4 a geodesic was defined as a curve that parallel-transports its own tangent vector. The tangent vector is defined as follows:

$$\vec{V} = \frac{d}{d\phi}\left(x_n^{\alpha}(\phi)\right)\vec{e}_{\alpha} = \frac{d\lambda}{d\phi}\frac{d}{d\lambda}\left(x_g^{\alpha}(\lambda)\right)\vec{e}_{\alpha} = \frac{d\lambda}{d\phi}\vec{U}.$$

The two tangent vectors of course point in the same direction since the paths of the two curves are identical. To find the rate of change of $\vec{V}$ along its curve, we differentiate with respect to ϕ:

$$\frac{d}{d\phi}\vec{V} = \frac{d}{d\phi}\left(\frac{d\lambda}{d\phi}\vec{U}\right) = \frac{d\lambda}{d\phi}\frac{d}{d\lambda}\left(\frac{d\lambda}{d\phi}\vec{U}\right) = \vec{U}\frac{d^2}{d\phi^2}\lambda + \left(\frac{d\lambda}{d\phi}\right)^2\underline{\frac{d}{d\lambda}\vec{U}} = \vec{U}\frac{d^2}{d\phi^2}\lambda.$$

The underlined quantity is zero because $\vec{U}$ is parallel-transported along the geodesic. So moving along the common path, the tangent vector $\vec{V}$ remains parallel to $\vec{U}$ but its magnitude will change if

$$\frac{d^2}{d\phi^2}\lambda \neq 0.$$

An affine parameter maintains the magnitude of the tangent vector.

SP 6.12 Recall in the solution to Exercise 6.39(c) we took the Lie derivative of the scalar

$$\pounds_{\vec{U}}[\tilde{\omega}(\vec{V})] = \vec{U}\cdot\nabla[\tilde{\omega}(\vec{V})] = U^{\alpha}\left(\omega_{\beta}V^{\beta}\right)_{,\alpha}$$
$$= U^{\alpha}\left(\omega_{\beta}V^{\beta}{}_{,\alpha} + V^{\beta}\omega_{\beta,\alpha}\right). \quad \text{used product rule} \quad (6.139)$$

We used the partial derivative for the scalar term. Convince yourself that we would have found the same thing, eventually, if we had used the covariant derivative instead.

SP 6.13 From the definition of the Einstein tensor in eqn. (6.82) show that it is symmetric, $G^{\alpha\beta} = G^{\beta\alpha}$.

Solution

The symmetry of $G^{\alpha\beta}$ follows from the symmetry of $R_{\alpha\beta}$ and $g_{\alpha\beta}$ as follows:

$$R^{\alpha\beta} = R_{\mu\nu}g^{\alpha\mu}g^{\beta\nu} = R_{\nu\mu}g^{\alpha\mu}g^{\beta\nu} \qquad\qquad \text{used eqn. (6.69)}$$

$$= R_{\nu\mu}g^{\mu\alpha}g^{\nu\beta} \qquad\qquad \text{symmetry of (inverse) metric tensor}$$

$$= R^{\beta\alpha}. \qquad\qquad\qquad\qquad\qquad\qquad\qquad (6.140)$$

SP 6.14   If you compare eqn. (6.34) found above in Exercise 6.9 with the formula for the divergence of a vector in spherical coordinates found on the Internet or in textbooks (e.g. Davis and Snider, 1979, Eq. (5.27)), you'll find something different:

$$\nabla \cdot \mathbf{F} = \frac{1}{r^2}\frac{\partial(r^2 F^r)}{\partial r} + \frac{1}{r\sin\theta}\frac{\partial(\sin\theta\, F^\theta)}{\partial\theta} + \frac{1}{r\sin\theta}\frac{\partial F^\phi}{\partial\phi}. \qquad (6.141)$$

What's gone wrong?!

SP 6.15   There is a subtlety about notation that arises regarding the derivative along a curve $d/d\lambda$ that deserves some consideration. Recall that the vector $\vec{V} = V^\alpha \vec{e}_\alpha$ is sometimes just written as V^α to "save time and space."[6] Usually it is very clear when there are implicit basis vectors but it can become troublesome when we take derivatives and suddenly it matters whether or not we have to differentiate the implicit basis vectors! For instance, in the geodesic equation,

$$\frac{d}{d\lambda}\left(\frac{dx^\alpha}{d\lambda}\right) + \Gamma^\alpha{}_{\mu\nu}\frac{dx^\mu}{d\lambda}\frac{dx^\nu}{d\lambda} = 0, \qquad \text{Schutz Eq. (6.51)} \qquad (6.142)$$

$dx^\alpha/d\lambda$ simply means the ordinary derivative of these four coordinate functions $x^\alpha = \{x^0(\lambda), x^1(\lambda), x^2(\lambda), x^3(\lambda)\}$ with respect to the parameter λ; no need to worry about covariant derivatives and basis vectors changing with position because all that is taken into account in the second term, the one with the Christoffel symbols. The geodesic equation eqn. (6.142) can be read as a statement about coordinates. But in other contexts we are differentiating basis vectors. For example, in the definition of parallel-transport

$$\frac{d}{d\lambda}\vec{V} = \nabla_{\vec{U}}\vec{V} = 0, \qquad \text{Schutz Eq. (6.48)} \qquad (6.143)$$

there are clearly basis vectors to be differentiated.

Sometimes it's less clear from the notation (albeit clear from the context). For example, in deriving the equation for the deviation ξ^α between two geodesics, an intermediate step was to take the covariant derivative of the deviation vector,

$$\nabla_V \nabla_V \xi^\alpha = \frac{d}{d\lambda}\left(\frac{d}{d\lambda}\xi^\alpha + \Gamma^\alpha{}_{\beta 0}\xi^\beta\right) + 0. \qquad \text{cf. Schutz Eq. (6.85)} \qquad (6.144)$$

Reintroduce the implicit basis vectors to make sense of eqn. (6.144).

---

[6] Unfortunately the notation is so standard that the relativist must get used to it but one wonders if this shorthand is actually a false economy.

Solution

Start with a single derivative,

$$\nabla_V(\xi^\alpha \vec{e}_\alpha) = \frac{d}{d\lambda}\left(\xi^\alpha \vec{e}_\alpha\right) = \vec{e}_\alpha \frac{d}{d\lambda}\xi^\alpha + \xi^\alpha \frac{d}{d\lambda}\vec{e}_\alpha$$

$$= \vec{e}_\alpha \frac{d}{d\lambda}\xi^\alpha + \xi^\alpha \frac{\partial \vec{e}_\alpha}{\partial x^\beta}\frac{dx^\beta}{d\lambda}$$

$$= \vec{e}_\alpha \frac{d}{d\lambda}\xi^\alpha + \xi^\alpha \Gamma^\sigma{}_{\alpha\beta}\vec{e}_\sigma \frac{dx^\beta}{d\lambda} \qquad\qquad \text{used eqn. (5.48) above}$$

$$= \vec{e}_\alpha \frac{d}{d\lambda}\xi^\alpha + \xi^\alpha \Gamma^\sigma{}_{\alpha\beta}\vec{e}_\sigma \delta^\beta_0 \qquad\qquad \text{tangent vector in } \vec{e}_0\text{-direction}$$

$$= \vec{e}_\alpha \frac{d}{d\lambda}\xi^\alpha + \xi^\alpha \Gamma^\sigma{}_{\alpha 0}\vec{e}_\sigma \qquad\qquad \text{summed over } \beta$$

$$= \vec{e}_\alpha \left(\frac{d}{d\lambda}\xi^\alpha + \xi^\beta \Gamma^\alpha{}_{\beta 0}\right). \qquad\qquad \text{relabeled dummy indices}$$

$$(6.145)$$

So when we include the implicit basis vector we get the extra term in $\Gamma^\alpha{}_{\beta 0}$. Now to arrive at eqn. (6.144) we differentiate again, giving

$$\nabla_V\nabla_V(\xi^\alpha \vec{e}_\alpha) = \frac{d}{d\lambda}\left[\vec{e}_\alpha\left(\frac{d}{d\lambda}\xi^\alpha + \xi^\beta\Gamma^\alpha{}_{\beta 0}\right)\right]$$

$$= \vec{e}_\alpha \frac{d}{d\lambda}\left(\frac{d}{d\lambda}\xi^\alpha + \xi^\beta\Gamma^\alpha{}_{\beta 0}\right) + \underline{\Gamma^\sigma{}_{\alpha\beta}\vec{e}_\sigma\frac{dx^\beta}{d\lambda}\left(\frac{d}{d\lambda}\xi^\alpha + \xi^\beta\Gamma^\alpha{}_{\beta 0}\right)}$$

$$= \vec{e}_\alpha \frac{d}{d\lambda}\left(\frac{d}{d\lambda}\xi^\alpha + \xi^\beta\Gamma^\alpha{}_{\beta 0}\right) + 0. \qquad\qquad (6.146)$$

Because $\Gamma^\alpha{}_{\beta\gamma} = 0$ at point A we can drop the underlined term above as Schutz has indicated with the final 0 on the RHS of eqn. (6.144). See SP6.9 for a cleaner derivation of the geodesic deviation.

SP 6.16   Argue that

$$\eta^\alpha{}_\beta \equiv \delta^\alpha{}_\beta \qquad\qquad \text{Schutz Eq. (3.60)} \qquad\qquad (6.147)$$

generalizes to curved spacetime in general coordinates, i.e.

$$g^\alpha{}_\beta \equiv \delta^\alpha{}_\beta. \qquad\qquad (6.148)$$

SP 6.17   Gauss' law on curved spacetime takes the form

$$\int V^\alpha{}_{;\alpha}\sqrt{-g}\,d^4x = \oint V^\alpha n_\alpha \sqrt{-g}\,d^3S, \qquad\qquad \text{Schutz Eq. (6.45)} \qquad\qquad (6.149)$$

where n_α is the unit normal to the hypersurface that encloses the volume of integration on the LHS, and g is the determinant of the metric. What metric does g correspond to on the LHS and the RHS of eqn. (6.149)?

Solution

On the LHS of eqn. (6.149) g is of course the determinant of the metric of the 4D spacetime. On the RHS the metric in question is the *induced metric* on the hypersurface of integration. Calculating the induced metric from the full metric is discussed in SP (7.7). Often γ_{ij} is used to denote the induced metric (e.g. Carroll, 2004; Baumgarte and Shapiro, 2010), in which case eqn. (6.149) could be written

$$\int V^{\alpha}_{;\alpha} \sqrt{-g}\, \mathrm{d}^4 x = \oint V^{\alpha} n_{\alpha} \sqrt{-\gamma}\, \mathrm{d}^3 S. \tag{6.150}$$

7 Physics in curved spacetime

In our new picture, there is no coordinate system [that] is inertial everywhere, i.e. in which $d^2x^i/dt^2 = 0$ for every particle for which $d\vec{U}/d\tau = 0$. Therefore we have to allow all coordinates on an equal footing.

Bernard Schutz, §7.3

It might help to tackle the supplementary problems first, see §7.2 below.

7.1 Exercises

7.1 (i) If

$$(nU^\alpha)_{;\alpha} = q\,R, \qquad \text{Schutz Eq. (7.3)} \qquad (7.1)$$

were the correct generalization of Schutz Eq. (7.1), $(nU^\alpha)_{,\alpha} = 0$, to a curved spactime, how would you interpret it? (ii) What would happen to the number of particles in a co-moving volume of fluid, as time evolves? (iii) In principle, can we distinguish experimentally between

$$(nU^\alpha)_{;\alpha} = 0, \qquad \text{Schutz Eq. (7.2)} \qquad (7.2)$$

and eqn. (7.1)?

Solution

(i) Recall that in the hypothetical eqn. (7.1) q was some constant and R the Ricci scalar. Based on kinematics, this equation states that, for $q > 0$ there would be a source of particles for positively curved space $R > 0$. This was shown in the solution to Exercise 4.20(a) above, which interpreted the divergence of the four-vector $N = n\vec{U}$ as the rate of generation of particles per unit volume. If $(q\,R)$ were negative there would be a sink of particles. Curvature of spacetime would somehow create or destroy particles.

(ii) What would happen to the number of particles in a co-moving volume of fluid, as time evolves?

Solution

(ii) Recall the solution to Exercise 4.20, where we derived eqn. (4.28):

$$\frac{\partial n\, U^0}{\partial t} = -\frac{\partial n\, U^x}{\partial x} - \frac{\partial n\, U^y}{\partial y} - \frac{\partial n\, U^z}{\partial z} + \varepsilon.$$

In the co-moving frame, U^0 is always unity, such that

$$\frac{\partial n}{\partial t} = -\frac{\partial n\, U^x}{\partial x} - \frac{\partial n\, U^y}{\partial y} - \frac{\partial n\, U^z}{\partial z} + \varepsilon. \tag{7.3}$$

The number of particles per unit volume n evolves in time, increasing according to the spatial convergence of the number flux $-N^i{}_{;i}$ and the source term, $\varepsilon = qR$.

(iii) Could we ever distinguish experimentally between eqn. (7.2) and eqn. (7.1)?

Solution

(iii) Yes one could, in principle, measure the terms on the RHS of eqn. (7.3) and thereby distinguish between eqn. (7.2) and eqn. (7.1). For instance, in a region of curved spacetime such that $R > 0$ and where initially $n = 0$, we would expect the number of particles per unit volume to remain nil according to eqn. (7.2) and yet should grow at a rate

$$\frac{dn}{dt} = qR \tag{7.4}$$

according to eqn. (7.1).

This theme reappears in SPs 7.12, 8.6, 10.12, 12.8 where an experimentalist searches for a practical setup with large R. They are indexed under "hopeful/frustrated experimentalist."

7.2 To first order in ϕ, compute $g^{\alpha\beta}$ for

$$ds^2 = -(1 + 2\phi)dt^2 + (1 - 2\phi)\left(dx^2 + dy^2 + dz^2\right). \quad \text{Schutz Eq. (7.8)} \tag{7.5}$$

[This is the line element for the *weak gravitational field* metric.]

Solution

By definition $g^{\alpha\beta}g_{\beta\gamma} = \delta^\alpha{}_\gamma$. For a diagonal metric like that in eqn. (7.5) the inverse is also diagonal with $g^{\alpha\alpha} = 1/g_{\alpha\alpha}$ (obvious after doing Exercise 6.3). For example, $g^{0\alpha} = g^{\alpha 0} = 0$ and

$$g^{00} = \frac{1}{-(1 + 2\phi)} = -(1 - 2\phi) + O(\phi^2). \qquad \text{used eqn. (B.3)} \qquad (7.6)$$

The rest of the calculation is similar, giving $g^{ij} = \delta^{ij}(1 + 2\phi) + O(\phi^2)$.

7.3   Calculate all the Christoffel symbols for the metric given by eqn. (7.5) to first order in $\phi(t, x, y, z)$. Assume ϕ is a general function of $t, x, y,$ and z.

Solution: This exercise requires a great deal of algebra but given the importance of this metric, a complete set of Christoffel symbols will prove to be very useful later. All non-zero Christoffel symbols can be found for this metric in the accompanying Maple™ worksheet but here we reduce them to first order in ϕ. First count the number of independent Christoffel symbols $\Gamma^\alpha{}_{\mu\nu}$ to calculate. For each α there are only ten independent terms because $\Gamma^\alpha{}_{\mu\nu} = \Gamma^\alpha{}_{\nu\mu}$ in any coordinate basis. Hereinafter we ignore redundant ones. Given the metric and inverse metric, see Exercise 7.2, we can calculate the Christoffel symbols using eqn. (6.17),

$$\Gamma^\alpha{}_{\mu\nu} = \frac{1}{2} g^{\alpha\beta}(g_{\beta\mu,\nu} + g_{\beta\nu,\mu} - g_{\mu\nu,\beta}). \qquad \text{eqn. (6.17)}$$

The calculation simplifies tremendously because $(g^{\alpha\beta})$ is diagonal. Thus we need only consider the $\beta = \alpha$ contribution in eqn. (6.17).

First consider $\Gamma^0{}_{\mu\nu}$ (the first equalities below introduces an alternative notation):

$$\Gamma^0{}_{00} = \Gamma^t{}_{tt} = \frac{1}{2} g^{tt} g_{tt,t} = \frac{-\phi_{,t}}{-(1 + 2\phi)} \simeq (1 - 2\phi)\,\phi_{,t} = \phi_{,t} + O(\phi^2). \qquad (7.7)$$

Similarly,

$$\Gamma^0{}_{01} = \Gamma^t{}_{tx} = \frac{1}{2} g^{tt} g_{tt,x} = \frac{-\phi_{,x}}{-(1 + 2\phi)} \simeq (1 - 2\phi)\,\phi_{,x} = \phi_{,x} + O(\phi^2). \qquad (7.8)$$

Note $x, y,$ and z play identical roles in the metric eqn. (7.5), so eqn. (7.8) implies:

$$\Gamma^0{}_{02} = \Gamma^t{}_{ty} = \phi_{,y} + O(\phi^2), \qquad \Gamma^0{}_{03} = \Gamma^t{}_{tz} = \phi_{,z} + O(\phi^2). \qquad (7.9)$$

Now consider:

$$\Gamma^0{}_{11} = \Gamma^t{}_{xx} = \frac{1}{2} g^{tt}(-g_{xx,t}) = \frac{-(-\phi_{,t})}{-(1 + 2\phi)} \simeq -(1 - 2\phi)\,\phi_{,t} = -\phi_{,t} + O(\phi^2).$$
$$(7.10)$$

And eqn. (7.10) implies:

$$\Gamma^0{}_{22} = \Gamma^t{}_{yy} = -\phi_{,t} + O(\phi^2), \qquad \Gamma^0{}_{33} = \Gamma^t{}_{zz} = -\phi_{,t} + O(\phi^2). \qquad (7.11)$$

For a general diagonal metric the Christoffel symbols vanish when all the indices are different:

$$\Gamma^0{}_{12} = \Gamma^t{}_{xy} = 0, \qquad \Gamma^0{}_{13} = \Gamma^t{}_{xz} = 0, \qquad \Gamma^0{}_{23} = \Gamma^t{}_{yz} = 0. \quad (7.12)$$

Now consider $\Gamma^1{}_{\mu\nu}$:

$$\Gamma^1{}_{00} = \Gamma^x{}_{tt} = \frac{1}{2} g^{xx}(-g_{tt,x}) = \frac{-(-\phi_{,x})}{(1-2\phi)} \simeq (1+2\phi)\phi_{,x} = \phi_{,x} + O(\phi^2), \quad (7.13)$$

and

$$\Gamma^1{}_{01} = \Gamma^x{}_{tx} = \frac{1}{2} g^{xx}(g_{xx,t}) = \frac{-\phi_{,t}}{(1-2\phi)} \simeq -(1+2\phi)\phi_{,t} = -\phi_{,t} + O(\phi^2). \quad (7.14)$$

And when the indices are all different (see above):

$$\Gamma^1{}_{02} = \Gamma^x{}_{ty} = 0, \qquad\qquad \Gamma^1{}_{03} = \Gamma^x{}_{tz} = 0. \quad (7.15)$$

Furthermore,

$$\Gamma^1{}_{11} = \Gamma^x{}_{xx} = \frac{1}{2} g^{xx}(g_{xx,x}) = \frac{-\phi_{,x}}{(1-2\phi)} \simeq -(1+2\phi)\phi_{,x} = -\phi_{,x} + O(\phi^2),$$

$$\Gamma^1{}_{12} = \Gamma^x{}_{xy} = \frac{1}{2} g^{xx}(g_{xx,y}) = \frac{-\phi_{,y}}{(1-2\phi)} \simeq -(1+2\phi)\phi_{,y} = -\phi_{,y} + O(\phi^2),$$

$$\Gamma^1{}_{13} = \Gamma^x{}_{xz} = \frac{1}{2} g^{xx}(g_{xx,z}) = \frac{-\phi_{,z}}{(1-2\phi)} \simeq -(1+2\phi)\phi_{,z} = -\phi_{,z} + O(\phi^2),$$

$$\Gamma^1{}_{22} = \Gamma^x{}_{yy} = \frac{1}{2} g^{xx}(-g_{yy,x}) = \frac{-(-\phi_{,x})}{(1-2\phi)} \simeq (1+2\phi)\phi_{,x} = \phi_{,x} + O(\phi^2). \quad (7.16)$$

$$\Gamma^1{}_{23} = \Gamma^x{}_{yz} = 0. \qquad\qquad \text{indices all different} \quad (7.17)$$

$$\Gamma^1{}_{33} = \Gamma^x{}_{zz} = \frac{1}{2} g^{xx}(-g_{zz,x}) = \frac{-(-\phi_{,x})}{(1-2\phi)} \simeq (1+2\phi)\phi_{,x} = \phi_{,x} + O(\phi^2). \quad (7.18)$$

The rest, $\Gamma^2{}_{\mu\nu}$ and $\Gamma^3{}_{\mu\nu}$, can be inferred from the above by noting that y and z play the same role as x. They can be represented succinctly through:

$$\Gamma^i{}_{00} = \phi_{,i} + O(\phi^2), \qquad\qquad \Gamma^i{}_{0j} = -\phi_{,t}\delta^i{}_j + O(\phi^2),$$
$$\Gamma^i{}_{jk} = \delta_{jk}\delta^{il}\phi_{,l} - \delta^i_j\phi_{,k} - \delta^i_k\phi_{,j} + O(\phi^2). \quad (7.19)$$

7.5 (a) For a perfect fluid, verify that the spatial components of,

$$T^{\alpha\beta}{}_{;\beta} = 0, \qquad\qquad \text{Schutz Eq. (7.6)} \quad (7.20)$$

in the Newtonian limit reduce to

$$\mathbf{v}_{,t} + (\mathbf{v} \cdot \nabla)\mathbf{v} + \nabla p/\rho + \nabla\phi = 0 \qquad \text{Schutz Eq. (7.38)} \quad (7.21)$$

for the metric eqn. (7.5). This is known as Euler's equation for non-relativistic fluid flow in a gravitational field. You will need to use eqn. (7.2), $(nU^\alpha)_{;\alpha} = 0$, to get this result.

Solution: A careful development here will pay off when in Exercises 7.5(b) and 7.5(c). The stress–energy tensor $T^{\alpha\beta}$ for a perfect fluid in curved spacetime is

$$T^{\alpha\beta} = [(\rho + p)U^\alpha U^\beta] + [pg^{\alpha\beta}], \qquad \text{Schutz Eq. (7.7)} \qquad (7.22)$$

the only modification from its flat-space counterpart eqn. (4.15) is the more general metric tensor. Substituting this into eqn. (7.20) above we get, without approximation:

$$\begin{aligned}
T^{\alpha\beta}_{\;;\beta} &= [(\rho + p)U^\alpha U^\beta]_{;\beta} + [pg^{\alpha\beta}]_{;\beta} = 0 \\
&= [(\rho + p)U^\alpha U^\beta]_{;\beta} + p_{;\beta}\, g^{\alpha\beta} && g^{\alpha\beta}_{\;;\beta} = 0 \\
&= [(\rho + p)\, U^\alpha U^\beta]_{;\beta} + p_{,\beta}\, g^{\alpha\beta} && p \text{ is a scalar} \\
&= U^\alpha[(\rho\, U^\beta)_{;\beta} + (p\, U^\beta)_{;\beta}] + (\rho + p)U^\beta U^\alpha_{\;;\beta} + p_{,\beta}\, g^{\alpha\beta}. && \text{expanded}
\end{aligned}$$
$$(7.23)$$

We keep the factor of ρ with U^β so that we can use Schutz Eq. (7.2) (see eqn. (7.2) above). In a general fluid $\rho = n(m + \Pi)$ is the total energy density (see Schutz Table 4.1 for a definition of symbols). But in the non-relativistic limit the rest mass energy will dominate $\rho \simeq nm = \rho_0$. Then

$$(\rho\, U^\beta)_{;\beta} \simeq m(n\, U^\beta)_{;\beta} = 0.$$

The pressure arises from the random motion of the particles, which provides them with negligible kinetic energy relative to the rest mass energy in the non-relativistic limit, so $p \ll \rho$. The terms in which the pressure is not differentiated are negligible:

$$pU^\alpha\, U^\beta_{\;;\beta} \ll \rho_0 U^\beta U^\alpha_{\;;\beta} \qquad pU^\beta U^\alpha_{\;;\beta} \ll \rho_0 U^\beta U^\alpha_{\;;\beta}.$$

So eqn. (7.23) above simplifies to

$$\begin{aligned}
T^{\alpha\beta}_{\;;\beta} &= U^\alpha U^\beta p_{,\beta} + \rho_0 U^\beta U^\alpha_{\;;\beta} + p_{,\beta}\, g^{\alpha\beta} \\
&= \rho_0\, U^\beta[U^\alpha_{\;,\beta} + \Gamma^\alpha_{\;\sigma\beta}U^\sigma] + p_{,\beta}\, (g^{\alpha\beta} + U^\alpha U^\beta). \quad \text{used Schutz Eq. (6.33)}
\end{aligned}$$
$$(7.24)$$

Now restrict attention to the $\alpha = i$ (spatial) components:

$$T^{i\beta}_{\;;\beta} = \rho_0\, U^\beta U^i_{\;,\beta} + \rho_0\, U^\beta \Gamma^i_{\;\sigma\beta}U^\sigma + p_{,\beta}\, (g^{i\beta} + U^i U^\beta). \qquad (7.25)$$

Consider the terms in eqn. (7.25) individually. The first term is the time derivative of the fluid momentum density. It is composed of the Eulerian time derivative part, obtained from $\beta = 0$, and the advection part, obtained from $\beta = j > 0$. To see this, expand the first term when $\beta = 0$:

$$\begin{aligned}
\rho_0\, U^0\, U^i_{\;,0} &= \rho_0\, U^0\, U^i_{\;,t} \\
&= \rho_0\, \gamma^2\, v^i_{\;,t}
\end{aligned}$$
$$(7.26)$$

where $\gamma = 1/\sqrt{1-v^2}$, v^i is the i-component of the three-velocity and v is its magnitude.

$$\rho_0 \gamma^2 v^i{}_{,t} \simeq \rho_0 v^i{}_{,t} \qquad \text{Newtonian limit}$$

$$= \rho_0 \frac{\partial \mathbf{v}}{\partial t}. \qquad \text{changed to three-vector notation.} \qquad (7.27)$$

And the advective part corresponds to this term with $\beta = j$:

$$\rho_0 \gamma^2 v^j v^i{}_{,j} \simeq \rho_0 v^j v^i{}_{,j} \qquad \text{Newtonian limit}$$

$$= \rho_0 (\mathbf{v} \cdot \nabla)\mathbf{v}. \qquad \text{changed to vector notation.} \qquad (7.28)$$

The next term in eqn. (7.25) contains the Christoffel symbols and can be written:

$$\rho_0 \Gamma^i{}_{\sigma\beta} U^\sigma U^\beta \simeq \rho_0 \Gamma^i{}_{00} U^0 U^0 \qquad \text{in Newtonian limit } U^0 \gg U^i$$

$$= \rho_0 \phi_{,i} U^0 U^0 \qquad \text{used eqn. (7.19) above}$$

$$= \rho_0 \phi_{,i} \gamma^2$$

$$\simeq \rho_0 \phi_{,i} \qquad \text{in Newtonian limit}$$

$$= \rho_0 \nabla\phi, \qquad \text{changed to vector notation} \qquad (7.29)$$

which is the gravitational force per unit volume of fluid. The final term in eqn. (7.25) is the pressure gradient force per unit volume,

$$p_{,\beta} (g^{i\beta} + U^i U^\beta) \simeq p_{,\beta}\, g^{i\beta} \qquad |U^i| \ll 1 \text{while } |g^{ii}| \sim 1$$

$$= p_{,i} (1 - 2\phi)^{-1} \qquad \text{used Schutz Eq. (7.20)}$$

$$\simeq p_{,i} (1 + 2\phi) \qquad \text{binomial series}$$

$$= p_{,i} \qquad \text{because } |\phi| \ll 1$$

$$= \nabla p. \qquad \text{changed to vector notation} \qquad (7.30)$$

Gathering eqns. (7.27)–(7.30) we construct the Euler equation applicable to perfect (inviscid) fluids in classical fluid mechanics (see for example Vallis, 2006, §1.7),

$$\rho_0 \frac{\partial \mathbf{v}}{\partial t} + \rho_0 (\mathbf{v} \cdot \nabla)\mathbf{v} = -\rho_0 \nabla\phi - \nabla p. \qquad (7.31)$$

7.5 (b) Examine the time component of eqn. (7.20) above under the same assumptions, and interpret each term.

Hint: The Newtonian approximation applied to the time component of the divergence of the perfect fluid stress–energy tensor leads to the classical fluid dynamics so-called *continuity equation* or equation of mass conservation, see (Misner et al., 1973, Eq. (39.15a)).

Going beyond the Newtonian approximation requires a systematic treatment keeping track of the order of accuracy of all the terms, as done in the post-Newtonian analysis. At this level of accuracy $T^{0\nu}_{\ ;\nu} = 0$ gives the first law of thermodynamics (Misner et al., 1973, Eq. (39.46)). In a more general context (not a perfect fluid) but to the same accuracy, Weinberg (1972) derives his Eq. (9.3.4), which includes a tidal forcing term, $T^{00}\partial\phi/\partial t$.

7.5 (c) Euler equation, eqn. (7.21) above, implies that a static fluid ($v = 0$) in a static Newtonian gravitational field obeys the equation of hydrostatic equilibrium,

$$\nabla p + \rho\nabla\phi = 0. \qquad\qquad \text{Schutz Eq. (7.39)} \qquad (7.32)$$

A metric tensor is said to be static if there exist coordinates in which $\vec{e}_0$ is timelike, $g_{i0} = 0$, and $g_{\alpha\beta,0} = 0$. Deduce from eqn. (7.20) that a static fluid ($U^i = 0$, $p_{,0} = 0$, etc.) obeys the relativistic equation of hydrostatic equilibrium,

$$p_{,i} + (\rho + p)\left[\frac{1}{2}\ln(-g_{00})\right]_{,i} = 0. \qquad \text{Schutz Eq. (7.40)} \qquad (7.33)$$

Hint: Consider trying SP7.1 below before tackling this question.

Solution: We start again with the divergence of the stress–energy tensor of a perfect fluid, eqn. (7.23) above and with $(\rho\, U^\beta)_{;\beta} = 0$ giving

$$T^{\mu\nu}_{\ ;\nu} = U^\mu(p\, U^\nu)_{;\nu} + (\rho + p)U^\nu U^\mu_{\ ;\nu} + p_{,\nu}\, g^{\mu\nu}. \qquad (7.34)$$

Consider individually the terms of eqn. (7.34) above starting with the last, applying the static condition when appropriate. The final term simplifies since the time derivatives vanish,

$$p_{,\nu}\, g^{\mu\nu} = p_{,i}\, g^{\mu i}. \qquad\qquad \text{static fluid} \qquad (7.35)$$

The second term is

$$\begin{aligned}
[(\rho + p)U^\nu]U^\mu_{\ ;\nu} &= [(\rho + p)U^0]U^\mu_{\ ;0} & \text{static fluid} \\
&= [(\rho + p)U^0](U^\mu_{\ ,t} + \Gamma^\mu_{\ \sigma 0}U^\sigma) & \text{used Schutz Eq. (6.33)} \\
&= (\rho + p)\, U^0\, \Gamma^\mu_{\ 00}\, U^0. & \text{static fluid} \quad (7.36)
\end{aligned}$$

To see that the first term vanishes, expand into two parts:

$$U^\alpha(pU^\beta)_{;\beta} = U^\alpha\, pU^\beta_{\ ;\beta} + U^\alpha U^\beta\, p_{,\beta}. \qquad (7.37)$$

The second part vanishes because for a static fluid $U^i = 0$ and so $U^\alpha U^\beta p_{,\beta} = U^0 U^0 p_{,t} = 0$. The first part also vanishes because

$$U^\beta_{\ ;\beta} = U^\beta_{\ ,\beta} + U^\beta(\ln(-g))_{,\beta} \qquad\qquad \text{used Schutz Eq. (6.41)}$$

$$= U^0{}_{,t} + U^0 (\ln(-g))_{,t} \qquad \text{everywhere static fluid}$$
$$= 0. \qquad \text{static metric}$$

The last line deserves comment. In the question Schutz introduced the idea of a static metric, hinting that the metric should be considered static in this question. It is, although you need the Einstein field equations, see Schutz Chapter 8, to fully appreciate this. Succinctly put, matter tells space how to bend and since the matter is static here so is the metric.

We conclude there is a (relativistic hydrostatic) balance between the second and third terms:

$$(\rho + p) U^0 \Gamma^\mu{}_{00} U^0 + p_{,i} g^{\mu i} = 0. \qquad (7.38)$$

Now simplify the Christoffel symbols,

$$\Gamma^\mu{}_{00} = \frac{1}{2} g^{\mu\beta} (-g_{00,\beta}) \qquad \text{used eqn. (5.65) \& static metric}$$
$$= \frac{1}{2} g^{\mu i} (-g_{00,i}). \qquad \text{static metric} \qquad (7.39)$$

Substitution in our hydrostatic balance equation eqn. (7.38) gives:

$$0 = (\rho + p) \frac{1}{2} g^{\mu i} (-g_{00,i}) U^0 U^0 + p_{,i} g^{\mu i}. \qquad (7.40)$$

And now for the tricky bit! In GR just as in SR $\vec{U} \cdot \vec{U} = -1$, so for a static fluid

$$\vec{U} \cdot \vec{U} = g_{\alpha\beta} U^\alpha U^\beta = U^0 U^0 g_{00} = 1 \cdot 1 \cdot (-1) = -1. \qquad (7.41)$$

Using this for $U^0 U^0$ in eqn. (7.40) and factoring $g^{\mu i}$ gives

$$0 = g^{\mu i} \left((\rho + p) \frac{1}{2} (-g_{00,i}) \left(\frac{1}{-g_{00}} \right) + p_{,i} \right) \qquad \text{sub eqn. (7.41) into eqn. (7.40)}$$
$$= (\rho + p) \frac{1}{2} (-g_{00,i}) \left(\frac{1}{-g_{00}} \right) + p_{,i} \qquad \text{true for all static } g^{\mu i}$$
$$= (\rho + p) \frac{1}{2} [\ln(-g_{00})]_{,i} + p_{,i}. \qquad \text{chain rule of differential calculus}$$

$$(7.42)$$

7.5   (d) This suggests that, at least for static situations, there is a close relation between g_{00} and $-\exp(2\phi)$, where ϕ is the Newtonian potential for a similar physical situation. Show that eqn. (7.5) and Exercise 7.4 are consistent with this.

Solution: First explore this "close relation." Consider the hydrostatic balance in the Newtonian limit, eqn. (7.32). We see that the pressure force on a unit volume of fluid arising from the pressure gradient, $-\nabla p$, is balanced by the weight of a unit volume of the fluid, $-\rho\nabla\phi$. Our result for the relativistic hydrostatic balance, eqn. (7.42) above, reveals that the term $\ln(-g_{00})/2 = \psi$ plays the role that the gravitational potential played in the Newtonian limit; it's the quantity the gradient of which balances the pressure gradient force per unit "mass." This is the sense in which g_{00} is related to the Newtonian gravitational potential. But the analogy is not exact because the weight of the fluid is augmented by the pressure in the relativistic case. And this is to be expected since pressure is a form of energy per unit volume and in relativity we learn that mass and energy are different forms of the same entity.

Now consider the relationship with the metric in eqn. (7.5) above, which we were told applies when the Newtonian potential $\phi = -GM/r$ is small, $|\phi| \ll 1$ in non-dimensional units. We can identify ϕ here with the "gravitational potential-like term" ψ because

$$
\begin{aligned}
-\exp(2\phi) &= -(1 + 2\phi + O(\phi)^2) & \text{Taylor series about } \phi = 0 \\
&\simeq -(1 + 2\phi) & \text{if } |\phi| \ll 1 \\
&= g_{00}. & \text{as in eqn. (7.5).} \quad (7.43)
\end{aligned}
$$

Or $\phi \simeq \ln(-g_{00})/2 = \psi$.

In Exercise 7.4 we were required to show that the geodesic equation, $\nabla_{\vec{p}}\,\vec{p} = 0$, in the Newtonian limit, which gave Schutz Eqs. (7.15) and (7.24), had only a dependence on g_{00}, and not the other components of the metric. These equations correspond respectively to the energy and momentum of a particle in a time-varying gravitational field. In the Newtonian limit we expect classical mechanics to apply, of course, from which we know that the energy depends upon the tidal forcing (time variation of the gravitational potential), and indeed we find the time derivative of ϕ on the RHS of Schutz Eq. (7.15):

$$
\frac{\mathrm{d}p^0}{\mathrm{d}\tau} = -m\frac{\partial\phi}{\partial\tau} \qquad \text{Schutz Eq. (7.15)}
$$

time derivate of energy $= -m \times$ time derivate of potential-like quantity.   (7.44)

Similarly classical momentum increases with time in proportion to the gradient of the gravitational potential and indeed in Schutz Eq. (7.24) we have

$$
\frac{\mathrm{d}p^i}{\mathrm{d}\tau} = -m\frac{\partial\phi}{\partial x^i} \qquad \text{Schutz Eq. (7.24)}
$$

time derivate of three-momentum $= -m \times$ gradient of potential-like quantity.
   (7.45)

7.7   Consider the following four different metrics, as given by their line elements:

(i) $\mathrm{d}s^2 = -\mathrm{d}t^2 + \mathrm{d}x^2 + \mathrm{d}y^2 + \mathrm{d}z^2$;

(ii) $\mathrm{d}s^2 = -\left(1 - \dfrac{2M}{r}\right)\mathrm{d}t^2 + \left(1 - \dfrac{2M}{r}\right)^{-1}\mathrm{d}r^2 + r^2(\mathrm{d}\theta^2 + \sin^2\theta)\mathrm{d}\phi^2$;

(iii) $\quad ds^2 = -\dfrac{\Delta - a^2 \sin^2\theta}{\rho^2} dt^2 - 2a \dfrac{2Mr \sin^2\theta}{\rho^2} dt\, d\phi$

$\quad\quad + \dfrac{(r^2 + a^2)^2 - a^2 \Delta \sin^2\theta}{\rho^2} \sin^2\theta\, d\phi^2 + \dfrac{\rho^2}{\Delta} dr^2 + \rho^2 d\theta^2;$

(iv) $\quad ds^2 = -dt^2 + R^2(t) \left[\dfrac{1}{1 - kr^2} dr^2 + r^2(d\theta^2 + \sin^2\theta d\phi^2) \right],$ $\quad$ (7.46)

where M and a are constants and we have introduced the shorthand notation $\Delta = r^2 - 2Mr + a^2$, $\rho^2 = r^2 + a^2 \cos^2\theta$. In (iv) k is a constant and $R(t)$ is an arbitrary function of t alone.

The first one should be familiar by now. We shall encounter the other three in later chapters. Their names are, respectively, the Schwarzschild, Kerr, and Robertson–Walker metrics.

7.7 (a) For each metric, find as many conserved components of p_α of a freely falling particle's four momentum as possible.

Solution: Refer to the comment after Schutz Eq. (7.29); for a given set of coordinates, if $g_{\mu\nu}$ is independent of $x^{\alpha*}$ (for a fixed index[1] $\alpha*$) then $p_{\alpha*}$ is conserved for free particles. By inspection of the metrics above we immediately conclude the following.
- For the Minkowski spacetime, all four components p_α are conserved.
- The Schwarzschild and Kerr spacetimes have conserved p_t and p_ϕ.
- For the Robertson–Walker spacetime p_ϕ is conserved.

7.7 (b) Use the results of Exercise 6.28 to transform the Minkowski metric (i) to the form

$$ds'^2 = -dt^2 + dr^2 + r^2(d\theta^2 + \sin^2\theta d\phi^2).$$

Use this to argue that (ii) and (iv) are spherically symmetric. Does this increase the number of conserved components of p_α?

Solution: The transformation from Cartesian coordinates x^α to spherical coordinates $x^{\alpha'}$ leaves the temporal components of the metric unchanged,

$$g_{0'\alpha'} = \dfrac{\partial x^\alpha}{\partial t'} \dfrac{\partial x^\beta}{\partial x^{\alpha'}} \eta_{\alpha\beta} = \eta_{0'\alpha'},$$

where we have used the fact that $\partial x^i / \partial t' = 0$, $\partial t / \partial t' = 1$. Hereafter drop primes on indices for sake of brevity. The spatial part of Minkowski space is 3D Euclidean space. In Exercise 6.28 we derived the metric of Euclidean space in spherical coordinates, as given in eqn. (6.30): $(g_{ij}) = \text{diag}(1, r^2, r^2 \sin^2\theta)$. It then follows that the Minkowski

---

[1] Some books (e.g. Poisson, 2004) add a '$*$' to indicate a fixed index.

space metric in spherical coordinates is:

$$ds'^2 = -dt^2 + dr^2 + r^2(d\theta^2 + \sin^2\theta d\phi^2).$$

Spherically symmetric spacetimes are defined in Schutz §10.1. A spacetime is spherically symmetric if coordinates can be found such that every point of the spacetime falls on a 2D submanifold with the metric of a sphere:

$$dl^2 = F(t,r)(d\theta^2 + \sin^2\theta d\phi^2) \tag{7.47}$$

where $F(t,r)$ is a constant for fixed (t,r). The metrics of both (ii) and (iv) above manifestly have this property.

Yes indeed the spherical symmetry of (ii) and (iv) implies that there are more conserved components of the four-momentum, $p_{\alpha*}$, in some as yet unspecified coordinate system. This crucially important point is taught in more advanced GR courses through the concept of Killing vector fields: fields of vectors that point in a direction of constant metric tensor, see Exercise 7.10 for more information. Here Schutz introduced the idea of conserved quantities through a more mathematically accessible idea: for a given set of coordinates, if $g_{\mu\nu}$ is independent of $x^{\alpha*}$ (for a fixed $\alpha*$) then $p_{\alpha*}$ is conserved for free particles. While beautifully simple, there is an inherent limitation in this formulation in that coordinate systems are arbitrary. For instance for spherically symmetric spacetimes, e.g. (i'), (ii), and (iv), the metric in the chosen coordinate systems is independent of ϕ, i.e. $g_{\mu\nu,\phi} = 0$, so p_ϕ will be conserved for free particles. Despite appearances (i.e. the $\sin^2\theta$ factor), the metric on the surface of a sphere centered at the origin is independent of position on the sphere because all points are identical. The chosen coordinates make the $g_{\mu\nu,\phi} = 0$ readily apparent, but there must be other directions for which the metric tensor does not change. For example consider a free particle in the equatorial plane. The p_ϕ component of $\tilde{p}$ is conserved. Now simply rotating the y and z axes by 90° about the x-axis (so the z'-axis points in the direction of negative y-axis), gives a new coordinate system with (same metric, different components) $g_{\mu'\nu',\phi'} = 0$, so the ϕ' component of $\tilde{p}$ is evidently conserved.

How many independent conserved quantites can we find like this? As in classical mechanics symmetries have associated conserved quantities. In GR any active transformation (a global mapping of the spacetime into itself) that leaves the metric unchanged (an isometry) provides a conserved quantity along geodesics. For spherically symmetric spacetimes rotations about the three independent spatial axes preserve the metric giving three independent conserved quantities along geodesics. A convincing demonstration benefits greatly from a coordinate independent approach using Killing vectors, see (§3.8 Carroll, 2004).

7.7   (c) For metrics (i'), (ii), (iii), and (iv), a geodesic that begins tangent to the equatorial plane stays on the equatorial plane (i.e. starts with $\theta = \pi/2$ and $p^\theta = 0$ and keeps $\theta = \pi/2$ and $p^\theta = 0$). For cases (i'), (ii), and (iii), use the equation

$\vec{p} \cdot \vec{p} = -m^2$ to solve for p^r in terms of m, and other conserved quantities, and known functions of position.

Solution: For the first two, (i') and (ii), the metric is diagonal and the solution proceeds very similarly:

$$-m^2 = \vec{p} \cdot \vec{p} = p^\alpha p^\beta g_{\alpha\beta} = (p^t)^2 g_{00} + (p^r)^2 g_{rr} + \cancelto{0}{(p^\theta)^2 g_{\theta\theta}} + (p^\phi)^2 g_{\phi\phi}$$

$$p^r = \pm \sqrt{\frac{-m^2 - (p^t)^2 g_{00} - (p^\phi)^2 g_{\phi\phi}}{g_{rr}}}. \tag{7.48}$$

Now we need to relate p^t and p^ϕ to the conserved quantities p_t and p_ϕ to fulfill the requirement of using "other conserved quantities." For diagonal metrics a single factor relates each component:

$$p_t = p^\alpha g_{\alpha t} = p^t g_{tt}$$

$$p_\phi = p^\phi g_{\alpha\phi} = p^\phi g_{\phi\phi}. \tag{7.49}$$

Substitute eqn. (7.49) into eqn. (7.48) to find

$$p^r = \pm \sqrt{\frac{-(p_t)^2/g_{00} - (p_\phi)^2/g_{\phi\phi} - m^2}{g_{rr}}}. \tag{7.50}$$

For Kerr, the non-diagonal metric (iii) above, two complications arise. There is an extra term in the product $\vec{p} \cdot \vec{p}$ resulting from $g_{t\phi} p^t p^\phi + g_{\phi t} p^\phi p^t = 2 g_{t\phi} p^t p^\phi$. So instead of eqn. (7.48) above we have

$$p^r = \pm \sqrt{\frac{-m^2 - (p^t)^2 g_{00} - (p^\phi)^2 g_{\phi\phi} - 2 p^t p^\phi g_{t\phi}}{g_{rr}}}. \tag{7.51}$$

The second complication is that the conserved quantities p_t and p_ϕ are not simply related to p^t and p^ϕ. To avoid inverting the whole 4×4 metric tensor, as one would need for $p^t = g^{t\alpha} p_\alpha$ and $p^\phi = g^{\phi\alpha} p_\alpha$, it is easier to write:

$$p_t = p^\sigma g_{t\sigma} = g_{tt} p^t + g_{t\phi} p^\phi$$

$$p_\phi = p^\sigma g_{\phi\sigma} = g_{\phi\phi} p^\phi + g_{t\phi} p^t. \tag{7.52}$$

We solve this 2×2 system (see eqn. (B.1) in Appendix B) to find p^t and p^ϕ in terms of the two conserved quantities p_t and p_ϕ:

$$p^t = -p_t \frac{g_{\phi\phi}}{g_{\phi t}^2 - g_{tt} g_{\phi\phi}} + p_\phi \frac{g_{t\phi}}{g_{\phi t}^2 - g_{tt} g_{\phi\phi}}$$

$$p^\phi = p_t \frac{g_{t\phi}}{g_{\phi t}^2 - g_{tt} g_{\phi\phi}} - p_\phi \frac{g_{tt}}{g_{\phi t}^2 - g_{tt} g_{\phi\phi}}. \qquad \text{used eqn. (B.1)} \tag{7.53}$$

Substitution of eqn. (7.53) into eqn. (7.51) above gives p^r in terms of the conserved quantities and the known metric component functions.

7.7   (d) For (iv), spherical symmetry implies that if a geodesic begins with $p^\theta = p^\phi = 0$, these remain zero. Use this to show from

$$m\frac{\mathrm{d}}{\mathrm{d}\tau}p_\alpha = \frac{1}{2}g_{\mu\nu,\alpha}\,p^\mu\,p^\nu \qquad \text{Schutz Eq. (7.29)} \qquad (7.54)$$

that when $k = 0$, p_r is conserved.

Solution: When $k = 0$ the Robertson–Walker metric simplifies to

$$(g_{\alpha\beta}) = \mathrm{diag}(-1, R^2(t), r^2 R^2(t), r^2\,\sin^2\theta\,R^2(t)). \qquad (7.55)$$

Writing out eqn. (7.54) for p_r and this metric we get:

$$
\begin{aligned}
m\frac{\mathrm{d}}{\mathrm{d}\tau}p_r &= \frac{1}{2}g_{\nu\alpha,r}\,p^\nu\,p^\alpha \\
&= \frac{1}{2}[g_{tt,r}\,(p^t)^2 + g_{rr,r}\,(p^r)^2 + g_{\theta\theta,r}\,(p^\theta)^2 + g_{\phi\phi,r}\,(p^\phi)^2] \\
&= \frac{1}{2}[2rR^2(t)\,(p^\theta)^2 + 2r\,\sin^2\theta\,R^2(t)\,(p^\phi)^2] \qquad\qquad \text{used eqn. (7.55)} \\
&= 0. \qquad\qquad\qquad\qquad\qquad\qquad\qquad\qquad\qquad \text{used } p^\theta = p^\phi = 0
\end{aligned}
$$

7.10   (a) Show that if a vector field ξ^α satisfies *Killing's equation*

$$\nabla_\alpha\xi_\beta + \nabla_\beta\xi_\alpha = 0, \qquad \text{Schutz Eq. (7.45)}$$

then along a geodesic, $p^\alpha\xi_\alpha =$ constant. This is a coordinate-invariant way of characterizing the conservation law we deduced from eqn. (7.54) above. We only have to know whether a metric admits Killing fields.

Solution: First we simply rearrange Killing's equation and note that it defines an antisymetric tensor $A_{\alpha\beta}$:

$$
\begin{aligned}
\nabla_\alpha\xi_\beta &= -\nabla_\beta\xi_\alpha \\
A_{\alpha\beta} &= -A_{\beta\alpha}. \qquad (7.56)
\end{aligned}
$$

Now we take the intrinsic derivative (recall terminology introduced in eqn. (6.136) of SP6.9) of the real number $(p^\alpha\xi_\alpha)$ along a geodesic, parameterized by proper time τ. We imagine that $p^\alpha = mu^\alpha$ is the four-momentum of some particle of rest mass m. We wish to show of course that this intrinsic derivative, $D/D\tau$, is nil.

$$
\begin{aligned}
\frac{D}{D\tau}(p^\alpha\xi_\alpha) &= m\frac{D}{D\tau}(u^\alpha\xi_\alpha) \qquad\qquad\qquad\qquad \text{rest mass } m \text{ is constant} \\
&= mu^\alpha\frac{D}{D\tau}(\xi_\alpha) \qquad \text{by definition of a geodesic have } \frac{Du^\alpha}{D\tau} = 0 \\
&= mu^\alpha u^\beta\nabla_\beta\xi_\alpha \qquad\qquad\qquad\qquad \text{notation of eqn. (6.143)} \\
&= 0, \qquad\qquad\qquad\qquad\qquad\qquad\qquad\qquad\qquad\qquad (7.57)
\end{aligned}
$$

because the second last line is the product of a symmetric tensor $(u^\alpha u^\beta)$ with an antisymmetric tensor $\nabla_\beta \xi_\alpha$.

7.10 (b) Find ten Killing fields of Minkowski spacetime.

Solution: In flat space the Christoffel symbols vanish and Killing's equation reduces to

$$\xi_{\alpha,\beta} + \xi_{\beta,\alpha} = 0. \tag{7.58}$$

Obviously a Killing vector field could be any constant vector field, e.g. $\vec{\xi} = \vec{e}_t$, or $\vec{\xi} = \vec{e}_x$, or either of the other two basis vectors, $\vec{\xi} = \vec{e}_y, \vec{e}_z$ or any linear combination of them

$$\vec{\xi} = a\vec{e}_t + b\vec{e}_x + d\vec{e}_y + f\vec{e}_z,$$

as long as the constants a, b, d, f don't change with position in spacetime. By inspection an obvious non-constant Killing (dual) vector field would be $\xi_\alpha = (-x, t, 0, 0)$, so $\vec{\xi} = x\vec{e}_t + t\vec{e}_x$. Similarly, $\xi_\alpha = (0, -y, x, 0)$, $\xi_\alpha = (0, -z, 0, x)$ and $\xi_\alpha = (0, 0, z, -y)$ are Killing in Minkowski. For future reference let

$$\vec{\xi}_A = y\vec{e}_x - x\vec{e}_y, \qquad\qquad \vec{\xi}_B = y\vec{e}_t + t\vec{e}_y. \tag{7.59}$$

How many linearly independent Killing vector fields are there? For a *maximally symmetric manifold* of dimension N there are $N(N+1)/2$ (Carroll, 2004, Eq. (3.189)), so in Minkowski space, which is an $N = 4$ maximally symmetric space, we expect 10 linearly independent Killing vectors. As stated in the solution to Exercise 7.7(b), each is related to a transformation and these are listed in Table 7.1. Here we're

Table 7.1 Killing vector fields of Minkowski spacetime

Killing one-form	Killing vector	Transformation
$(-x, t, 0, 0)$	$x\vec{e}_t + t\vec{e}_x$	boost along x-axis
$(-y, 0, t, 0)$	$y\vec{e}_t + t\vec{e}_y$	boost along y-axis
$(-z, 0, 0, t)$	$z\vec{e}_t + t\vec{e}_z$	boost along z-axis
$(0, 0, z, -y)$	$z\vec{e}_y - y\vec{e}_z$	rotation about x-axis
$(0, z, 0, -x)$	$z\vec{e}_x - x\vec{e}_z$	rotation about y-axis
$(0, y, -x, 0)$	$y\vec{e}_x - x\vec{e}_y$	rotation about z-axis
$(-1, 0, 0, 0)$	$\vec{e}_t$	translation along t-axis
$(0, 1, 0, 0)$	$\vec{e}_x$	translation along x-axis
$(0, 0, 1, 0)$	$\vec{e}_y$	translation along y-axis
$(0, 0, 0, 1)$	$\vec{e}_z$	translation along z-axis

broaching the deeper aspects of the exercise. Killing vector fields are related to conserved quantities, which are related to symmetries, which are related to active transformations that preserve the metric (see solution to Exercise 7.7(c) above). So the 10 Killing vector fields of Minkowski are in fact linked to transformations that preserve the metric. Recall that Lorentz transformations can be defined as the transformations that preserve the metric, eqn. (3.79). There are six "types" of Lorentz transformation; in Cartesian coordinates they corespond to a boost along one of the three x^{i*} axes and a rotation about one of the three x^{i*} axes for fixed $i*$, see first six rows of Table 7.1. They are elements of the *Lorentz group* explored in Exercise 3.33. The remaining four Killing vector fields correspond to the translations along the four coordinate axes, final four rows of Table 7.1. These translations together with the Lorentz transformations form the Poincaré group. A *general Lorentz transformation* or arbitrary Poincaré transformation can be constructed from products of these 10 (Rindler, 2006, §2.7).

7.10   (c) Show that if $\vec{\xi}$ and $\vec{\eta}$ are Killing fields, then so is $\alpha\,\vec{\xi} + \beta\,\vec{\eta}$ for constant α and β.

Solution: The covariant derivative is a linear operator, so it follows immediately that a linear combination of Killing vector fields $\vec{\xi}$ and $\vec{\eta}$ also satisfies the Killing equation:

$$\nabla_\mu(\alpha\xi_\nu + \beta\eta_\nu) + \nabla_\nu(\alpha\xi_\mu + \beta\eta_\mu) = \alpha(\nabla_\mu\xi_\nu + \nabla_\nu\xi_\mu) + \beta(\nabla_\mu\eta_\nu + \nabla_\nu\eta_\mu)$$
$$= 0. \tag{7.60}$$

We have assumed that the constants α and β are independent of space and time; it's the same linear combination everywhere and always!

7.10   (d) Show that Lorentz transformations of the fields in (b) simply produce linear combinations as in (c).

Solution: In Execise. 7.10(b) we found a set of rather obvious Killing vector fields for Minkowski spacetime. Of course, on the one hand, a given vector field $\vec{\xi}$ can be expressed in more than one Lorentz frame,

$$\vec{\xi} = \xi^\alpha \vec{e}_\alpha = \xi^{\alpha'} \vec{e}_{\alpha'},$$

and the components and basis vectors are related by a Lorentz transformation

$$\xi^{\alpha'} = \Lambda^{\alpha'}{}_\alpha \xi^\alpha \qquad\qquad \vec{e}_{\alpha'} = \Lambda^\alpha{}_{\alpha'} \vec{e}_\alpha. \tag{7.61}$$

On the other hand, in the new Lorentz frame there will be a *new* set of "obvious" Killing vector fields, analogous to those in Table 7.1 but in the new Lorentz frame, e.g. $\vec{\xi}' = y'\vec{e}_{x'} - x'\vec{e}_{y'}$ is different from but analogous to the second element in

Table 7.1. The question here is to relate the new Killing vectors to the original 10, and in particular show that their relationship has the form of a linear combination.

Abstraction is helpful here. Note that eqn. (7.58) is a linear, first-order, homogeneous differential equation. We expect, and indeed proved in Exercise 7.10(c), that a linear combination of Killing vectors is itself a Killing vector; the Killing vectors form a linear (vector) space. Because Minkowski space is maximally symmetric there are 10 linearly independent solutions, see Exercise 7.10(b); these form a basis for this linear space. The "new" Killing vectors must be in the solution space (they are Killing vectors of the spacetime, regardless of the coordinates we used to find them), and so must be represented by a linear combination of the 10 (basis) vector fields in Table 7.1. That completes the proof.

Let's work through one example to see how this works. Consider $\vec{\xi}' = y'\vec{e}_{x'} - x'\vec{e}_{y'}$, where $x^{\alpha'}$ is a coordinate system on Minkowski spacetime related to the coordinates of Exercise 7.10(b) by, say, a boost $x^{\alpha'} = \Lambda^{\alpha'}_{\ \alpha}(v)x^\alpha$ in the x-direction i.e.,

$$\begin{pmatrix} t' \\ x' \\ y' \\ z' \end{pmatrix} = \begin{pmatrix} \gamma & -v\gamma & 0 & 0 \\ -v\gamma & \gamma & 0 & 0 \\ 0 & 0 & 1 & 0 \\ 0 & 0 & 0 & 1 \end{pmatrix} \begin{pmatrix} t \\ x \\ y \\ z \end{pmatrix} \implies \begin{array}{l} \vec{e}_{t'} = \gamma\vec{e}_t + v\gamma\vec{e}_x \\ \vec{e}_{x'} = v\gamma\vec{e}_t + \gamma\vec{e}_x \\ \vec{e}_{y'} = \vec{e}_y \\ \vec{e}_{z'} = \vec{e}_z. \end{array} \quad (7.62)$$

Applying this coordinate transformation we find $x' = \gamma(x - vt)$ and $y' = y$. Substituting into $\vec{\xi}'$ we find:

$$\begin{aligned} \vec{\xi}' = y'\vec{e}_{x'} - x'\vec{e}_{y'} &= y(v\gamma\vec{e}_t + \gamma\vec{e}_x) - \gamma(x - vt)\vec{e}_y \\ &= v\gamma(y\vec{e}_t + t\vec{e}_y) + \gamma(y\vec{e}_x - x\vec{e}_y) \\ &= v\gamma\vec{\xi}_B + \gamma\vec{\xi}_A. \qquad\qquad \text{used eqn. (7.59)} \quad (7.63) \end{aligned}$$

Here v and γ are constants and $\vec{\xi}_A$ and $\vec{\xi}_B$ are two of the obvious Killing vector fields found in Exercise 7.10(b). So for this example the new Killing vector field $\vec{\xi}'$ can be written as a linear combination of two other Killing vector fields. This is to be expected because the 10 Killing vector fields form a basis of the linear space.

7.10 (e) If you did Exercise 7.7, use the results of Exercise 7.7(a) to find Killing vectors of metrics (ii)–(iv).

Solution: Note that we are asked to use the results on conserved quantities, so the implied symmetries, to find the Killing vector fields as opposed to solving Killing's equation eqn. (7.56). In general it would be hard to know when one has found all the solutions. Fortunately the metrics (ii)–(iv) are very well-studied and we can appeal to the literature.

For the Schwarzschild spacetime, the conserved quantities found in Exercise 7.7(a) were p_t (resulting from time translational symmetry) and p_ϕ (resulting from spherical symmetry about the spatial origin). Table 7.1 rows four, five and six, lists the three Killing vector fields associated with invariance for rotations about the three spatial Cartesian axes. Because Schwarzschild also has spherical symmetry it enjoys the same Killing vector fields. We can transform these into the spherical coordinates of (ii) using relations in Appendix B giving:

$$\vec{Q} = \vec{e}_t$$

$$\vec{R} = \vec{e}_\phi$$

$$\vec{S} = \left(\frac{\partial\theta}{\partial x}S^x + \frac{\partial\theta}{\partial z}S^z\right)\vec{e}_\theta + \left(\frac{\partial\phi}{\partial x}S^x + \frac{\partial\phi}{\partial z}S^z\right)\vec{e}_\phi = \cos\phi\ \vec{e}_\theta - \cot\theta\ \sin\phi\ \vec{e}_\phi$$

$$\vec{T} = \left(\frac{\partial\theta}{\partial x}T^x + \frac{\partial\theta}{\partial z}T^z\right)\vec{e}_\theta + \left(\frac{\partial\phi}{\partial x}T^x + \frac{\partial\phi}{\partial z}T^z\right)\vec{e}_\phi = -\sin\phi\ \vec{e}_\theta - \cot\theta\ \cos\phi\ \vec{e}_\phi.$$

$$(7.64)$$

And these are the *only* Killing vector fields for Schwarzschild (Carroll, 2004, §5.4).

For the Kerr spacetime, we found two conserved quantities, p_t and p_ϕ. But now spherical symmetry does *not* apply so we do not gain the two further Killing vector fields $\vec{S}$ and $\vec{T}$ above. The two obvious Killing vector fields are just

$$\vec{Q} = \vec{e}_t \qquad \text{and} \qquad \vec{R} = \vec{e}_\phi. \qquad (7.65)$$

And these are the *only* Killing vector fields for Kerr (Carroll, 2004, §6.6) (although the Kerr metric does have something called a Killing tensor).

The Killing vector fields for the complete Robertson–Walker spacetime are quite complicated to find, see (Weinberg, 1972, Chapters 13 and 14) or (Aldrovandi et al., 2007, Appendix A). Putting special restrictions on $R(t)$ (so that the space-time becomes the so-called de Sitter spacetime) the spacetime becomes maximally symmetric and has 10 independent Killing vector fields (Aldrovandi et al., 2007, Eq. (22)), but for general $R(t)$ there are just six. The procedure to find these is outlined in SP7.13.

Restricting ourselves to the special case considered in Exercise 7(d), wherein $k = 0$ (the so-called flat-space model), things simplify considerably. The spacetime is spherically symmetric so we have the three vectors $\vec{R}, \vec{S}$, and $\vec{T}$ of eqn. (7.64) above. Note that the spacetime is not stationary; it is *not* invariant to time-translations, so $\vec{e}_t$ is not a Killing vector. On the other hand the flat-space Robertson–Walker spacetime conserved p_r for initial conditions $p_\theta = p_\phi = 0$. But p_r could be pointing along any one of three independent directions, so we expect three associated Killing vectors. (This becomes really obvious if you transform $g_{\alpha\beta}$ of (iv), with $k = 0$, to Cartesian coordinates.) This immediately gives three Killing vector fields: $\vec{U} = \vec{e}_x, \vec{V} = \vec{e}_y, \vec{W} = \vec{e}_z$. For completeness we transform them to the spherical coordinates of (iv) using eqn. (B.12):

$$\vec{U} = \frac{\partial x^\alpha}{\partial x}U^x\vec{e}_\alpha = \sin\theta\ \cos\phi\ \vec{e}_r + \frac{\cos\theta\ \cos\phi}{r}\ \vec{e}_\theta - \frac{\sin\phi}{r\sin\theta}\vec{e}_\phi$$

$$\vec{V} = \frac{\partial x^{\alpha}}{\partial y} V^{y} \vec{e}_{\alpha} = \sin\theta \ \sin\phi \ \vec{e}_{r} + \frac{\cos\theta \ \sin\phi}{r} \vec{e}_{\theta} + \frac{\cos\phi}{r\sin\theta} \vec{e}_{\phi}$$

$$\vec{W} = \frac{\partial x^{\alpha}}{\partial z} W^{z} \vec{e}_{\alpha} = \cos\theta \ \vec{e}_{r} - \frac{\sin\theta}{r} \vec{e}_{\theta}. \tag{7.66}$$

7.2 Supplementary problems

SP 7.1   Recall we learned in SR that the four velocity of a stationary particle was the speed of light in the direction of time, cf. §2.2, so that

$$\vec{U} \cdot \vec{U} = \eta_{\alpha\beta} U^{\alpha} U^{\beta} = U^{0} U^{0} \eta_{00}$$
$$= 1 \cdot 1 \cdot (-1) = -1. \qquad\qquad \text{eqn. (2.40)}$$

Now in GR the metric has changed from $\eta_{\alpha\beta}$ to the general $g_{\alpha\beta}$, but do we keep the magnitude of the $\vec{U} \cdot \vec{U} = -1$ implying that the components of U^{α} change accordingly? Or do we keep the components of U^{α} as in Minkowski space, and find a different normalization? I recommend you do this problem before tackling Exercise 7.5(c).

Solution

We keep the condition that $\vec{U} \cdot \vec{U} = -1$. This is a frame-invariant expression that we want to generalize to curved spacetime.

$$\vec{U} \cdot \vec{U} = g_{\alpha\beta} U^{\alpha} U^{\beta} \qquad\qquad \text{in general}$$
$$\underset{\text{LIF}}{=} \eta_{\alpha\beta} U^{\alpha} U^{\beta} = -1 \qquad\qquad \text{in LIF}$$
$$\underset{\text{MCRF}}{=} \eta_{00} U^{0} U^{0} = -1. \qquad\qquad \text{in MCRF} \tag{7.67}$$

A local inertial frame (LIF, see Schutz §5.1) is one for which SR holds locally, and therefore the metric is $\eta_{\alpha\beta}$. So the same normalization $\vec{U} \cdot \vec{U} = -1$ holds in both SR and GR. The MCRF is a LIF that has been Lorentz boosted such that it's momentarily stationary with respect to the fluid at a given event.

SP 7.2   In SR the dot product between two four-vectors was given in Schutz Eq. (3.1)

$$\vec{A} \cdot \vec{B} = A^{\alpha} B^{\beta} \eta_{\alpha\beta}.$$

Generalize this to curved spacetime with metric $g_{\alpha\beta}$.

SP 7.3   Show that $T^{\nu}_{\ \mu;\nu} = 0$ can be derived from the conservation law eqn. (7.20) above, $T^{\mu\nu}_{\ \ ;\nu} = 0$.

Solution

We simply multiply eqn. (7.20) by the metric tensor to lower the index:

$$0 = T^{\sigma\nu}_{\ \ ;\nu} \qquad\qquad\qquad\qquad \text{eqn. (7.20)}$$

$$= T^{\nu\sigma}_{\ \ ;\nu} \qquad\qquad \text{stress–energy tensor symmetric, Schutz §4.5}$$

$$0 = g_{\mu\sigma} T^{\nu\sigma}_{\ \ ;\nu} \qquad\qquad\qquad \text{multiplied both sides by } \mathbf{g}$$

$$= (g_{\mu\sigma} T^{\nu\sigma})_{;\nu} \qquad\qquad\qquad\qquad \text{used eqn. (6.18)}$$

$$= T^{\nu}_{\ \mu;\nu}. \qquad\qquad\qquad\qquad\qquad (7.68)$$

SP 7.4   This and the next four problems aim to deepen our understanding of the physical meaning of the metric and proper time in curved spacetime. An ideal clock, i.e. a clock not affected by acceleration, moves along a world line $x^\alpha(\tau)$ in an arbitrary curved spacetime, where τ is the proper time measured by the clock. Using the information given in Schutz §7.1 argue that the relationship between coordinate time $t \equiv x^0$ and proper time τ is

$$\frac{d\tau}{dt} = \sqrt{-g_{00} - 2g_{0i}\frac{dx^i}{dt} - g_{ij}\frac{dx^i}{dt}\frac{dx^j}{dt}}. \qquad (7.69)$$

SP 7.5   In a stationary spacetime one can find coordinates such that,

$$\frac{\partial g_{\alpha\beta}}{\partial t} = 0,$$

where $\vec{e}_t$ is timelike, i.e. $\vec{e}_t \cdot \vec{e}_t < 0$. One can adjust the local (ideal) clocks, at fixed spatial coordinates, such that they indicate this global coordinate time. In physical terms, the clock-rate synchronization and rate adjustment procedure is a simple generalization of that used in SR (Rindler, 2006, see §9.2). In mathematical terms we simply let

$$\tau = f(x^i)\, t,$$

where τ is the unadjusted, ideal, stationary clock time. Find the function of the spatial coordinates $f(x^i)$.

Solution

The clocks are stationary in the chosen coordinate system (they have fixed spatial coordinates) so eqn. (7.69) simplifies to $d\tau/dt = \sqrt{-g_{00}}$. So we simply choose $f(x^i) = \sqrt{-g_{00}(x^i)}$ so that

$$\tau = \sqrt{-g_{00}(x^i)}\, t \qquad \Rightarrow \qquad d\tau = \sqrt{-g_{00}(x^i)}\, dt.$$

Chapter 9 of Rindler (2006) is well worth reading for gaining a deeper appreciation of the meaning of the metric and its relationship to physical experiments.

SP 7.6 The Schwarzschild metric, see (ii) in Exercise 7.7, describes the geometry of the spacetime of the vacuum surrounding a non-rotating, spherical mass. It will be discussed in detail in Chapters 10 and 11. You might expect, by analogy with Kepler's circular orbits in Newtonian mechanics, that there are circular geodesics and this turns out to be true (but with some surprising conditions attached as we will see in Chapter 11). Without loss of generality, consider the equatorial plane $\theta = \pi/2$. Then a circular orbit will have both $dr = 0$ and $d\theta = 0$. We will see in Chapter 11, see Schutz Eqs. (11.20) and (11.22), that

$$\dot{\phi} := \frac{d\phi}{d\tau} = \frac{1}{r^2} \left(\frac{Mr}{1 - 3M/r} \right)^{1/2}. \tag{7.70}$$

First convince yourself that for large orbits, far from the mass $r \gg M$, eqn. (7.70) reduces to Kepler's third law (assume that $\tau \simeq t$ in this limit). Use eqn. (7.70) to show that the relation between the time read by a clock in a circular orbit of radial coordinate r and the Schwarzschild coordinate time is

$$\frac{d\tau}{dt} = \left(1 - \frac{3M}{r} \right)^{1/2}, \qquad \text{when } r > 3M. \tag{7.71}$$

This concept is also explored in Exercise 11.7.

SP 7.7 Consider an aribtrary event $\mathcal{A}$ in a general spacetime, and a neighboring event $\mathcal{B}$ that is simultaneous with $\mathcal{A}$ in some reference frame x^μ. Find the distance measured by a standard ruler between $\mathcal{A}$ and $\mathcal{B}$ in terms of the difference in coordinates between events, dx^i. This distance is called *proper distance* in Chapter 9, consistent with the term "proper volume element" introduced in Schutz Eq. (6.18), see eqn. (6.102) above, for the volume element physically measured by rods and clocks at some event $\mathcal{P}$, see also "proper length" in SR. Proper distance is sometimes called "ruler distance" to emphasize its relation to distance measured with *local* rulers (Rindler, 2006, §11.5).

Solution

A good strategy is to introduce a local inertial frame with Cartesian coordinates $x^{\mu'}$ that is instantaneously at rest with event $\mathcal{A}$, a MCRF. For a sufficiently close event $\mathcal{B}$ this frame will also be essentially stationary at $\mathcal{B}$. Distance measurements are easy in the inertial frame; the metric is just that of Minkowski and the squared proper distance is just

$$d\sigma^2 = \eta_{i'j'} dx^{i'} dx^{j'} = (dx')^2 + (dy')^2 + (dz')^2. \tag{7.72}$$

Now we must relate eqn. (7.72) to the coordinates and metric, $g_{\mu\nu}$, in the given reference frame. From differential calculus we have

$$dx^{i'} = \frac{\partial x^{i'}}{\partial x^\mu} dx^\mu = \frac{\partial x^{i'}}{\partial x^i} dx^i, \tag{7.73}$$

where the second equality arises from the fact that points of fixed spatial coordinates in the MCRF do not vary with time t (because they're co-moving!). Substituting eqn. (7.73) into eqn. (7.72) we find

$$
d\sigma^2 = \eta_{i'j'} \frac{\partial x^{i'}}{\partial x^i} \frac{\partial x^{j'}}{\partial x^j} dx^i dx^j
$$

$$
= \left(\eta_{\mu'\nu'} \frac{\partial x^{\mu'}}{\partial x^i} \frac{\partial x^{\nu'}}{\partial x^j} - \eta_{0'0'} \frac{\partial t'}{\partial x^i} \frac{\partial t'}{\partial x^j} \right) dx^i dx^j
$$

$$
= \left(g_{ij} + \frac{\partial t'}{\partial x^i} \frac{\partial t'}{\partial x^j} \right) dx^i dx^j \equiv \gamma_{ij} dx^i dx^j. \qquad \text{used } \eta_{0'0'} = -1 \quad (7.74)
$$

The thing in parentheses in the last line of eqn. (7.74), we called it γ_{ij}, is the thing we're trying to find. It is called the *induced metric* on the hypersurface, where here the hypersurface is the 3D volume defined by fixing the temporal coordinate to that of events $\mathcal{A}$ and $\mathcal{B}$ in our 4D spacetime. We can find the $\partial t'/\partial x^i$ terms by considering

$$
g_{0i} = \frac{\partial x^{\mu'}}{\partial t} \frac{\partial x^{\nu'}}{\partial x^i} \eta_{\mu'\nu'} = \frac{\partial t'}{\partial t} \frac{\partial x^{\nu'}}{\partial x^i} \eta_{0'\nu'} \qquad \text{used MCRF not moving}
$$

$$
= \frac{\partial t'}{\partial t} \frac{\partial t'}{\partial x^i} \eta_{0'0'} = -\frac{\partial t'}{\partial t} \frac{\partial t'}{\partial x^i}. \qquad \text{used } \eta_{0'i'} = 0 \quad (7.75)
$$

Now we can use the result eqn. (7.69) if we interpret the coordinate time in Minkowski spacetime, here t', as the proper time τ. In particular, for stationary clocks eqn. (7.69) gives

$$
\frac{\partial t'}{\partial t} = \frac{\partial \tau}{\partial t} = \sqrt{-g_{00}}. \tag{7.76}
$$

Substituting eqn. (7.76) into eqn. (7.75) gives

$$
\frac{\partial t'}{\partial x^i} = -\frac{g_{0i}}{\sqrt{-g_{00}}}. \tag{7.77}
$$

And substituting eqn. (7.77) into eqn. (7.74) above we conclude that the induced metric can be written

$$
\gamma_{ij} = g_{ij} - \frac{g_{0i} g_{0j}}{g_{00}}. \tag{7.78}
$$

The proper distance then between events $\mathcal{A}$ at x^μ and $\mathcal{B}$ at $x^\mu + dx^i$ is

$$
d\sigma = \left(\gamma_{ij} dx^i dx^j \right)^{1/2}. \tag{7.79}
$$

This solution was based upon the presentation by Möller (1952, §89). For more on induced metrics see Sean Carroll's textbook (Carroll, 2004, Appendix D).

SP 7.8   Argue that the relation between radial coordinate r and proper distance l in the r-direction in the Schwarzschild metric, Exercise 7.7 (ii), is

$$
\frac{dl}{dr} = \left(1 - \frac{2M}{r} \right)^{-1/2},
$$

when $r > 2M$. Discuss the situation at $r = 2M$. This problem emphasizes that r in Schwarzschild coordinates is not just the radius.

SP 7.9 Killing vector fields were introduced in Exercise 7.10 and provide a coordinate independent way of expressing the conservation law that if all components of a metric are independent of a given coordinate variable $x^{\mu*}$, i.e. $g_{\alpha\beta,\mu*} = 0$ for fixed $\mu*$, then $p_{\mu*}$ is conserved. In general it is not easy to find all the Killing vector fields of a given spacetime, but some can be obvious. Show that if the metric has the property $g_{\alpha\beta,\mu*} = 0$ for fixed $\mu*$ then the vector $K^\alpha = \delta^\alpha{}_{\mu*}$ for fixed $\mu*$ is a Killing vector field.

Solution

We must show that the (dual) vector $K_\alpha = g_{\alpha\sigma}\delta^\sigma{}_{\mu*} = g_{\alpha\mu*}$ satisfies Killing's equation:

$$0 = \nabla_\alpha K_\beta + \nabla_\beta K_\alpha \qquad\qquad \text{Killing's equation}$$

$$= K_{\beta,\alpha} - \Gamma^\sigma{}_{\beta\alpha}K_\sigma + K_{\alpha,\beta} - \Gamma^\sigma{}_{\alpha\beta}K_\sigma$$

$$= K_{\beta,\alpha} + K_{\alpha,\beta} - 2\Gamma^\sigma{}_{\alpha\beta}K_\sigma \qquad\qquad \text{used symmetry of } \Gamma^\sigma{}_{\alpha\beta}$$

$$= g_{\beta\mu*,\alpha} + g_{\alpha\mu*,\beta} - 2\Gamma^\sigma{}_{\alpha\beta}g_{\sigma\mu*} \qquad\qquad \text{substituted } K_\alpha$$

$$= g_{\beta\mu*,\alpha} + g_{\alpha\mu*,\beta} - 2g_{\sigma\mu*}\frac{1}{2}g^{\sigma\lambda}[g_{\lambda\alpha,\beta} + g_{\lambda\beta,\alpha} - g_{\alpha\beta,\lambda}] \qquad \text{used eqn. (6.17)}$$

$$= g_{\beta\mu*,\alpha} + g_{\alpha\mu*,\beta} - \delta^\lambda_{\mu*}[g_{\lambda\alpha,\beta} + g_{\lambda\beta,\alpha} - g_{\alpha\beta,\lambda}] \qquad \text{summed over } \sigma$$

$$= g_{\alpha\beta,\mu*}. \qquad\qquad \text{summed over } \lambda$$

$$\qquad\qquad (7.80)$$

So Killing's equation is satisfied under the condition stipulated, i.e. $g_{\alpha\beta,\mu*} = 0$ for fixed $\mu*$. We can interpret this physically as saying that the metric does not change in the direction of the Killing vector.

SP 7.10 A centrifuge is a common laboratory device that exploits the equivalence between the gravitational field and centripetal acceleration to achieve high rates of sedimentation. An ultracentrifuge can achieve accelerations of 2 million $g \approx 2 \times 10^7$ m s^{-2}. We can model the centrifuge as a rotating disk in Minkowski space. Show that the metric in a reference frame rotating with the disk at constant angular velocity ω can be written as:

$$ds^2 = -(1 - r^2\omega^2)dt^2 + 2\omega r^2 dt\, d\theta + r^2 d\theta^2 + dz^2, \qquad (7.81)$$

where θ is the angular coordinate of a polar coordinate system, with z along the axis of rotation, r is the radial coordinate from the axis of rotation. Find the Ricci scalar at a radius of $r = 10$ cm, assuming $r\omega^2 = 2 \times 10^7$ m s^{-2}.

> **Hint**
>
> Some thoughtful reflection saves copious computation.

SP 7.11 The four-acceleration vector is defined by eqn. (2.48) as the derivative of the four-velocity with respect to proper time, $\vec{a} \equiv d\vec{U}/d\tau$. In §7.3, Schutz reminds us that acceleration is a coordinate independent quantity. (a) Derive an expression for the acceleration by differentiating $\vec{U} = (dx^\mu/d\tau)\vec{e}_\mu$. (b) Set the acceleration to zero to derive the geodesic equation in the form of eqn. (6.132):

$$\frac{d^2 x^\mu}{d\tau^2} + \Gamma^\mu{}_{\alpha\beta} \frac{dx^\alpha}{d\tau} \frac{dx^\beta}{d\tau} = 0, \tag{7.82}$$

where we have chosen the affine parameter of the geodesic to be proper time τ. (c) Show that the acceleration vector is a geometric quantity since it transforms as required.

Solution

(a) The acceleration is given by

$$\frac{d\vec{U}}{d\tau} \equiv \frac{d}{d\tau}\left(\frac{dx^\mu}{d\tau}\vec{e}_\mu\right) = \frac{d^2 x^\mu}{d\tau^2}\vec{e}_\mu + \frac{dx^\mu}{d\tau}\frac{d\vec{e}_\mu}{d\tau}. \tag{7.83}$$

Focus on the last term:

$$\frac{d\vec{e}_\mu}{d\tau} = \frac{dx^\nu}{d\tau}\frac{\partial \vec{e}_\mu}{\partial x^\nu} = \frac{dx^\nu}{d\tau}\Gamma^\alpha{}_{\mu\nu}\vec{e}_\alpha. \qquad \text{used eqn. (5.48)} \tag{7.84}$$

Relabeling indices so we can factor $\vec{e}_\alpha$ we have

$$\frac{d\vec{U}}{d\tau} = \left(\frac{d^2 x^\alpha}{d\tau^2} + \frac{dx^\mu}{d\tau}\frac{dx^\nu}{d\tau}\Gamma^\alpha{}_{\mu\nu}\right)\vec{e}_\alpha. \tag{7.85}$$

(b) Setting $\vec{a} = 0$ we immediately recover the geodesic equation in the desired form

$$\frac{d^2 x^\alpha}{d\tau^2} + \frac{dx^\mu}{d\tau}\frac{dx^\nu}{d\tau}\Gamma^\alpha{}_{\mu\nu} = 0.$$

(c) We wish to show that one can consistently apply a general coordinate transformation to $\vec{a}$ as defined above. That is,

$$\vec{a} = a^\alpha \vec{e}_\alpha = a^{\alpha'} \vec{e}_{\alpha'} = a^{\alpha'} \vec{e}_\alpha \frac{\partial x^\alpha}{\partial x^{\alpha'}}. \tag{7.86}$$

Equation eqn. (7.86) will be true if:

$$\left(\frac{d^2 x^\alpha}{d\tau^2} + \frac{dx^\mu}{d\tau}\frac{dx^\nu}{d\tau}\Gamma^\alpha{}_{\mu\nu}\right) = \left(\frac{d^2 x^{\alpha'}}{d\tau^2} + \frac{dx^{\mu'}}{d\tau}\frac{dx^{\nu'}}{d\tau}\Gamma^{\alpha'}{}_{\mu'\nu'}\right)\frac{\partial x^\alpha}{\partial x^{\alpha'}}$$

$$= \left[\frac{d}{d\tau}\left(\frac{dx^\sigma}{d\tau}\frac{\partial x^{\alpha'}}{\partial x^\sigma}\right) + \frac{dx^{\mu'}}{d\tau}\frac{dx^{\nu'}}{d\tau}\Gamma^{\alpha'}{}_{\mu'\nu'}\right]\frac{\partial x^\alpha}{\partial x^{\alpha'}}. \tag{7.87}$$

Note that we used a σ for the first dummy index, instead of α because α was already used. That first term on the far RHS of eqn. (7.87) above splits into two contributions:

$$\frac{d}{d\tau}\left(\frac{dx^\sigma}{d\tau}\frac{\partial x^{\alpha'}}{\partial x^\sigma}\right) = \frac{d^2x^\sigma}{d\tau^2}\frac{\partial x^{\alpha'}}{\partial x^\sigma} + \frac{dx^\sigma}{d\tau}\frac{d}{d\tau}\frac{\partial x^{\alpha'}}{\partial x^\sigma}.\tag{7.88}$$

The first term above, when multiplied by $\partial x^\alpha/\partial x^{\alpha'}$, gives the first term on the LHS of eqn. (7.87). For the Christoffel symbols in eqn. (7.87) above, recall from SP5.11 that

$$\Gamma^{\alpha'}{}_{\mu'\nu'} = \frac{\partial x^{\alpha'}}{\partial x^\sigma}\frac{\partial x^\mu}{\partial x^{\mu'}}\frac{\partial x^\nu}{\partial x^{\nu'}}\Gamma^\sigma{}_{\mu\nu} - \frac{\partial^2 x^{\alpha'}}{\partial x^\mu \partial x^\nu}\frac{\partial x^\mu}{\partial x^{\mu'}}\frac{\partial x^\nu}{\partial x^{\nu'}}.\tag{7.89}$$

And the first term on the RHS of eqn. (7.89), when multiplied by $\partial x^\alpha/\partial x^{\alpha'}$, gives the second term on the LHS of eqn. (7.87). To establish the equality in eqn. (7.87) it remains to show that the second term of the RHS of eqn. (7.88) above

$$\frac{dx^\sigma}{d\tau}\frac{d}{d\tau}\frac{\partial x^{\alpha'}}{\partial x^\sigma} = \frac{dx^\sigma}{d\tau}\frac{dx^\gamma}{d\tau}\frac{\partial^2 x^{\alpha'}}{\partial x^\gamma \partial x^\sigma},\tag{7.90}$$

cancels the contribution arising from the second term of eqn. (7.89) above:

$$-\frac{dx^{\mu'}}{d\tau}\frac{dx^{\nu'}}{d\tau}\frac{\partial^2 x^{\alpha'}}{\partial x^\mu \partial x^\nu}\frac{\partial x^\mu}{\partial x^{\mu'}}\frac{\partial x^\nu}{\partial x^{\nu'}} = -\frac{dx^\mu}{d\tau}\frac{dx^\nu}{d\tau}\frac{\partial^2 x^{\alpha'}}{\partial x^\mu \partial x^\nu}.\tag{7.91}$$

Clearly these do cancel, which establishes the equality in eqn. (7.87) above.

SP 7.12 A hopeful experimentalist wishes to perform an experiment to test the role of the Ricci scalar on particle flux divergence; see Exercise 7.1(iii). He proposes to exploit the energy density associated with a photon gas to create a curved spacetime inside a vacuum chamber. After reading Chapter 8, wherein we learn the Einstein field equations, see eqn. (8.86), it will be clear that the Ricci scalar is proportional to the trace of the stress–energy tensor. Calculate the trace of the stress–energy tensor $T^\alpha{}_\alpha$ for a photon gas, and thereby explain why he needs a new experimental design. See also SP8.6, SP10.12, and SP12.8.

Hint

Treat the photon gas as a perfect fluid. See Exercise 4.22 for the equation of state.

SP 7.13 Recall that in the solution to Exercise 7.10(e) we claimed the Killing vector fields for the complete Robertson–Walker spacetime are quite complicated to find. Here we give a flavour of the calculations involved.

(a) Write out the 10 coupled differential equations associated with Killing's equation eqn. (7.56) for the Robertson–Walker spacetime, (iv) of Exercise 7.7.

(b) Except for the special case of a de Sitter universe, the time component is always zero, $\xi_0 = 0$, and the spatial components are time-independent, $\xi_{i,t} = 0$ (see e.g. Aldrovandi et al., 2007). Using this and the equations you will find in (a) leads to the following equations for the other three components:

$$\xi_r = \frac{1}{\sqrt{1-kr^2}}\left[K_3\cos\theta + (K_2\sin\phi + K_1\cos\phi)\sin\theta\right] \tag{7.92}$$

$$\xi_\theta = \sqrt{1-kr^2}\left[(K_1\cos\phi + K_2\sin\phi)\cos\theta - K_3\sin\theta\right] + r^2\left[L_1\sin\phi - L_2\cos\phi\right] \tag{7.93}$$

$$\xi_\phi = \sqrt{1-kr^2}(K_2\cos\phi - K_1\sin\phi)\sin\theta$$
$$+ r^2\left[\sin\theta\ \cos\theta(L_1\cos\phi + L_2\sin\phi) - \sin^2\theta\ L_3\right], \tag{7.94}$$

where K_n and L_n, with $n \in \{1,2,3\}$, are six real parameters. Setting five parameters to zero and one to, say $+1$, one can generate all six Killing vector fields. Verify one of these components.

Solution

(a) In general Killing's equation is

$$0 = \nabla_\alpha\xi_\beta + \nabla_\beta\xi_\alpha \qquad \text{Killing's equation, eqn. (7.56)}$$
$$= \xi_{\alpha,\beta} + \xi_{\beta,\alpha} - 2\Gamma^\sigma{}_{\alpha\beta}\ \xi_\sigma. \tag{7.95}$$

Using Christoffel symbols for the Robertston–Walker metric eqn. (B.34) we obtain 10 coupled linear PDEs for the 4 (one-form) components of the Killing fields:

$$0 = \xi_{t,t} \tag{7.96}$$
$$0 = \xi_{t,r} + \xi_{r,t} - 2H\xi_r \tag{7.97}$$
$$0 = \xi_{t,\theta} + \xi_{\theta,t} - 2H\xi_\theta \tag{7.98}$$
$$0 = \xi_{t,\phi} + \xi_{\phi,t} - 2H\xi_\phi \tag{7.99}$$
$$0 = \xi_{r,\theta} + \xi_{\theta,r} - \frac{2}{r}\xi_\theta \tag{7.100}$$
$$0 = \xi_{r,\phi} + \xi_{\phi,r} - \frac{2}{r}\xi_\phi \tag{7.101}$$
$$0 = \xi_{\theta,\phi} + \xi_{\phi,\theta} - 2\cot\theta\ \xi_\phi \tag{7.102}$$
$$0 = (1-kr^2)\xi_{r,r} - kr\xi_r - R^2H\xi_t \tag{7.103}$$
$$0 = \xi_{\theta,\theta} + r(1-kr^2)\xi_r - R^2Hr^2\xi_t \tag{7.104}$$
$$0 = \xi_{\phi,\phi} + \sin\theta\ \cos\theta\ \xi_\theta + r(1-kr^2)\sin^2\theta\ \xi_r - R^2Hr^2\sin^2\theta\ \xi_t. \tag{7.105}$$

where $H \equiv \dot{R}/R$ (the Hubble parameter, of great importance in cosmology).

(b) When $\xi_t = 0$, eqn. (7.103) separates and we can integrate to find ξ_r:

$$\frac{d\xi_r}{\xi_r} = \frac{kr}{1-kr^2}dr$$

$$\xi_r(r,\theta,\phi) = \frac{C(\theta,\phi)}{\sqrt{1-kr^2}}, \tag{7.106}$$

where $C(\theta, \phi)$ is an integration "constant" with respect to r. To solve for the θ-dependence of C we can take $\partial/\partial\theta$ of eqn. (7.100) above and use eqn. (7.104) to eliminate $\xi_{\theta,\theta}$. Some fortunate cancellation leaves:

$$\frac{\partial^2 C}{\partial\theta^2} + C = 0, \tag{7.107}$$

with obvious general solution:

$$C(\theta, \phi) = B(\phi)\cos\theta + A(\phi)\sin\theta. \tag{7.108}$$

Now, take a deep breath and differentiate eqn. (7.101) by ϕ, from which we can eliminate the ξ_ϕ-dependence using eqn. (7.105):

$$\frac{\partial^2 \xi_r}{\partial\phi^2} - \left(\frac{\partial}{\partial r} - \frac{2}{r}\right)\left[\sin\theta\,\cos\theta\,\xi_\theta + r(1 - kr^2)\sin^2\theta\,\xi_r\right] = 0. \tag{7.109}$$

Unfortunately we've introduced an unwanted ξ_θ-dependence, but with a stroke of luck we can immediately eliminate this using eqn. (7.100):

$$\frac{\partial^2 \xi_r}{\partial\phi^2} + \sin\theta\,\cos\theta\,\frac{\partial\xi_r}{\partial\theta} - \left(\frac{\partial}{\partial r} - \frac{2}{r}\right)\left[+r(1 - kr^2)\sin^2\theta\,\xi_r\right] = 0. \tag{7.110}$$

Substituting what we know about ξ_r up to this point, eqn. (7.106) and eqn. (7.108),

$$\xi_r = \frac{1}{\sqrt{1 - kr^2}}\left(B(\phi)\cos\theta + A(\phi)\sin\theta\right),$$

into eqn. (7.110) above we find

$$(A'' + A)\sin\theta + B''\cos\theta = 0. \tag{7.111}$$

For this to be possible for all values of θ the coefficients of $\sin\theta$ and $\cos\theta$ must vanish separately. Integrating $B'' = 0$ gives $B(\phi) = B_2\phi + B_1$ but applying periodicity the linear dependence must drop out ($B_2 = 0$). The solution for A is clearly $A = A_1\sin\phi + A_2\cos\phi$. And we arrive at our goal, the radial component of our Killing one-form:

$$\xi_r = \frac{1}{\sqrt{1 - kr^2}}\left[B_1\cos\theta + (A_1\sin\phi + A_2\cos\phi)\sin\theta\right]. \tag{7.112}$$

Obvious relabeling of constants verifies eqn. (7.92) above. Similar calculations lead to the other components.

SP 7.14 Argue that for free particles in the Newtonian limit we expect the gravitational potential $\phi = O(v^2)$ where v is a typical velocity.

Hint

Consider a test particle in a circular Keplerian orbit.

8 The Einstein field equations

[The Einstein Field Equations] Eq. (8.10) should be regarded as a system of ten coupled differential equations. ... These ten, then, are not independent, and the ten Einstein equations are really only six independent differential equations for the six functions among the ten $g_{\alpha\beta}$ that characterize the geometry independently of the coordinates.

Bernard Schutz, §8.2

... the first step in the solution of any problem in GR must be an attempt to construct coordinates that will make the calculation simplest.

Bernard Schutz, §8.3

8.1 Exercises

8.1 Show that

$$\phi = -\frac{Gm}{r} \qquad\qquad \text{Schutz Eq. (8.2)} \qquad\qquad (8.1)$$

is a solution to

$$\nabla^2 \phi = 4\pi G\rho \qquad\qquad \text{Schutz Eq. (8.1)} \qquad\qquad (8.2)$$

using the following method. Assume the point particle to be at the origin, $r = 0$, and to produce a spherically symmetric field. Then use Gauss' law on a sphere of radius r to conclude

$$\frac{d\phi}{dr} = Gm/r^2.$$

Deduce eqn. (8.1) from this. (Consider the behavior at infinity.)

Solution: Gauss' law in 3D space is

$$\int_{\Omega} \nabla \cdot (\nabla\phi) dV = \int_{d\Omega} \hat{n} \cdot (\nabla\phi) dA,$$

where Ω is a volume, dV a differential volume element, $d\Omega$ is the surface bounding Ω, $\hat{n}$ is the outward pointing unit normal vector, and dA is an area element on the bounding surface. As suggested, we consider a point particle at the origin of a spherical coordinate system. Eqn. (8.2) applies and we can integrate this equation over a spherical volume centered at the origin:

$$\int_\Omega \nabla \cdot (\nabla\phi)\mathrm{d}V = \int_\Omega 4\pi G\rho\mathrm{d}V, \qquad \text{integrate eqn. (8.2)}$$

$$= 4\pi G \int_\Omega \rho\,\mathrm{d}V = 4\pi Gm, \tag{8.3}$$

where m is the mass of the particle. We have assumed that the particle is isolated and in a vacuum so that ϕ is only due to this particle. The LHS of eqn. (8.3) above gives, using the surface of a sphere of radius r for $\mathrm{d}\Omega$:

$$\int_\Omega \nabla \cdot (\nabla\phi)\mathrm{d}V = \int_{\mathrm{d}\Omega} \hat{n} \cdot (\nabla\phi)\mathrm{d}A, \qquad \text{used Gauss' theorem}$$

$$= \int_{\mathrm{d}\Omega} \frac{\mathrm{d}\phi}{\mathrm{d}r}\mathrm{d}A = \frac{\mathrm{d}\phi}{\mathrm{d}r}4\pi r^2, \tag{8.4}$$

where by spherical symmetry the integrand is constant on the surface of the sphere. Combining eqns. (8.3) and (8.4) gives the first-order ODE indicated in the question:

$$\frac{\mathrm{d}\phi}{\mathrm{d}r} = \frac{Gm}{r^2}.$$

Solve by integrating both sides and imposing $\phi(\infty) = 0$,

$$\int_r^\infty \frac{\mathrm{d}\phi}{\mathrm{d}r'}\mathrm{d}r' = \int_r^\infty \frac{Gm}{r'^2}\mathrm{d}r'$$

$$[\phi]_r^\infty = \left[-\frac{Gm}{r'}\right]_r^\infty$$

$$\phi(r) = -\frac{Gm}{r}, \tag{8.5}$$

which is eqn. (8.1).

8.3 (a) Calculate in geometrized units:

(i) Newtonian potential of the Sun at the surface of the Sun.

Solution: Note that gravitational potential has dimensions of energy per unit mass or velocity squared. In geometrized units this corresponds to units of the speed of light squared. Eqn. (8.1) gives

$$\phi = -\frac{GM}{r} = -\frac{1 \cdot 1476 \text{ m}}{6.96 \times 10^8 \text{ m}} = -2.12 \times 10^{-6}. \tag{8.6}$$

Note we used 1476 m for the radius of the Sun in geometric units from Schutz Table 8.1, in which case $G = 1$. Alternatively, to be difficult, one could work in SI units; use M in [kg], G in [J m kg^{-2}] and r in [m]. Then we have to divide this energy per unit mass, GM/r, by c^2 to obtain dimensionless (fractions of the speed of light squared) units:

$$\phi = -\frac{G}{c^2}\frac{M}{r}. \tag{8.7}$$

But then we recognize

$$\frac{G}{c^2} = 7.425 \times 10^{-28} \text{m kg}^{-1} \qquad\qquad \text{Schutz Eq. (8.9)} \qquad (8.8)$$

as the conversion factor to transform mass to length in SI units, which justifies the calculation in geometric units in eqn. (8.6).

(ii) Newtonian potential of the Sun at the radius of the Earth's orbit.

Solution: Eqn. (8.1) gives

$$\phi = -\frac{GM}{r} = -\frac{1 \cdot 1476 \text{ m}}{1.496 \times 10^{11} \text{ m}} = -9.87 \times 10^{-9}. \qquad (8.9)$$

(iii) Newtonian potential of Earth at the surface of the Earth.

Solution: Eqn. (8.1) gives

$$\phi = -\frac{GM}{r} = -\frac{1 \cdot 4.434 \times 10^{-3} \text{ m}}{6.371 \times 10^{6} \text{ m}} = -6.96 \times 10^{-10}. \qquad (8.10)$$

The Earth has a mass of less than 4.5 mm!

(iv) Speed of Earth in its orbit around the Sun.

Solution: We assume a circular Keplerian orbit, (for explanation see also Exercise 8.3(c) below) for which the velocity is

$$v = \sqrt{\frac{GM}{r}} = \sqrt{\frac{1 \cdot 1476 \, [\text{m}]}{1.496 \times 10^{11} \, [\text{m}]}} = 0.993 \times 10^{-4}. \qquad (8.11)$$

8.3 (b) You should have found that your answer to (ii) was larger than to (iii). Why, then, do we on Earth feel Earth's gravitational pull much more than the Sun's?

Solution: The force of gravity per unit mass is determined by the gradient of the gravitational potential. While the Sun's potential is larger at the surface of the Earth, its potential gradient is much less than that of the Earth's own gravitational potential.

8.3 (c) Show that a circular orbit around a body of mass M has an orbital velocity, in Newtonian theory, of $v^2 = -\phi$, where ϕ is the Newtonian potential.

Solution: (Recall bold face indicates a traditional three-vector.) The centripetal acceleration of a body in a circular orbit of radius r is

$$\mathbf{\Omega} \times (\mathbf{\Omega} \times r\,\hat{\mathbf{r}}) = -r\Omega^2\,\hat{\mathbf{r}} = -\frac{v^2}{r}\,\hat{\mathbf{r}}, \qquad (8.12)$$

as can be found in basic texts on classical mechanics (e.g. Kibble and Berkshire, 2004, Eq. (5.15)). In classical Newtonian mechanics this centripetal acceleration is provided by the gravitational force per unit mass,

$$-\frac{GM}{r^2}\,\hat{\mathbf{r}} = \frac{\phi}{r}\,\hat{\mathbf{r}}. \quad \text{used Newton's law of gravity and eqn. (8.1) above} \qquad (8.13)$$

Equating eqns. (8.12) and (8.13), i.e. using Newton's second law of motion, and solving for v^2 gives $v^2 = -\phi$.

8.5 (a) Show that if $h_{\alpha\beta} = \xi_{\alpha,\beta} + \xi_{\beta,\alpha}$ then the Riemann tensor to first order in $h_{\alpha\beta}$

$$R_{\alpha\beta\mu\nu} = \frac{1}{2}(h_{\alpha\nu,\beta\mu} + h_{\beta\mu,\alpha\nu} - h_{\alpha\mu,\beta\nu} - h_{\beta\nu,\alpha\mu}) \quad \text{Schutz Eq. (8.25)} \quad (8.14)$$

vanishes.

Solution: We simply substitute the given expression for $h_{\alpha\beta}$ into eqn. (8.14) and find that all six terms cancel. (Underline, overline etc. are used to line up the terms that cancel.)

$$R_{\alpha\beta\mu\nu} = \frac{1}{2}(h_{\alpha\nu,\beta\mu} + h_{\beta\mu,\alpha\nu} - h_{\alpha\mu,\beta\nu} - h_{\beta\nu,\alpha\mu}) \qquad \text{eqn. (8.14)}$$

$$= \frac{1}{2}(\xi_{\alpha,\nu\beta\mu} + \overline{\xi_{\nu,\alpha\beta\mu}} + \underline{\xi_{\beta,\mu\alpha\nu}} + \xi_{\mu,\beta\alpha\nu} - \xi_{\alpha,\mu\beta\nu} - \xi_{\mu,\alpha\beta\nu} - \underline{\xi_{\beta,\nu\alpha\mu}} - \overline{\xi_{\nu,\beta\alpha\mu}})$$

$$= 0, \qquad (8.15)$$

where we have used the fact that the order of partial derivatives does not matter.

8.5 (b) Argue from this [i.e. the results of Exercise 8.5(a)] that eqn. (8.14) is gauge invariant.

Solution: Eqn. (8.14) for the Riemann tensor has the form of a linear differential operator on the metric terms $h_{\alpha\beta}$, so $R_{\alpha\beta\mu\nu} = L(h_{\alpha\beta})$. So in general if we added something to $h_{\alpha\beta}$, say $h_{\alpha\beta} \to h_{\alpha\beta} + \Delta h_{\alpha\beta}$ then the Riemann tensor would be augmented

$$R_{\alpha\beta\mu\nu} \to L(h_{\alpha\beta} + \Delta h_{\alpha\beta}) = R_{\alpha\beta\mu\nu} + L(\Delta h_{\alpha\beta}).$$

But the effect of a gauge transformation is to modify $h_{\alpha\beta}$ only by the terms $\Delta h_{\alpha\beta} = -\xi_{\alpha,\beta} - \xi_{\beta,\alpha}$. In other words, a gauge transformation changes the metric terms from

$h^{(\text{old})}_{\alpha\beta}$ to $h^{(\text{new})}_{\alpha\beta}$ with

$$h^{(\text{new})}_{\alpha\beta} = h^{(\text{old})}_{\alpha\beta} - \xi_{\alpha,\beta} - \xi_{\beta,\alpha},$$

see Schutz Eq. (8.24). Furthermore as we saw in Exercise 8.5(a) above, $L(\xi_{\alpha,\beta} + \xi_{\beta,\alpha}) = 0$. Applying a gauge transformation will not change the Riemann curvature tensor approximated with eqn. (8.14); this means that $R_{\alpha\beta\mu\nu}$ is gauge invariant.

8.5   (c) Relate this to Exercise 7.10.

Solution: In general a gauge transformation $x^\alpha \to x^\alpha + \xi^\alpha(x^\alpha)$ certainly does, as we just mentioned in the solution to Exercise 8.5(b), affect the metric quantities $h_{\alpha\beta}$. But if the vector field $\xi^\alpha(x^\alpha)$ were to satisfy Killing's equation in Minkowski space, see eqn. (7.58) above, then the metric coefficients would be unchanged:

$$h^{(\text{new})}_{\alpha\beta} = h^{(\text{old})}_{\alpha\beta} - \xi_{\alpha,\beta} - \xi_{\beta,\alpha} = h^{(\text{old})}_{\alpha\beta}.$$

This makes sense as follows. Recall we are treating $h_{\alpha\beta}$ as a tensor field living in Minkowski spacetime. And the Killing vector fields in general are the vectors that point in the direction along which the metric does not change. They are related to infinitesimal active transformations we could perform on spacetime that leave the metric invariant (move all points in spacetime a tiny bit, while keeping the observers fixed, and observers see the same metric field). The gauge transformation ξ^α could be interpreted as a tiny active transformation of spacetime. When ξ^α satisfies the Minkowski spacetime Killing's equation, the metric perturbation fields $h_{\alpha\beta}$ are unchanged.

8.7   (a) Prove that

$$\bar{h}^\alpha{}_\alpha = -h^\alpha{}_\alpha. \qquad\qquad \text{Schutz Eq. (8.30)} \qquad (8.16)$$

Solution: We start with the definition of $\bar{h}^{\alpha\beta}$ in Schutz Eq. (8.29). It is clear from Schutz Eqs. (8.26) and (8.27) that we can lower the index using $\eta_{\alpha\beta}$. (If this bothers you, see SP8.2 below.)

$$\bar{h}^\alpha{}_\alpha = \bar{h}^{\alpha\beta}\eta_{\alpha\beta} = \left(h^{\alpha\beta} - \frac{1}{2}\eta^{\alpha\beta}h\right)\eta_{\alpha\beta}$$

$$= h^\alpha{}_\alpha - \frac{1}{2}\delta^\alpha_\alpha h = h - \frac{4}{2}h = -h \equiv -h^\alpha{}_\alpha. \qquad (8.17)$$

8.7   (b) Prove

$$h^{\alpha\beta} = \bar{h}^{\alpha\beta} - \frac{1}{2}\eta^{\alpha\beta}\bar{h}. \qquad\qquad \text{Schutz Eq. (8.31)} \qquad (8.18)$$

Solution: Starting with the definition

$$\bar{h}^{\alpha\beta} := h^{\alpha\beta} - \frac{1}{2}\eta^{\alpha\beta}h \qquad \text{Schutz Eq. (8.29)} \qquad (8.19)$$

the result follows almost immediately:

$$h^{\alpha\beta} = \bar{h}^{\alpha\beta} + \frac{1}{2}\eta^{\alpha\beta}h \qquad \text{rearranged eqn. (8.19)}$$

$$= \bar{h}^{\alpha\beta} - \frac{1}{2}\eta^{\alpha\beta}\bar{h}. \qquad \text{used eqn. (8.16)} \qquad (8.20)$$

8.9 (a) Show from

$$G_{\alpha\beta} = -\frac{1}{2}\left[\bar{h}_{\alpha\beta,\mu}{}^{,\mu} + \eta_{\alpha\beta}\bar{h}_{\mu\nu}{}^{,\mu\nu} - \bar{h}_{\alpha\mu,\beta}{}^{,\mu} - \bar{h}_{\beta\mu,\alpha}{}^{,\mu} + O(h^2_{\alpha\beta})\right] \quad \text{Schutz Eq. (8.32)}$$
$$(8.21)$$

that G_{00} and G_{0i} do not contain second time derivatives of any $\bar{h}_{\alpha\beta}$. Thus only the *six* equations, $G_{ij} = 8\pi T_{ij}$, are true dynamical equations. Relate this to the discussion at the end of §8.2. The equations $G_{0\alpha} = 8\pi T_{0\alpha}$ are called *constraint* equations because they are relations among the initial data for the other six equations, which prevent us choosing all these data freely.

Solution: We start with eqn. (8.21), writing the second term with a derivative in the same form as the others, and dropping the second-order terms:

$$G_{\alpha\beta} = -\frac{1}{2}\left[\bar{h}_{\alpha\beta,\mu}{}^{,\mu} + \eta_{\alpha\beta}\bar{h}_{\mu\nu}{}^{,\mu\nu} - \bar{h}_{\alpha\mu,\beta}{}^{,\mu} - \bar{h}_{\beta\mu,\alpha}{}^{,\mu}\right] \qquad \text{from eqn. (8.21)}$$

$$= -\frac{1}{2}\left[\bar{h}_{\alpha\beta,\mu}{}^{,\mu} + \eta_{\alpha\beta}\eta^{\sigma\mu}\bar{h}_{\mu\nu,\sigma}{}^{,\nu} - \bar{h}_{\alpha\mu,\beta}{}^{,\mu} - \bar{h}_{\beta\mu,\alpha}{}^{,\mu}\right]. \qquad (8.22)$$

Consider first the G_{00} term by setting $\alpha = \beta = 0$,

$$G_{00} = -\frac{1}{2}\left[\bar{h}_{00,\mu}{}^{,\mu} + \eta_{00}\eta^{\sigma\mu}\bar{h}_{\mu\nu,\sigma}{}^{,\nu} - \bar{h}_{0\mu,0}{}^{,\mu} - \bar{h}_{0\mu,0}{}^{,\mu}\right]$$

$$= -\frac{1}{2}\left[\bar{h}_{00,\mu}{}^{,\mu} - \eta^{\mu\sigma}\bar{h}_{\sigma\nu,\mu}{}^{,\nu} - \bar{h}_{0\mu,0}{}^{,\mu} - \bar{h}_{0\mu,0}{}^{,\mu}\right]. \quad \text{relabeled indices, } \eta_{00} = -1$$
$$(8.23)$$

The indices μ, ν, and σ are all dummy indices on the RHS of eqn. (8.23) above. Focus on second time derivative terms on the RHS of eqn. (8.23) by setting $\mu = 0$ and $\nu = 0$. The RHS of eqn. (8.23) becomes:

$$-\frac{1}{2}\left[\cancel{\bar{h}_{00,0}{}^{,0}} - \eta^{0\sigma}\bar{h}_{\sigma0,0}{}^{,0} - \cancel{\bar{h}_{00,0}{}^{,0}} - \bar{h}_{00,0}{}^{,0}\right]$$

$$= -\frac{1}{2}\left[\bar{h}_{00,0}{}^{,0} - \bar{h}_{00,0}{}^{,0}\right] = 0. \qquad \text{used } \eta^{0i} = 0 \text{ and } \eta^{00} = -1$$
$$(8.24)$$

Consider now the G_{0i} term by setting $\alpha = 0, \beta = i$ in eqn. (8.22) above:

$$G_{0i} = -\frac{1}{2}\left[\bar{h}_{0i,\mu}{}^{,\mu} + \eta_{0i}\eta^{\sigma\mu}\bar{h}_{\mu\nu,\sigma}{}^{,\nu} - \bar{h}_{0\mu,i}{}^{,\mu} - \bar{h}_{i\mu,0}{}^{,\mu}\right]$$

$$= -\frac{1}{2}\left[\bar{h}_{0i,\mu}{}^{,\mu} - \bar{h}_{0\mu,i}{}^{,\mu} - \bar{h}_{i\mu,0}{}^{,\mu}\right]. \qquad\qquad \text{used } \eta_{0i} = 0 \quad (8.25)$$

The middle term on the RHS above cannot give a second time derivative ($i \in \{1, 2, 3\}$). The first and third terms from the RHS give the only second time derivatives, when $\mu = 0$, but they cancel each other in that case because $\bar{h}_{i0} = \bar{h}_{0i}$.

8.9   (b) The field equations of linearized theory,

$$\Box\, \bar{h}^{\mu\nu} = -16\pi\, T^{\mu\nu}, \qquad\qquad \text{Schutz Eq. (8.42)} \qquad (8.26)$$

contains second time derivatives even when μ or ν is zero. Does this contradict (a)? Why?

Solution: The field equations eqn. (8.26) represent 10 PDEs for the 10 components $\bar{h}_{\alpha\beta}$. (Because $\bar{h}_{\alpha\beta} = \bar{h}_{\beta\alpha}$, six of the $4 \times 4 = 16$ are not counted.) And, as Schutz alludes to in this exercise, each of these contains a second time derivative because the D'Alembertian is

$$\Box = -\frac{\partial^2}{\partial t^2} + \nabla^2.$$

At first sight you might say, well there's no problem because we can always manipulate this system of PDEs to eliminate some terms. That's certainly the case but it doesn't help. That's because each one of the 10 equations in eqn. (8.26) has its own unique term:

$$\bar{h}^{\mu\nu}{}_{,00} = \frac{\partial^2}{\partial t^2}\bar{h}^{\mu\nu}.$$

So a straightforward linear combination of these 10 PDEs does *not* permit us to eliminate one of the $\partial^2/\partial t^2$ terms!

We need different equations. And these arise from the gauge conditions. For eqn. (8.26) only holds in the Lorenz gauge, for which

$$\bar{h}^{\mu\nu}{}_{,\nu} = 0. \qquad\qquad \text{Schutz Eq. (8.33)} \qquad (8.27)$$

Differentiating this gauge condition with respect to t gives us four equations (one for each value of μ)

$$0 = \bar{h}^{\mu\nu}{}_{,\nu 0} = \bar{h}^{\mu 0}{}_{,00} + \bar{h}^{\mu i}{}_{,i0},$$

each of which allows us to eliminate a second time derivative term like the LHS of:

$$\bar{h}^{\mu 0}{}_{,00} = -\bar{h}^{\mu i}{}_{,i0} \qquad\qquad (8.28)$$

from eqn. (8.26). Albeit a bit messy, in principle we could do this, and it would highlight the fact that we have only six linearly independent equations involving second time derivatives. The 10 equations of eqn. (8.26) are nicely compact but we should bear in mind that they are not linearly independent.

8.11 When we write Maxwell's equations in special-relativistic form, we identify the scalar potential ϕ and three-vector potential A_i (signs defined by $E_i = -\phi_{,i} - A_{i,0}$) as components of a one-form $A_0 = -\phi, A_i$ (one-form) $= A_i$ (three-vector). A gauge transformation is the replacement $\phi \to \phi - \partial f/\partial t, A_i \to A_i + f_{,i}$. This leaves the electric and magnetic fields unchanged. The Lorenz gauge is a gauge in which $\partial \phi/\partial t + \nabla_i A_i = 0$. Write both the gauge transformation and the Lorenz gauge condition in four-tensor notation. Draw the analogy with similar equations in linearized gravity.

Solution: With some trial and error one arrives at the following (it's hard to go wrong here):

$$A_\mu^{(\text{new})} = A_\mu^{(\text{old})} + f_{,\mu} \qquad\qquad \text{gauge transformation} \qquad (8.29)$$

$$A_\mu^{,\mu} = 0. \qquad\qquad\qquad\qquad \text{Lorenz gauge condition} \qquad (8.30)$$

It's perhaps instructive to see the Lorenz gauge condition eqn. (8.30) expanded:

$$A_\mu^{,\mu} = A_{\mu,\nu}\, \eta^{\mu\nu} = -A_{0,t} + A_{1,x} + A_{2,y} + A_{3,z}$$

$$= -\frac{\partial(-\phi)}{\partial t} + \frac{\partial A_1}{\partial x} + \frac{\partial A_y}{\partial y} + \frac{\partial A_z}{\partial z}. \qquad (8.31)$$

The gauge transformation eqn. (8.29) is analogous to

$$\bar{h}_{\alpha\beta}^{(\text{new})} = \bar{h}_{\alpha\beta}^{(\text{old})} - \xi_{\alpha,\beta} - \xi_{\beta,\alpha} + \eta_{\alpha\beta}\xi^\sigma_{,\sigma}. \qquad \text{Schutz Eq. (8.34)} \qquad (8.32)$$

And the gauge condition eqn. (8.30) is analogous to eqn. (8.27), hence the name "Lorenz gauge."

8.13 The inequalities

$$|T^{00}| \gg |T^{0i}| \gg |T^{ij}| \qquad (8.33)$$

for a Newtonian system are illustrated in Exercises 8.2(c). Devise physical arguments to justify them in general.

Solution: T^{00} is the total energy density, or total energy per unit volume including the rest mass energy, the kinetic energy, etc. as discussed in Chapter 4. T^{0i} is the density of i-direction momentum, or the flux of energy in the i-direction. This latter interpretation is more helpful here. For in a Newtonian system the ratio T^{0i}/T^{00}

scales like v^i, the component of three-velocity in the x^i-direction. Since $|v^i| \ll 1$ in a Newtonian system, the first inequality in eqn. (8.33) follows. Interpreting T^{0i} as the density of i-direction momentum, and T^{ij} as the flux of i-direction momentum in the j-direction, we can make a similar argument to that above. That is, in a Newtonian system the ratio T^{ij}/T^{0i} scales like v^j. Since $|v^j| \ll 1$ in a Newtonian system, the second inequality in eqn. (8.33) follows.

This is especially clear for the case of a perfect fluid in nearly flat space,

$$T^{0i} = (\rho + p)\gamma^2 v^i + pg^{0i}, \qquad \text{used eqn. (4.15)}$$

where $\gamma = 1/\sqrt{1 - v^2}$, v is the magnitude of the three-velocity. For a Newtonian system, $p \ll \rho \approx \rho_0$ and $g^{0i} \approx \eta^{0i} = 0$, so $T^{0i} \approx \rho_0 v^i$ in keeping with our classical notion of i-direction momentum per unit volume or flux of rest mass in i-direction. Taking the ratio we get

$$\left| \frac{T^{0i}}{T^{00}} \right| \approx \left| \frac{\rho_0 v^i}{\rho_0} \right| = |v^i| \ll 1.$$

Again using eqn. (4.15),

$$T^{ij} = (\rho + p)\gamma^2 u^i u^j + pg^{ij} \approx \rho_0 u^i u^j.$$

Hence

$$\left| \frac{T^{ij}}{T^{0i}} \right| \approx \left| \frac{\rho_0 u^i u^j}{\rho_0 u^i} \right| = |u^j| \ll 1.$$

More generally T^{ij} are the stresses between neighboring elements of the general fluid or other material. These stresses cannot be much larger than $\rho_0 u^i$ for otherwise they would create velocities that would violate the Newtonian system approximation. So the second inequality in eqn. (8.33) must be satisfied.

8.15 We have argued that we should use convenient coordinates to solve the weak-field problem (or any other!), but that any physical results should be expressible in coordinate-free language. From this point of view our demonstration of the Newtonian limit is as yet incomplete, since in Chapter 7 we merely showed that the metric eqn. (7.5) led to Newton's law $dp/dt = -m\nabla\phi$. But surely this is a coordinate-dependent equation, involving coordinate time and position. It is certainly not a valid four-dimensional tensor equation. Fill in this gap in our reasoning by showing that we can make physical measurements to verify that the relativistic predictions match the Newtonian ones. (For example, what is the relation between the proper time one orbit takes and its proper circumference?)

Solution: Let's add to the challenge here and imagine we have a strongly relativistic source, but assume it's stationary, spherically symmetric, and of course we're "at a safe distance" so that $\nabla^2 \bar{h}^{\mu\nu} = 0$ applies, Schutz Eq. (8.51). As Schutz pointed out in

Chapter 7, the Newtonian limit of the geodesic equation for the metric eqn. (7.5) led to

$$\frac{dp^i}{d\tau} = -m\, \phi_{,j}\delta^{ij}, \qquad\qquad \text{Schutz Eq. (7.24)} \qquad (8.34)$$

where p^i is the four-momentum of a particle of rest mass m. Assuming higher order effects are too small to measure, GR predicts Keplerian motion. The LHS of eqn. (8.34) is the rate of change of momentum, which for a circular orbit we can write as

$$\frac{dp^i}{d\tau} = m\frac{du^i}{d\tau} \approx -mr\left(\frac{d\theta}{d\tau}\right)^2. \qquad (8.35)$$

The RHS of eqn. (8.34) is the gradient of a potential. Recall the argument in Schutz §8.5 leading up to Schutz Eq. (8.57) that concluded we can identify the ϕ of the metric eqn. (7.5) with the far-field potential from a possibly relativistic source, and we can define the constant to be the total mass M of this source,

$$\phi = -\frac{GM}{r}, \qquad \text{far field of a relativistic source.} \qquad \text{Schutz Eq. (8.59)} \qquad (8.36)$$

So equating eqns. (8.34) and (8.35) we find

$$-mr\left(\frac{d\theta}{d\tau}\right)^2 = -m\frac{d\phi}{dr} = -m\frac{GM}{r^2} \qquad \text{used eqn. (8.36)}$$

$$d\theta = \sqrt{\frac{GM}{r^3}}\,d\tau \qquad\qquad \text{rearranged}$$

$$2\pi = \sqrt{\frac{GM}{r^3}}\,T, \qquad\qquad \text{assumed circular orbit} \qquad (8.37)$$

where T is the period of the circular orbit of "radius" r. The radial coordinate r is not, for a strongly relativistic source, the proper distance to the center of symmetry, see SP7.8. But the circumference of the orbit, i.e. the proper distance measured around the circular orbit, is $C = 2\pi r$. Making this substitution in eqn. (8.37) we can express the result in terms of more directly measurable quantites:

$$\frac{C^3}{T^2} = 2\pi GM. \qquad (8.38)$$

Here both C and T could in principle be measured by standard rulers and clocks. However, M, would not have a simple Newtonian interpretation. For example, if the source could be cut up into little pieces and each brought home to the laboratory for analysis, we would not find M to be the sum of the masses of the pieces.

8.17   (a) A small planet orbits a static neutron star in a circular orbit whose proper circumference is 6×10^{11} m. The orbital period takes 200 days of the planet's proper time. Estimate the mass M of the star. (b) Five satellites are placed into circular orbits around a static black hole. The proper circumferences and proper periods of their orbits are given in the table below. Use the method of (a) to estimate the hole's mass. Explain the trend of the results you get for the satellites.

Proper circumference	2.5×10^6 m	6.3×10^6 m	6.3×10^7 m	3.1×10^8 m	6.3×10^9 m
Proper period	8.4×10^{-3} s	0.055 s	2.1 s	23 s	2.1×10^3 s

Solution:
  (a) We can use the formula eqn. (8.38) from Exercise 8.15, which gives $M \approx 1.73 \times 10^{30}$ kg for the mass of the star, or about 0.87 solar masses. In geometric units, $M \approx 1281$ m. (See Maple™ worksheet for computations.)
  (b) See Maple™ worksheet for computations.

Proper circumference	2.5×10^6 m	6.3×10^6 m	6.3×10^7 m	3.1×10^8 m	6.3×10^9 m
Proper period	8.4×10^{-3} s	0.055 s	2.1 s	23 s	2.1×10^3 s
Estimated M	266 $M_\odot$	99.1 $M_\odot$	68.0 $M_\odot$	67.5 $M_\odot$	68.0 $M_\odot$
M/r	1	0.1	0.01	2×10^{-3}	1×10^{-4}

The black hole is, of course, a strongly relativistic source. The Kepler formula for the ratio involving the period and circumference are valid in so far as the orbit is large enough that the orbiting body remains in nearly flat space. For the shortest orbit, we found $M/r \sim 1$, so the mass estimate is not actually valid. As the proper circumference increases M/r falls and the estimated mass appears to be tending to a limit near 68 solar masses.

8.19 In this exercise we shall compute the first correction to the Newtonian solution caused by a source that rotates. In Newtonian gravity, the angular momentum of the source does not affect the field: two sources with the same $\rho(x^i)$ but different angular momenta have the same field. Not so in relativity, since *all* components of $T^{\mu\nu}$ generate the field. This is our first example of a *post-Newtonian effect*, an effect that introduces an aspect of general relativity that is not present in Newtonian gravity.

  (a) Suppose a spherical body of uniform density ρ and radius R rotates rigidly about the x^3-axis (z-axis) with constant angular velocity Ω. Write down the components $T^{0\nu}$ in a Lorentz frame at rest with respect to the center of mass of the body, assuming ρ, Ω, and R are independent of time. For each component, work to the lowest non-vanishing order in ΩR.

Solution: We can use the stress–energy tensor for dust, $T^{\mu\nu} = \rho U^\mu U^\nu$, cf. eqn. (4.7), where ρ is the rest mass density. If the source is a fluid this implies that we are ignoring the pressure $p \ll \rho$ and if the source is solid we are ignoring internal stresses, see solution to Exercise 8.13 above. The Lorentz frame approximation implies we are ignoring terms of order $O(\phi)$ in the metric $g_{\mu\nu} = \eta_{\mu\nu} + O(\phi)$. This is consistent with ignoring terms $O(v^2)$, see SP7.14.

In the Minkowski metric $U^\mu = \gamma(v)(1, v^i)$, recall eqn. (2.39), where here the three-velocity v^i is that due to the solid-body rotation about the z-axis of our Lorentz frame, $v^i = \Omega \epsilon^i{}_{jk} \delta^j_3 x^k$. This gives for the energy density within the body $0 \le r = \sqrt{x^2 + y^2 + z^2} \le R$:

$$T^{00} = \rho\gamma^2(v) = \rho\left(1 + a^2\Omega^2 R^2 + a^4\Omega^4 R^4 \cdots\right) \qquad \text{used Taylor series in } v^2$$

$$= \rho(1 + O(R^2\Omega^2)) \approx \rho, \tag{8.39}$$

where $a \equiv \sqrt{x^2 + y^2}/R$ is a geometric factor. We wrote the magnitude of the three-velocity v as this geometric factor a, which is at most unity, times the quantity $R\Omega$, which is fixed by the parameters of the problem. Thus we can refer to terms proportional to v^n as $O(R\Omega)^n$. For the off-diagonal terms,

$$T^{0i} = \rho\gamma^2(v)v^i$$

$$= \rho\left(1 + a^2\Omega^2 R^2 + a^4\Omega^4 R^4 \cdots\right)\Omega \epsilon^i{}_{jk}\delta^j_3 x^k \qquad \text{used Taylor series in } v^2$$

$$= \rho\left(\Omega \epsilon^i{}_{jk}\delta^j_3 x^k + O(R^3\Omega^3)\right) \approx \rho\Omega \epsilon^i{}_{jk}\delta^j_3 x^k. \tag{8.40}$$

In summary,

$$T^{0\mu} = (T^{00}, T^{0x}, T^{0y}, T^{0z}) = (\rho, -y\Omega\rho, x\Omega\rho, 0). \tag{8.41}$$

For future reference we note that T^{ij} vanish when keeping only terms to order $O(v)$:

$$T^{ij} = \rho\gamma^2(v)v^i v^j = \rho\, O(v^2) \approx 0. \tag{8.42}$$

8.19 (b) The general solution to the equation $\nabla^2 f = g$, which vanishes at infinity, is the generalization of eqn. (8.1),

$$f(\vec{r}) = -\frac{1}{4\pi} \int \frac{g(\vec{r}')}{|\vec{r} - \vec{r}'|} d^3\vec{r}', \tag{8.43}$$

which reduces to eqn. (8.1) when g is non-zero in a very small region. Use this to solve the weak-field Einstein equations, eqn. (8.26), for $\bar{h}^{00}$ and $\bar{h}^{0j}$ for the source described in (a). Obtain the solutions only outside the body, and only to the lowest non-vanishing order in r^{-1}, where r is the distance from the body's center. Express the result for $\bar{h}^{0j}$ in terms of the body's annular momentum. Find the metric tensor within this approximation, and transform it to spherical coordinates.

Solution: The D'Alembertian reduces to the Laplacian in the weak-field Einstein equations because of the symmetry of the source: a spherical body of uniform density rotating rigidly has $\partial/\partial t = 0$. Even without this symmetry, for slow rotation we could invoke the approximation in Schutz Eq. (8.44):

$$\Box \, \bar{h}^{\mu\nu} = -16\pi \, T^{\mu\nu} \qquad \text{Schutz Eq. (8.42)}$$

$$(\nabla^2 + O(R^2\Omega^2\nabla^2))\bar{h}^{\mu\nu} = -16\pi \, T^{\mu\nu}. \qquad \text{used Schutz Eq. (8.44)} \qquad (8.44)$$

So we can use eqn. (8.43) to invert the Laplacian and solve for $\bar{h}^{\mu\nu}$. The first component is straightforward:

$$\bar{h}^{00}(\vec{r}) = \frac{-1}{4\pi} \int \frac{-16\pi \, T^{00}(\vec{r'})}{|\vec{r} - \vec{r'}|} \mathrm{d}^3\vec{r'} = 4 \int \frac{\rho(\vec{r'})}{|\vec{r} - \vec{r'}|} \mathrm{d}^3\vec{r'} = 4\frac{M}{r}, \qquad (8.45)$$

where M is the mass of the spherical body, and r is the distance from the point $\vec{r}$ to the center of the body. The integral here is over all of three-space, but the integrand is non-zero only within the body. The final equality follows from identifying the integral with that associated with the Newtonian gravitational potential.

The off-diagonal components are more involved:

$$\bar{h}^{0x}(\vec{r}) = \frac{-1}{4\pi} \int \frac{-16\pi \, T^{0x}(\vec{r'})}{|\vec{r} - \vec{r'}|} \mathrm{d}^3\vec{r'} = 4 \int \frac{-y(\vec{r'})\Omega\rho(\vec{r'})}{|\vec{r} - \vec{r'}|} \mathrm{d}^3\vec{r'}$$

$$= -4\Omega\rho \int_{r' \leq R} \frac{y'}{|\vec{r} - \vec{r'}|} \mathrm{d}^3\vec{r'}, \qquad (8.46)$$

where $y(\vec{r'}) = y'$ denotes the y-component of the position vector $\vec{r'}$. The integral in eqn. (8.46) is challenging even for Maple™ – see accompanying worksheet! But we can approximate the integral when $|\vec{r}| = r \gg R$. Writing out the denominator in Cartesian components helps identify a key simplification:

$$\frac{1}{|\vec{r} - \vec{r'}|} = [(x' - x)^2 + (y' - y)^2 + (z' - z)^2]^{-1/2}$$

$$= [x'^2 + x^2 - 2xx' + y'^2 + y^2 - 2y'y + z'^2 + z^2 - 2z'z]^{-1/2}$$

$$= [\vec{r'} \cdot \vec{r'} + \vec{r} \cdot \vec{r} - 2\vec{r'} \cdot \vec{r}]^{-1/2}$$

$$= r^{-1}\left[1 - 2\frac{\vec{r'} \cdot \vec{r}}{r^2} + O\left(\frac{R^2}{r^2}\right)\right]^{-1/2}$$

$$= r^{-1}\left[1 + \frac{\vec{r'} \cdot \vec{r}}{r^2} + O\left(\frac{R^2}{r^2}\right)\right]. \qquad \text{used eqn. (B.8)}$$

$$(8.47)$$

Substitute eqn. (8.47) into eqn. (8.46) and dropping the $O(R^2/r^2)$ term:

$$\bar{h}^{0x}(\vec{r}) = -4\frac{\Omega\rho}{r} \int_{r' \leq R} y'\left[1 + \frac{\vec{r'} \cdot \vec{r}}{r^2}\right] \mathrm{d}^3\vec{r'}$$

$$= -4\frac{\Omega\rho}{r} \int_{r' \leq R} y'\left[1 + \frac{x'x + y'y + z'z}{r^2}\right] \mathrm{d}x'\,\mathrm{d}y'\,\mathrm{d}z'$$

$$= -4\frac{\Omega\rho y}{r^3} \int_{r' \leq R} y'^2 \mathrm{d}x'\,\mathrm{d}y'\,\mathrm{d}z'. \qquad \text{used symmetry!}$$

$$(8.48)$$

Only even powers in y' have a non-zero contribution to the integral. The second moment in y' on the sphere is straightforward to calculate:

$$\int_{r' \leq R} y'^2 \mathrm{d}x' \, \mathrm{d}y' \, \mathrm{d}z' = \int_{-R}^{R} y'^2 \pi (R^2 - y'^2) \, \mathrm{d}y' \quad \text{divide into disks of constant } y'$$

$$= \frac{4}{15} \pi R^5. \tag{8.49}$$

Substituting eqn. (8.49) into eqn. (8.48) above gives

$$\bar{h}^{0x}(\vec{r}) = -\frac{16}{15} \pi R^5 \Omega \rho \frac{y}{r^3}. \tag{8.50}$$

Similarly

$$\bar{h}^{0y}(\vec{r}) = \frac{16}{15} \pi R^5 \Omega \rho \frac{x}{r^3} \qquad \qquad \bar{h}^{0z} = 0. \tag{8.51}$$

The angular momentum of a rotating spherical ball is the momentum of inertia of the spherical ball, $2MR^2/5$ times its angular velocity Ω:

$$J = \Omega \frac{2}{5} MR^2 = \Omega \frac{8}{15} \pi \rho R^5. \qquad \text{derived in Exercise 10.19} \tag{8.52}$$

So we can write:

$$\bar{h}^{0i} = \left(-\frac{2y}{r^3} J, \frac{2x}{r^3} J, 0 \right). \tag{8.53}$$

To find the metric tensor, $g_{\mu\nu} = \eta_{\mu\nu} + h_{\mu\nu}$ we need also $\bar{h}^{ij}$ components. However these all vanish because the corresponding T^{ij} vanish to $O(v)$, see eqn. (8.42) above. The trace is

$$\bar{h} = \bar{h}^{\mu}{}_{\mu} = \bar{h}^{\mu\nu} \eta_{\mu\nu} = -\bar{h}^{00} = -4 \frac{M}{r}. \qquad \text{used eqn. (8.45) above} \tag{8.54}$$

Lowering the indices in eqn. (8.18), we find

$$h_{\mu\nu} = \bar{h}_{\mu\nu} - \frac{1}{2} \eta_{\mu\nu} \bar{h}, \tag{8.55}$$

which gives the metric coefficients:

$$h_{0x} = \bar{h}_{0x} = -\bar{h}^{0x} = \frac{2y}{r^3} J \qquad h_{0y} = \bar{h}_{0y} = -\bar{h}^{0y} = -\frac{2x}{r^3} J \qquad h_{0z} = 0$$

$$h_{00} = \bar{h}_{00} + \frac{1}{2} \bar{h} = \frac{2M}{r} \qquad h_{ij} = -\frac{1}{2} \eta_{ij} \bar{h} = 2 \frac{M}{r} \eta_{ij}. \tag{8.56}$$

And from these, via

$$g_{\alpha\beta} = \eta_{\alpha\beta} + h_{\alpha\beta} \qquad \qquad \text{Schutz Eq. (8.12)} \tag{8.57}$$

we obtain the metric

$$\mathrm{d}s^2 = -\left(1 - \frac{2M}{r} \right) \mathrm{d}t^2 + \frac{4y}{r^3} J \, \mathrm{d}t \, \mathrm{d}x - \frac{4x}{r^3} J \, \mathrm{d}t \, \mathrm{d}y + \left(1 + \frac{2M}{r} \right) (\mathrm{d}x^2 + \mathrm{d}y^2 + \mathrm{d}z^2).$$

$$\tag{8.58}$$

To convert this metric to spherical coordinates we need the derivatives of the Cartesian coordinates with respect to spherical coordinate, see Appendix B.3,

$$ds^2 = -\left(1 - \frac{2M}{r}\right)dt^2 - 4J\frac{\sin^2\theta}{r}\,dt\,d\phi + \left(1 + \frac{2M}{r}\right)(dr^2 + r^2\,d\theta^2 + r^2\sin^2\theta\,d\phi^2).$$

(8.59)

8.19 (c) Because the metric is independent of t and the azimuthal angle ϕ, particles orbiting this body will have p_0 and p_ϕ constant along their trajectories (see Schutz §7.4). Consider a particle of non-zero rest-mass in a circular orbit of radius r in the equatorial plane. To lowest order in Ω, calculate the difference between its orbital period in the positive sense (i.e., rotating in the sense of the central body's rotation) and in the negative sense. (Define the period to be the coordinate time taken for one orbit of $\Delta\phi = 2\pi$.)

Hint: Write the period in terms of the temporal and azimuthal components of the four-momentum of the orbiting particle, p^t and p^ϕ. Then write p^t and p^ϕ in terms of the conserved quantites, total energy $E \equiv -p_t \equiv -p_0$, and angular momentum $L \equiv p_\phi$. Assume that rest mass dominates the energy, and use the classical expression for angular momentum of a particle in a Keplerian orbit.

Solution: The angular velocity of a massive particle, measured in coordinate time, is

$$\frac{\partial\phi}{\partial t} = \frac{\partial\phi}{\partial\tau}\frac{\partial\tau}{\partial t} = \frac{mU^\phi}{mU^t} = \frac{p^\phi}{p^t}.$$

(8.60)

So the period we seek is

$$T = 2\pi\left|\frac{\partial t}{\partial\phi}\right| = 2\pi\left|\frac{p^t}{p^\phi}\right|.$$

(8.61)

However, we want to work with the conserved quantities, which are the corresponding covariant components: $p_t \equiv -E$ and $p_\phi \equiv L$. Using the metric to lower the indices we obtain a 2×2 system for the pair $(-E, L)$ that we can easily solve:

$$-E \equiv p_t = g_{\alpha t}p^\alpha = g_{tt}p^t + g_{t\phi}p^\phi$$
$$L \equiv p_\phi = g_{\alpha\phi}p^\alpha = g_{t\phi}p^t + g_{\phi\phi}p^\phi.$$

(8.62)

We immediately find

$$\begin{pmatrix} p^t \\ p^\phi \end{pmatrix} = \begin{pmatrix} g_{tt} & g_{t\phi} \\ g_{t\phi} & g_{\phi\phi} \end{pmatrix}^{-1}\begin{pmatrix} -E \\ L \end{pmatrix} = \begin{pmatrix} g_{\phi\phi} & -g_{t\phi} \\ -g_{t\phi} & g_{tt} \end{pmatrix}\begin{pmatrix} -E \\ L \end{pmatrix}\frac{1}{\det}, \quad \text{used eqn. (B.1)}$$

(8.63)

where $\det = g_{tt}g_{\phi\phi} - g_{t\phi}^2$. This gives

$$p^t = \frac{-g_{\phi\phi}E - g_{t\phi}L}{g_{tt}g_{\phi\phi} - g_{t\phi}^2} \qquad p^\phi = \frac{g_{t\phi}E + g_{tt}L}{g_{tt}g_{\phi\phi} - g_{t\phi}^2}. \qquad (8.64)$$

For the ratio p^t/p^ϕ the det cancels, and dividing through by $-L$ gives:

$$\frac{p^t}{p^\phi} = \frac{-g_{\phi\phi}E - g_{t\phi}L}{g_{t\phi}E + g_{tt}L} = \frac{(1 + \frac{2M}{r})r^2\frac{E}{L} - \frac{2J}{r}}{\frac{2J}{r}\frac{E}{L} + (1 - \frac{2M}{r})}. \qquad \text{used eqn. (8.59)} \qquad (8.65)$$

So we need E and L. These turn out to be well approximated by their Newtonian counterparts, see SP8.7. In fact we can approximate $E = m$, simply the rest mass of the orbiting particle. The angular momentum of a particle of mass m in a circular Keplerian orbit is $L = mrv^\phi = mr\sqrt{M/r} = m\sqrt{Mr}$.

An important lesson of this exercise is to learn to handle a delicate equation like eqn. (8.65) with some care. The leading order balance, as you might have guessed and we will show below, is just Kepler's third law. The rest are tiny general relativistic "post-Newtonian" corrections. Our challenge is to strip down the equation as much as possible, so that it's easier to handle, while retaining the tiny corrections that lead to the difference in orbital period for prograde and retrograde orbits. First we exploit the smallness of the off-diagonal metric terms via a binomial series approximation to bring the denominator to the numerator:

$$\frac{p^t}{p^\phi} \approx \left((1 + \frac{2M}{r})r^2\frac{E}{L} - \frac{2J}{r}\right)\left(1 + \frac{2M}{r} - \frac{2J}{r}\frac{E}{L}\right). \qquad \text{used eqn. (8.65)} \qquad (8.66)$$

Now the subtle part. The orbiting body can rotate in one of two possible senses, either with J, a *prograde orbit*, or in the opposite sense, a *retrograde orbit*. In either case the orbital period will be a positive number so we must take the absolute value in eqn. (8.66) to find the period, $T = 2\pi|p^t/p^\phi|$. We're looking for a small change in T due to the change in sign of L. It's safe to ignore the small term $2M/r$ in $(1 + 2M/r)$ but we need to *keep the small terms that change sign with L*. So we write

$$T = 2\pi\left|\frac{p^t}{p^\phi}\right| \approx 2\pi\left|\left(r^2\frac{E}{L} - \frac{2J}{r}\right)\left(1 - \frac{2J}{r}\frac{E}{L}\right)\right|. \qquad \text{used eqn. (8.59)} \qquad (8.67)$$

It's helpful to recognize the first term as the leading order term, $2\pi r^2 E/L \approx 2\pi r/v^\phi$, where v^ϕ is the orbiting three-velocity. This is Kepler's third law in disguise:

$$T \approx 2\pi\left|r^2\frac{E}{L}\right| = 2\pi r^2\frac{m}{m\sqrt{Mr}} \qquad \text{leading order balance in eqn. (8.67)}$$

$$\frac{T^2}{r^3} = \frac{4\pi^2}{M}. \qquad \text{squared, rearranged} \qquad (8.68)$$

All the others are the small post-Newtonian corrections. The effect of the change in sign of L is more transparent if we multiply by the sign of L, so that the leading order term remains positive and the small corrections change sign. (Because we're taking the absolute value in the end, we can multiply by a $+1$ or -1 to no effect.) Ignoring squares of small terms eqn. (8.67) can be written

$$T = 2\pi \left| \mathrm{sgn}(L) \left(r^2 \frac{E}{L} - \frac{2J}{r} \right) \left(1 - \frac{2J}{r} \frac{E}{L} \right) \right| \qquad |\mathrm{sgn}(L)| \times \text{eqn. (8.67)}$$

$$\approx 2\pi \left| r^2 \frac{E}{|L|} - \frac{2J}{r} \mathrm{sgn}(L) - r^2 \left(\frac{E}{L} \right)^2 \frac{2J}{r} \mathrm{sgn}(L) \right|. \qquad \text{expanded}$$

$$(8.69)$$

There are two terms in eqn. (8.69) that change sign with L, but the second one is much bigger. Twice this term gives the difference in period between pro and retrograde orbits:

$$\Delta T = 2 \times 2\pi r^2 \left(\frac{E}{L} \right)^2 \frac{2J}{r} = 4\pi r^2 \left(\frac{m}{m\sqrt{Mr}} \right)^2 \frac{2J}{r} = \frac{8\pi J}{M}. \qquad (8.70)$$

It's clear that prograde is the shorter one because the correction involved is proportional to $(-J \, \mathrm{sgn}(L))$ so that J and L having the same sign gives a negative correction term.

8.19   (d) From this devise an experiment to measure the angular momentum J of the central body. We take the central body to be the Sun ($M = 2 \times 10^{30}$ kg, $R = 7 \times 10^8$ m, $\Omega = 3 \times 10^{-6}$ s^{-1}) and the orbiting particle Earth ($r = 1.5 \times 10^{11}$ m). What would be the difference in the year between positive and negative orbits?

Solution: We simply launch two clocks in counter-rotating circular orbits in the equatorial plane and measure the difference in period ΔT. To a good approximation the two clocks agree with coordinate time. Then the calculations of Exercise 8.19(c) apply and we can use eqn. (8.70). A measurement of ΔT, and Kepler's law eqn. (8.68) for M, allows us to solve for J. For the Sun eqn. (8.70) gives ΔT:

$$\Delta T = 8\pi \frac{2}{5} \frac{MR^2\Omega}{c^2 M} = 1.6 \times 10^{-4} \text{ s}, \qquad (8.71)$$

where the c^2 was necessary to obtain an answer in seconds.

8.2 Supplementary problems

SP 8.1   Recall from Exercise 8.8(c) that $g_{\alpha\beta} R = \eta_{\alpha\beta} \, \eta^{\mu\nu} R_{\mu\nu} + O(h^2)$. Use this and the result from Exercise 8.8(b),

$$R_{\alpha\beta} = \frac{1}{2}(h_{\beta\mu,\alpha}{}^{,\mu} + h_{\alpha\mu,\beta}{}^{,\mu} - h_{,\alpha\beta} - h_{\alpha\beta,\mu}{}^{,\mu}) + O(h_{\alpha\beta}^2) \qquad (8.72)$$

to show that

$$R = h_{v\sigma}{}^{,v\sigma} - h_{,\mu}{}^{,\mu} + O(h^2).$$

Solution

This is a straightforward plug and calculate exercise. The underline will be used to track the evolution of the second term,

$$R = \eta^{\mu v} \frac{1}{2} \left(h_{v\sigma,\mu}{}^{,\sigma} + \underline{h_{\mu\sigma,v}{}^{,\sigma}} - h_{,\mu v} - h_{\mu v,\sigma}{}^{,\sigma} \right) + O(h^2) \qquad \text{contracted eqn. (8.72)}$$

$$= \frac{1}{2} \left(h_{v\sigma}{}^{,v\sigma} - h_{,v}{}^{,v} - h_{,\sigma}{}^{,\sigma} \right) + \frac{1}{2} \eta^{\mu v} \underline{h_{\mu\sigma,v}{}^{,\sigma}} + O(h^2) \qquad \text{summed on } \mu \text{ except in 2nd term}$$

$$= \frac{1}{2} \left(h_{v\sigma}{}^{,v\sigma} - h_{,v}{}^{,v} - h_{,\sigma}{}^{,\sigma} \right) + \frac{1}{2} \underline{h_{\mu\sigma}{}^{;\mu\sigma}} + O(h^2) \qquad \text{summed on } v \text{ in 2nd term}$$

$$= \left(h_{v\sigma}{}^{,v\sigma} - h_{,\mu}{}^{,\mu} \right) + O(h^2). \qquad \text{combined like terms}$$
$$\text{(8.73)}$$

The last line of course involved relabeling indices.

SP 8.2 Using the definitions Schutz Eq. (8.26) $h^\mu{}_\beta \equiv \eta^{\mu\alpha} h_{\alpha\beta}$ and Schutz Eq. (8.27) $h^{\mu v} \equiv \eta^{v\beta} h^\mu{}_\beta$ to convince yourself that eqn. (8.18) holds for the indices simply lowered to the covariant position:

$$h_{\alpha\beta} = \bar{h}_{\alpha\beta} - \frac{1}{2} \eta_{\alpha\beta} \bar{h}. \tag{8.74}$$

SP 8.3 Using eqn. (8.72) and eqn. (8.73) and the definition of the Einstein tensor, one immediately obtains

$$G_{\alpha\beta} = \frac{1}{2}(h_{\beta\mu,\alpha}{}^{,\mu} + h_{\alpha\mu,\beta}{}^{,\mu} - h_{,\alpha\beta} - h_{\alpha\beta,\mu}{}^{,\mu}) - \frac{1}{2}\eta_{\alpha\beta} \left(h_{v\sigma}{}^{,v\sigma} - h_{,\mu}{}^{,\mu} \right) + O(h_{\alpha\beta}^2).$$

Use this and eqn. (8.18) to obtain eqn. (8.21) for $G_{\alpha\beta}$ in the weak-field approximation. If the contravariant position of the indices in eqn. (8.18) bothers you, see SP8.2!

Solution

We repeatedly use eqn. (8.18) with lowered indices, i.e. eqn. (8.74) above, and $h = -\bar{h}$, to replace terms in $h_{\alpha\beta}$ and h with their corresponding trace-reverse terms. The number of terms nearly doubles, but many cancel. To help line up the terms that cancel we use underlines and overlines.

$$G_{\alpha\beta} = \frac{1}{2}\left(h_{\beta\mu,\alpha}{}^{,\mu} + h_{\alpha\mu,\beta}{}^{,\mu} - h_{\alpha\beta,\mu}{}^{,\mu} - h_{,\alpha\beta}\right) - \frac{1}{2}\eta_{\alpha\beta}\left[h_{\nu\sigma}{}^{,\nu\sigma} - h_{,\mu}{}^{,\mu}\right] + O(h_{\alpha\beta}^2)$$

$$= \frac{1}{2}\left(\bar{h}_{\beta\mu,\alpha}{}^{,\mu} + \bar{h}_{\alpha\mu,\beta}{}^{,\mu} - \bar{h}_{\alpha\beta,\mu}{}^{,\mu} + \bar{h}_{,\alpha\beta} - \frac{1}{2}\left\{\eta_{\beta\mu}\bar{h}_{,\alpha}{}^{,\mu} + \eta_{\alpha\mu}\bar{h}_{,\beta}{}^{,\mu} - \overline{\eta_{\alpha\beta}\bar{h}_{,\mu}{}^{,\mu}}\right\}\right)$$

$$- \frac{1}{2}\eta_{\alpha\beta}\left[\bar{h}_{\nu\sigma}{}^{,\nu\sigma} - \frac{1}{2}\overline{\eta_{\nu\sigma}\bar{h}^{,\nu\sigma}} + \overline{\bar{h}_{,\mu}{}^{,\mu}}\right] + O(h_{\alpha\beta}^2)$$

$$= -\frac{1}{2}[\bar{h}_{\alpha\beta,\mu}{}^{,\mu} + \eta_{\alpha\beta}\bar{h}_{\mu\nu}{}^{,\mu\nu} - \bar{h}_{\alpha\mu,\beta}{}^{,\mu} - \bar{h}_{\beta\mu,\alpha}{}^{,\mu}] + O(h_{\alpha\beta}^2). \tag{8.75}$$

In the last line we relabeled $\sigma \to \mu$ and used symmetry to change the order to get the second term. These are the weak gravitational field equations before the Lorenz gauge condition is envoked.

SP 8.4   Starting with eqn. (8.32) derive,

$$\bar{h}^{(\text{new})\mu\nu}{}_{,\nu} = \bar{h}^{(\text{old})\mu\nu}{}_{,\nu} - \xi^{\mu,\nu}{}_{,\nu}. \qquad\qquad \text{Schutz Eq. (3.35)} \qquad (8.76)$$

SP 8.5   The dimensions of the interval $(ds)^2$ are length squared. The coordinate variables can have dimensions of length, e.g. for the pseudo-Carstesian coordinates $\{ct, x, y, z\}$, but sometimes can be unitless, such as the angular coordinates $\{\theta, \phi\}$ of spherical coordinates. Define an operator D that gives us the dimensions of a quantity, so we can write $D(ds^2) = length^2$.

(a) Argue that the dimensions of the basis vectors $\vec{e}_\alpha$ are the square root of the cooresponding metric tensor component,

$$D(\vec{e}_\alpha) = \sqrt{D(g_{\alpha\alpha})} = D(\sqrt{|g_{\alpha\alpha}|}) = \frac{length}{D(x^\alpha)}. \tag{8.77}$$

(b) Find the dimensions of the Christoffel symbol $\Gamma^\sigma{}_{\alpha\beta}$ in terms of the dimensions of the corresponding coordinates $x^\sigma, x^\alpha, x^\beta$.

(c) Show that the definition of the Riemann curvature tensor in terms of the Christoffel symbols, eqn. (6.52), is dimensionally consistent. What are the dimensions of $R^\alpha{}_{\beta\mu\nu}$?

(d) What are the dimensions of $R_{\alpha\beta}$, expressed in terms of $D(x^\alpha)$ and $D(x^\beta)$? What are the dimensions of R? Verify that $G_{\alpha\beta}$ is dimensionally consistent.

(e) Argue that the dimensions of the Einstein field equations are consistent. For the dimensions of the stress–energy tensor use the expression for a perfect fluid.

Solution

(a) One strategy is to use the expression eqn. (5.63), $\vec{e}_\alpha \cdot \vec{e}_\beta = g_{\alpha\beta}$, from which it immediately follows that $D(\vec{e}_\alpha) = \sqrt{D(g_{\alpha\alpha})}$. Then from $g_{\alpha\beta}dx^\alpha dx^\beta = ds^2$ we have immediately that

$$D(\vec{e}_\alpha) = \sqrt{D(g_{\alpha\alpha})} = \frac{length}{D(x^\alpha)}, \tag{8.78}$$

which we will use many times below.

(b) From the expression for the Christoffel symbol,

$$\frac{\partial \vec{e}_\alpha}{\partial x^\beta} = \Gamma^\mu_{\alpha\beta} \, \vec{e}_\mu, \qquad\qquad \text{eqn. (5.48)}$$

it immediately follows that

$$D(\Gamma^\mu_{\alpha\beta}) = \frac{D(\vec{e}_\alpha)}{D(\vec{e}_\mu) \, D(x^\beta)}$$

$$= \frac{D(x^\mu)}{D(x^\alpha x^\beta)}. \qquad\qquad \text{used eqn. (8.78)} \tag{8.79}$$

(c) The Riemann tensor $R^\alpha{}_{\beta\mu\nu}$ is given as the sum of four terms in eqn. (6.52),

$$R^\alpha{}_{\beta\mu\nu} := \Gamma^\alpha{}_{\beta\nu,\mu} - \Gamma^\alpha{}_{\beta\mu,\nu} + \Gamma^\alpha{}_{\sigma\mu}\Gamma^\sigma{}_{\beta\nu} - \Gamma^\alpha{}_{\sigma\nu}\Gamma^\sigma{}_{\beta\mu},$$

and we must show that they all have the same dimensions. The first two terms in eqn. (6.52) have the same dimensions because only the placement of the μ and ν indices are interchanged:

$$D(R^\alpha{}_{\beta\mu\nu}) = D(\Gamma^\alpha{}_{\beta\mu,\nu}) = D(\Gamma^\alpha{}_{\beta\nu,\mu}) = \frac{D(x^\alpha)}{D(x^\beta x^\mu x^\nu)}. \qquad \text{used eqn. (8.79)} \tag{8.80}$$

Similarly the dimensions of the third and fourth terms in eqn. (6.52) are the same, and furthermore equal to that of the first two terms:

$$D(\Gamma^\alpha{}_{\sigma\mu}\Gamma^\sigma{}_{\beta\nu}) = \frac{D(\cancel{\vec{e}_\sigma})}{D(\vec{e}_\alpha)\,D(x^\mu)} \frac{D(\vec{e}_\beta)}{D(\cancel{\vec{e}_\sigma})\,D(x^\nu)} = D(\Gamma^\alpha{}_{\sigma\nu}\Gamma^\sigma{}_{\beta\mu})$$

$$= \frac{D(x^\alpha)}{D(x^\beta x^\mu x^\nu)} = D(\Gamma^\alpha{}_{\beta\mu,\nu}). \qquad\qquad \text{used eqn. (8.80)} \tag{8.81}$$

(d) In taking the contraction on the first and third indices of the Riemann tensor we find the dimensions of the Ricci tensor components:

$$D(R^\sigma{}_{\alpha\sigma\beta}) = D(R_{\alpha\beta}) = \frac{D(\cancel{x^\sigma})}{D(x^\alpha) \cdot D(\cancel{x^\sigma}) \cdot D(x^\beta)} \qquad \text{used eqn. (8.80)}$$

$$= \frac{1}{D(x^\alpha)D(x^\beta)}. \tag{8.82}$$

For the Ricci scalar we need the dimensions of the inverse metric,

$$g^{\alpha\beta} = \frac{1}{D(\vec{e}_\alpha \cdot \vec{e}_\beta)} = \frac{D(x^\alpha x^\beta)}{length^2}. \tag{8.83}$$

Substituting (8.83) into the following

$$D(R) = D(R^{\beta}_{\ \beta}) = D(g^{\beta\alpha}R_{\alpha\beta}) = \frac{1}{length^2}. \qquad \text{used eqn. (8.82)} \qquad (8.84)$$

We find the Ricci scalar always has dimensions of inverse length squared *independent of coordinate system* (as one expects for a scalar quantity).

For the Einstein tensor we lower the indices in eqn. (6.82) giving $G_{\alpha\beta} \equiv R_{\alpha\beta} - g_{\alpha\beta}R/2$. To verify that it is dimensionally consistent we note that

$$D(R\ g_{\alpha\beta}) = \frac{1}{length^2}\frac{length^2}{D(x^\alpha x^\beta)} \qquad \text{used eqns. (8.83) and (8.84)}$$

$$= \frac{1}{D(x^\alpha)\,D(x^\beta)}, \qquad (8.85)$$

in agreement with eqn. (8.82). So we can consistently form the Einstein tensor in a given coordinate system but the units will depend upon the coordinate system.

(e) At first sight this question seems like a straightforward extension of the previous ones, but it requires first coming up with the right dimensional constants in the Einstein field equations, which were expressed in geometric units in

$$G_{\alpha\beta} = 8\pi T_{\alpha\beta}. \qquad \text{Schutz Eq. (8.10)} \qquad (8.86)$$

It helps to start with a simple situation, where all the coordinate variables x^μ have dimension of length. Then the LHS of eqn. (8.86), as we've just seen, has dimensions of $length^{-2}$. Consider the $G_{00} = 8\pi T_{00}$ component of eqn. (8.86). Recall from Chapter 4 that T_{00} is the energy density, which is apparent for instance from some relativistic contemplation of eqn. (4.5). How do we express energy density in units of $length^{-2}$? Dividing by c^2 will convert to units of density of *mass*. Then eqn. (8.8) tells us to multiply by G/c^2 to convert *mass* to *length*. So $T_{00}G/c^4$ has units of *length* per unit $length^3$, or $length^{-2}$, consistent with G_{00}. This tells us G/c^4 is the hidden factor in the field equations eqn. (8.86), which can be written

$$G_{\alpha\beta} = 8\pi\frac{G}{c^4}T_{\alpha\beta}, \qquad (8.87)$$

cf. (Hobson et al., 2006, Eq. (8.14)) or (Rindler, 2006, Eqs. (14.10, 14.11)). What if we had used, say, the T_{0x} component instead? T_{0x} is the energy flux density in the x-direction. One arrives at the same result, eqn. (8.87), because energy flux density is energy density times velocity and velocities are dimensionless in geometric units where time is measured in units of length (or velocities are fractions of the speed of light). The dimensions of G/c^4 are

$$D\left(\frac{G}{c^4}\right) = \frac{time^2}{mass \cdot length}. \qquad (8.88)$$

Next we establish the dimensional consistency of the stress–energy tensor for a perfect fluid. Substituting eqn. (7.22) for the stress–energy tensor on the RHS of the field equations eqn. (8.86) gives

$$G_{\alpha\beta} = 8\pi \left[(\rho + p)U_\alpha U_\beta + p\,g_{\alpha\beta} \right]. \qquad \text{used eqn. (7.22)} \qquad (8.89)$$

For the pressure to have the same dimensions as density we must divide by velocity (recall for instance Bernoulli's equation (Landau and Lifshitz, 1966, Eq. (10.4))), so there is an implicit $c^2 = 1$ dividing the first p in eqn. (8.89) above:

$$T_{\alpha\beta} = \left(\rho + \frac{p}{c^2} \right) U_\alpha U_\beta + p\,g_{\alpha\beta}. \qquad (8.90)$$

The dimensions of the four-velocity are

$$D(U_\alpha) = D \left(g_{\alpha\beta} \frac{dx^\beta}{d\tau} \right) = \frac{length^2}{D(x^\alpha)D(x^\beta)} \frac{D(x^\beta)}{time}$$

$$= \frac{length^2}{D(x^\alpha) \cdot time}. \qquad (8.91)$$

Substituting eqn. (8.91) into eqn. (8.90) we find

$$D \left([\rho + \frac{p}{c^2}]U_\alpha U_\beta \right) = \frac{mass}{length^3} \frac{length^2}{D(x^\alpha) \cdot time} \frac{length^2}{D(x^\beta) \cdot time}$$

$$= \frac{mass \cdot length}{D(x^\alpha)D(x^\beta) \cdot time^2}. \qquad (8.92)$$

Expressing force in terms of mass and acceleration one finds for the final term in eqn. (8.90)

$$D \left(p\,g_{\alpha\beta} \right) = \frac{mass}{length \cdot time^2} \frac{length^2}{D(x^\alpha)\,D(x^\beta)}$$

$$= \frac{mass \cdot length}{D(x^\alpha)D(x^\beta) \cdot time^2}, \qquad (8.93)$$

in agreement with eqn. (8.92) and verifying the dimensional consistency of eqn. (8.90). Multiplying by eqn. (8.88) we find the RHS of the Einstein equations eqn. (8.87) has dimensions of $1/D(x^\alpha)D(x^\beta)$ in agreement with the dimensions of $R_{\alpha\beta}$, see eqn. (8.82), and thus $G_{\alpha\beta}$.

SP 8.6 A hopeful experimentalist wishes to perform an experiment to test the role of the Ricci scalar on particle flux divergence; see Exercise 7.1(iii). To simplify the calculations she wants to work in a vacuum, so $n = 0$ initially, but needs a non-zero Ricci scalar there. She proposes to exploit the energy density associated with a strong electromagnetic field to create a curved spacetime inside a vacuum chamber. The stress–energy tensor resulting from an electromagnetic field can be written, see (Hobson et al., 2006, Exer. 8.3, §19.12):

$$T^{\alpha\beta} = \frac{1}{\mu_0} \left(F^{\alpha\rho} F^\beta_{\ \rho} - \frac{1}{4} g^{\alpha\beta} F_{\rho\sigma} F^{\rho\sigma} \right), \qquad (8.94)$$

where μ_0 is the permeability of free space, and $F^{\alpha\beta}$ is the electromagnetic field tensor, also called the Faraday (Misner et al., 1973, Eq. (3.5)). In pseudo-Cartesian coordinates the non-zero components are $F^{0i} = -F^{i0} = E^i$, $F^{xy} = -F^{yx} = B^z$, $F^{yz} = -F^{zy} = B^x$, $F^{zx} = -F^{xz} = B^y$, see Exercise 4.25. Calculate the Ricci scalar for a general electromagnetic field, and thereby explain why she needs a new experimental design. Use this result to anticipate the trace of the stress–energy tensor of a photon gas. See also SP7.12, SP10.12, and SP12.8.

SP 8.7 Consider a particle in a circular orbit around a rotating spherical body using the same level of approximation used in Exercise 8.19, so that the metric eqn. (8.59) applies. Show that the energy of the particle is the sum of the rest mass plus Newtonian gravitational potential plus kinetic energy. Show that the angular momentum is that expected from a classical particle in a Keplerian orbit.

Solution

We can find E from the equation $\vec{p} \cdot \vec{p} = -m^2$, or equivalently from dividing ds^2 by $-d\tau^2$. Either way we obtain:

$$-Ep^t + Lp^\phi = -m^2$$

$$-E\left(-g_{\phi\phi}E - g_{t\phi}L\right) + L\left(g_{t\phi}E + g_{tt}L\right) = -m^2(g_{tt}g_{\phi\phi} - g_{t\phi}^2). \quad \text{used eqn. (8.64)}$$

$$(8.95)$$

We might expect the terms involving the off-diagonal metric components to be small (the solution to SP8.8 provides a more detailed argument) so that

$$g_{t\phi}L \ll g_{\phi\phi}E. \tag{8.96}$$

Likewise the following quantity on the RHS of eqn. (8.95) simplifies (again see SP8.8):

$$\det = g_{tt}g_{\phi\phi} - g_{t\phi}^2 \approx g_{tt}g_{\phi\phi}. \tag{8.97}$$

Applying approximations eqn. (8.96) and eqn. (8.97) we find that eqn. (8.95) simplifies

$$E^2 = -m^2 g_{tt} - \frac{g_{tt}}{g_{\phi\phi}}L^2$$

$$E = m\left(1 - \frac{2M}{r} - \frac{g_{tt}}{r^2 m^2}L^2\right)^{1/2}$$

$$E = m\left(1 - \frac{M}{r} + \frac{1}{2}\frac{L^2}{r^2 m^2}\right). \quad \text{used eqn. (B.6)} \tag{8.98}$$

The RHS of eqn. (8.98) is clearly the sum of respectively the rest mass, the Newtonian gravitational potential energy, and the kinetic energy.

The angular momentum can be determined from the r component of the geodesic equation. (Appendix B.4.5 provides the Christoffel symbols.) For a circular orbit there are only two non-zero four-velocity components:

$$m^2 \left(\frac{\mathrm{d}^2 r}{\mathrm{d}\tau^2} + \Gamma^r_{\mu\nu} \frac{\mathrm{d}x^\mu}{\mathrm{d}\tau} \frac{\mathrm{d}x^\nu}{\mathrm{d}\tau} \right) = 0$$

$$0 + \Gamma^r_{tt}(p^t)^2 + \Gamma^r_{\phi\phi}(p^\phi)^2 + 2\Gamma^r_{t\phi}p^t p^\phi = 0 \qquad \text{circular orbit}$$

$$\frac{M}{r(r+2M)} - \frac{r(M+r)}{r+2M}\left(\frac{p^\phi}{p^t}\right)^2 - 2\frac{J}{r(r+2M)}\left(\frac{p^\phi}{p^t}\right) = 0. \quad \text{used eqn. (B.30)}$$

$$(8.99)$$

From eqn. (8.65) we can simplify this ratio because the diagonal metric terms dominate:

$$\frac{p^\phi}{p^t} = \frac{g_{t\phi}E + g_{tt}L}{-g_{\phi\phi}E - g_{t\phi}L} \approx -\frac{g_{tt}L}{g_{\phi\phi}E}. \tag{8.100}$$

The solution to SP8.8 provides a more detailed justification of eqn. (8.100). Substituting eqn. (8.100) into eqn. (8.99) we find, after cancelling $(r+2M)$:

$$\frac{M}{r} - r(M+r)\left(\frac{g_{tt}L}{g_{\phi\phi}E}\right)^2 + \frac{2J}{r}\left(\frac{g_{tt}L}{g_{\phi\phi}E}\right) = 0. \tag{8.101}$$

The third term in eqn. (8.101) can be dropped; it is proportional to $2J/r = g_{t\phi}$ which can be arbitrarily small for small Ω. (Detailed justification is deferred to SP8.8.) So dropping this term in favor of the first two in eqn. (8.101) finally we arrive at:

$$0 = \frac{M}{r} - r(M+r)\left(\frac{g_{tt}L}{g_{\phi\phi}E}\right)^2$$

$$L = \sqrt{\frac{M}{(M+r)r^2} \frac{r^2(1+2M/r)E}{1-2M/r}} \qquad \text{rearranged}$$

$$L \approx \sqrt{Mr}E \approx m\sqrt{Mr}. \qquad \text{keeping largest terms} \tag{8.102}$$

This is just what we find for a particle in a circular Keplerian orbit. From the balance between Newtonian gravitational and centripetal acceleration $M/r^2 = v^2/r$, the angular moment is $L = mrv = m\sqrt{Mr}$.

SP 8.8 In SP8.7 we made use of the approximation that the diagonal metric terms dominated when finding the energy E and angular momentum L of a body orbiting a slowly rotating spherical ball. In particular we argued,

$$g_{t\phi}L \ll g_{\phi\phi}E \tag{8.103}$$

and

$$g_{t\phi}^2 \ll g_{tt}g_{\phi\phi} \tag{8.104}$$

and later, see eqn. (8.100) above, that

$$\frac{p^\phi}{p^t} = \frac{g_{t\phi}E + g_{tt}L}{-g_{\phi\phi}E - g_{t\phi}L} \approx -\frac{g_{tt}L}{g_{\phi\phi}E}. \tag{8.105}$$

Defend these approximations in more detail for the case of a non-relativistic orbiting body.

9 Gravitational radiation

The action of gravitational waves is sometimes characterized as a stretching of space. Eq. (9.24) makes it clear what this means: as the wave passes through, the proper separations of free objects that are simply sitting at rest change with time.

Bernard Schutz, §9.1

... gravitational waves from the Big Bang originated when the Universe was perhaps only 10^{-25} s old; they are our earliest messengers from the beginning of our Universe, and they should carry the imprint of unknown physics at energies far higher than anything we can hope to reach in accelerators on the Earth.

Bernard Schutz, §9.2

9.1 Exercises

9.1 A function $f(s)$ has derivative $f'(s) = \mathrm{d}f/\mathrm{d}s$. Prove that $\partial f(k_\mu x^\mu)/\partial x^\nu = k_\nu f'(k_\mu x^\mu)$. Use this to prove

$$\bar{h}^{\alpha\beta}{}_{,\mu} = ik_\mu \bar{h}^{\alpha\beta} \qquad\qquad \text{Schutz Eq. (9.4)} \qquad (9.1)$$

$$\eta^{\mu\nu} \bar{h}^{\alpha\beta}{}_{,\mu\nu} = -k_\mu k_\nu \bar{h}^{\alpha\beta} = 0. \qquad\qquad (9.2)$$

> Hint: These results follow from the chain rule of differential calculus with $s = k_\mu x^\mu$.

9.3 Let $\bar{h}^{\alpha\beta}(t, x^i)$ be any solution of the vacuum weak-field Einstein equations

$$\Box \bar{h}^{\alpha\beta} = \left(-\frac{\partial^2}{\partial t^2} + \nabla^2\right)\bar{h}^{\alpha\beta} = 0 \qquad\qquad \text{Schutz Eq. (9.1)} \qquad (9.3)$$

that has the property $\int \mathrm{d}x^\alpha |\bar{h}^{\mu\nu}|^2 < \infty$, for the integral over any particular x^α holding other coordinates fixed. Define the Fourier transform of $\bar{h}^{\alpha\beta}(t, x^i)$ as

$$\bar{H}^{\alpha\beta}(\omega, k^i) = \int \bar{h}^{\alpha\beta}(t, x^i) \exp[i(\omega t - k_j x^j)]\, \mathrm{d}t\, \mathrm{d}^3\mathbf{x}. \qquad (9.4)$$

Show, by transforming eqn. (9.3), that $\bar{H}^{\alpha\beta}(\omega, k^i)$ is zero except for those values of ω and k^i that satisfy the dispersion relation

$$\omega^2 = |\mathbf{k}|^2. \qquad\qquad \text{Schutz Eq. (9.10)} \qquad (9.5)$$

By applying the inverse transform, write $\bar{h}^{\alpha\beta}(t, x^i)$ as a superposition of plane waves.

Solution: To Fourier transform eqn. (9.3) we multiply each term by $\exp[i(\omega t - k_j x^j)]$ and integrate over all space and time:

$$0 = \int \exp[i(\omega t - k_j x^j)] \left(-\frac{\partial^2}{\partial t^2} + \nabla^2 \right) \bar{h}^{\alpha\beta} \, dt \, d^3\mathbf{x}.$$

Suppose we do the time integral first. Using integration by parts and assuming the function vanishes at infinity we find that each term simplifies:

$$\int - \exp[i(\omega t - k_j x^j)] \frac{\partial^2}{\partial t^2} \bar{h}^{\alpha\beta} \, dt \, d^3\mathbf{x} = (i\omega) \int \exp[i(\omega t - k_j x^j)] \frac{\partial}{\partial t} \bar{h}^{\alpha\beta} \, dt \, d^3\mathbf{x}.$$

$$(9.6)$$

Note that we have assumed the boundary term at infinity vanishes,

$$\left[\int - \exp[i(\omega t - k_j x^j)] \frac{\partial}{\partial t} \bar{h}^{\alpha\beta} \, d^3\mathbf{x} \right]_{t=-\infty}^{t=+\infty} = 0. \qquad (9.7)$$

A more rigorous treatment would involve defending this assumption, see SP9.6. Integrating again by parts and again dropping the boundary term gives

$$-(i\omega)^2 \int \exp[i(\omega t - k_j x^j)] \bar{h}^{\alpha\beta} \, dt \, d^3\mathbf{x} = \omega^2 \bar{H}^{\alpha\beta}(\omega, k^i). \quad \text{used eqn. (9.4)} \quad (9.8)$$

Similarly the other terms can be written in terms of their Fourier transform and combine to give:

$$(\omega^2 - k_x^2 - k_y^2 - k_z^2) \bar{H}^{\alpha\beta}(\omega, k^i) = (\omega^2 - |\mathbf{k}|^2) \bar{H}^{\alpha\beta}(\omega, k^i) = 0. \qquad (9.9)$$

We immediately see that when $\omega^2 \neq |\mathbf{k}|^2$ then $\bar{H}^{\alpha\beta}(\omega, k^i) = 0$, as claimed. The only non-zero $\bar{H}^{\alpha\beta}(\omega, k^i)$ are those that agree with the dispersion relation eqn. (9.5). And these correspond to plane waves. The inverse transform is (e.g. Hassani, 1999, Eq.(8.23))[1]

$$\bar{h}^{\alpha\beta}(t, x^i) = \frac{1}{(2\pi)^4} \int \bar{H}^{\alpha\beta}(\omega, k^i) \exp[-i(\omega t - k_j x^j)] \, d\omega \, d^3\mathbf{k}, \qquad (9.10)$$

which represents a sum of plane waves. The conclusion is that any solution $\bar{h}^{\alpha\beta}$ to the weak-field vacuum Einstein equations is a sum of plane waves.

9.5   (a) Show that $A_{\alpha\beta}^{(\text{NEW})}$, given by

$$A_{\alpha\beta}^{(\text{NEW})} = A_{\alpha\beta}^{(\text{OLD})} - i B_\alpha k_\beta - i B_\beta k_\alpha + i \eta_{\alpha\beta} B^\mu k_\mu, \quad \text{Schutz Eq. (9.17)} \quad (9.11)$$

satisfies the gauge condition $A^{\alpha\beta} k_\beta = 0$ if $A_{\alpha\beta}^{(\text{OLD})}$ does.

---

[1] Recall there is flexibility in defining the Fourier transform and inverse transform pair in terms of where the factors of 2π and -1 go.

Solution: First raise both indices in eqn. (9.11):

$$A^{\mu\nu}_{(\text{NEW})} = \eta^{\mu\alpha}\eta^{\nu\beta}A^{(\text{NEW})}_{\alpha\beta} = A^{\mu\nu}_{(\text{OLD})} - iB^\mu k^\nu - iB^\nu k^\mu + i\eta^{\mu\nu}B^\sigma k_\sigma. \qquad (9.12)$$

Now contract with k_ν:

$$A^{\mu\nu}_{(\text{NEW})}k_\nu = A^{\mu\nu}_{(\text{OLD})}k_\nu - iB^\mu \overset{0}{\cancel{k^\nu k_\nu}} - iB^\nu k^\mu k_\nu + i\eta^{\mu\nu}B^\sigma k_\sigma k_\nu \qquad k_\nu \text{ is null}$$

$$= A^{\mu\nu}_{(\text{OLD})}k_\nu - \cancel{i(B^\nu k_\nu)k^\mu} + \cancel{i(B^\sigma k_\sigma)k^\mu}$$

$$= A^{\mu\nu}_{(\text{OLD})}k_\nu. \qquad (9.13)$$

So $A^{\alpha\beta}_{(\text{NEW})}k_\beta = 0$ if, and in fact only if, $A^{\alpha\beta}_{(\text{OLD})}k_\beta = 0$.

9.5   (b) Use

$$A^\alpha{}_\alpha = 0 \qquad\qquad \text{Schutz Eq. (9.18)} \qquad (9.14)$$

for $A^{(\text{NEW})}_{\alpha\beta}$ to constrain B_μ.

Solution: We take the trace of eqn. (9.11) by contracting with $\eta^{\beta\alpha}$, giving

$$A^{(\text{NEW})\alpha}{}_\alpha = \eta^{\beta\alpha}A^{(\text{NEW})}_{\alpha\beta} = A^{(\text{OLD})\alpha}{}_\alpha - iB_\alpha k^\alpha - iB^\alpha k_\alpha + i\eta_\alpha{}^\alpha B^\sigma k_\sigma. \qquad (9.15)$$

Recall $\eta_\sigma{}^\sigma = \delta^\sigma_\sigma = 4$ because there are four spacetime dimensions. Of course $B^\alpha k_\alpha = B_\alpha k^\alpha$, see SP3.1. Finally the traceless condition, Schutz Eq. (9.18), applies to both old and new $A_{\alpha\beta}$, so the eqn. (9.15) reduces to

$$A^{(\text{NEW})\beta}{}_\beta = 0 = A^{(\text{OLD})\beta}{}_\beta + i2B^\sigma k_\sigma. \qquad (9.16)$$

This gives us an equation for $B^\sigma k_\sigma$ that we will use in Exercise 9.5(d).

9.5   (c) Show that

$$A_{\alpha\beta}U^\beta = 0, \qquad\qquad \text{Schutz Eq. (9.19)} \qquad (9.17)$$

for $A^{(\text{NEW})}$ imposes only three constraints on B^μ, not the four that we might expect from the fact that the free index α can take any values from 0 to 3. Do this by showing that the particular linear combination $k^\alpha(A_{\alpha\beta}U^\beta)$ vanishes for any B^μ.

Solution: First let's write the equation for B_α in a more transparent form. We contract $A^{(\text{NEW})}_{\alpha\beta}$ in eqn. (9.11) with U^β and set to zero:

$$U^\beta A^{(\mathrm{NEW})}_{\alpha\beta} = U^\beta A^{(\mathrm{OLD})}_{\alpha\beta} - iU^\beta B_\alpha k_\beta - iU^\beta B_\beta k_\alpha + iU^\beta \eta_{\alpha\beta} B^\sigma k_\sigma \qquad U^\beta \times \text{ eqn. (9.11)}$$

$$0 = U^\beta A^{(\mathrm{OLD})}_{\alpha\beta} - iU^\beta B_\alpha k_\beta - iU^\beta B_\beta k_\alpha + iU^\beta \eta_{\alpha\beta} B^\sigma k_\sigma \qquad \text{imposed eqn. (9.17)}$$

$$0 = iU^\beta A^{(\mathrm{OLD})}_{\alpha\beta} + \left(U^\sigma k_\sigma \delta^\beta_\alpha + k_\alpha U^\beta - U_\alpha k^\beta\right) B_\beta \qquad \times \text{ i, factored } B_\beta$$

$$a_\alpha{}^\beta B_\beta = b_\alpha, \tag{9.18}$$

where $a_\alpha{}^\beta \equiv (U^\sigma k_\sigma)\delta_\alpha{}^\beta + k_\alpha U^\beta - U_\alpha k^\beta$ and $b_\alpha = -iU^\beta A^{(\mathrm{OLD})}_{\alpha\beta}$. Clearly eqn. (9.18) represents a set of four equations for B_β, but they are not linearly independent. The linear combinations $k^\alpha a_\alpha{}^\beta = 0$ on the LHS and $k^\alpha b_\alpha = 0$ on the RHS because

$$k^\alpha a_\alpha{}^\beta = k^\alpha \left((U^\sigma k_\sigma)\delta_\alpha{}^\beta + k_\alpha U^\beta - U_\alpha k^\beta\right)$$
$$= k^\beta(U^\sigma k_\sigma) - k^\beta(U_\alpha k^\alpha) = 0, \qquad k_\alpha \text{ is null} \tag{9.19}$$

and

$$k^\alpha b_\alpha = -k^\alpha iU^\beta A^{(\mathrm{OLD})}_{\alpha\beta} = 0. \qquad \text{used Schutz Eq. (9.12)} \tag{9.20}$$

9.5 (d) Using (b) and (c), solve for B_μ as a function of k_μ, $A^{(\mathrm{OLD})}_{\alpha\beta}$, and U^μ. These determine B^μ: there is no further gauge freedom.

Solution: Eq. (9.18) above gives

$$\left((U^\sigma k_\sigma)\delta_\alpha{}^\beta + k_\alpha U^\beta - U_\alpha k^\beta\right) B_\beta = -iU^\beta A^{(\mathrm{OLD})}_{\alpha\beta}. \tag{9.21}$$

Collect knowns on the RHS. Summing over β, the LHS contains terms with respectively B_α, $U^\beta B_\beta$, and $k^\beta B_\beta$. The latter is given by eqn. (9.16) in Exercise 9.5 (b) above and can be put on the RHS, giving

$$(U^\sigma k_\sigma)B_\alpha + k_\alpha U^\beta B_\beta = -iU^\beta A^{(\mathrm{OLD})}_{\alpha\beta} + \frac{i}{2}A^{(\mathrm{OLD})\beta}_{\quad\ \beta} U_\alpha. \tag{9.22}$$

Contracting with U^α, the two terms on the LHS are equal, giving:

$$2(U^\sigma k_\sigma)U^\alpha B_\alpha = U^\alpha\left(-iU^\beta A^{(\mathrm{OLD})}_{\alpha\beta} + \frac{i}{2}A^{(\mathrm{OLD})\beta}_{\quad\ \beta} U_\alpha\right)$$

$$U^\alpha B_\alpha = \frac{1}{2(U^\sigma k_\sigma)}\left(-iU^\alpha U^\beta A^{(\mathrm{OLD})}_{\alpha\beta} - \frac{i}{2}A^{(\mathrm{OLD})\beta}_{\quad\ \beta}\right). \tag{9.23}$$

Rather than substitute this into eqn. (9.22), we simply note that the second term on the LHS of eqn. (9.22) is "known" by eqn. (9.23), and can be moved to the RHS so eqn. (9.22) becomes

$$(U^\sigma k_\sigma)B_\alpha = -k_\alpha U^\beta B_\beta - iU^\beta A^{(\mathrm{OLD})}_{\alpha\beta} + \frac{i}{2}A^{(\mathrm{OLD})\beta}_{\quad\ \beta} U_\alpha$$

$$B_\alpha = \frac{1}{U^\sigma k_\sigma}\left(-k_\alpha U^\beta B_\beta - iU^\beta A^{(\mathrm{OLD})}_{\alpha\beta} + \frac{i}{2}A^{(\mathrm{OLD})\beta}_{\quad\ \beta} U_\alpha\right). \tag{9.24}$$

Recall $\vec{U}$ was some constant timelike vector we chose. So from $A_{\alpha\beta}^{(OLD)}$ eqns. (9.23) and (9.24) allow us to calculate B_α for each wavenumber k_α.

9.5   (e) Show that it is possible to choose ξ^β in

$$h_{\alpha\beta}^{(NEW)} = h_{\alpha\beta}^{(OLD)} - \xi_{\alpha,\beta} - \xi_{\beta,\alpha} \qquad \text{Schutz Eq. (9.15)} \qquad (9.25)$$

to make any superposition of plane waves satisfy eqns. (9.14) and (9.17), so that these are generally applicable to gravitational waves of any sort.

Solution: We generalize

$$\bar{h}^{\alpha\beta} = A^{\alpha\beta} \exp(ik_\sigma x^\sigma) \qquad \text{Schutz Eq. (9.2)} \qquad (9.26)$$

to a superposition of plane waves (cf. eqn. (9.10) in Exercise 9.3 above):

$$\bar{h}_{\alpha\beta}(\vec{x}) = \frac{1}{(2\pi)^4} \int \bar{H}_{\alpha\beta}(\vec{k}) \exp(ik_\sigma x^\sigma) d^4\vec{k}. \qquad (9.27)$$

And similarly generalize Schutz Eq. (9.14) $\xi_\alpha = B_\alpha \exp(ik_\alpha x^\alpha)$ to the superposition:

$$\xi_\alpha(\vec{x}) = \frac{1}{(2\pi)^4} \int \Xi_\alpha(\vec{k}) \exp(ik_\sigma x^\sigma) d^4\vec{k}. \qquad (9.28)$$

Substitution of eqn. (9.28) into eqn. (9.25) and using eqn. (8.19), or simply using Schutz Eq. (9.16) directly, gives:

$$\bar{h}_{\alpha\beta}^{(NEW)}(\vec{x}) = \frac{1}{(2\pi)^4} \int \bar{H}_{\alpha\beta}^{(NEW)}(\vec{k}) \exp(ik_\sigma x^\sigma) d^4\vec{k}$$
$$= \frac{1}{(2\pi)^4} \int \left(\bar{H}_{\alpha\beta}^{(OLD)}(\vec{k}) - ik_\beta \Xi_\alpha(\vec{k}) - ik_\alpha \Xi_\beta(\vec{k}) + i\eta_{\alpha\beta} \Xi^\gamma(\vec{k})k_\gamma \right) \exp(ik_\sigma x^\sigma) d^4\vec{k}. \qquad (9.29)$$

Taking the Fourier transform of eqn. (9.29) above gives

$$\bar{H}_{\alpha\beta}^{(NEW)}(\vec{k}) = \bar{H}_{\alpha\beta}^{(OLD)}(\vec{k}) - ik_\beta \Xi_\alpha(\vec{k}) - ik_\alpha \Xi_\beta(\vec{k}) + i\eta_{\alpha\beta} \Xi^\gamma(\vec{k})k_\gamma. \qquad (9.30)$$

But eqn. (9.30) above is identical with eqn. (9.11), with $\bar{H}_{\alpha\beta}$ playing the role of $A_{\alpha\beta}$ and Ξ_α playing the role of B_α. So all the results of Exercise 9.5(a)–(d) carry over to the case of a superposition of waves of the form eqn. (9.27) simply by choosing ξ_α as in eqn. (9.28) above.

9.5   (f) Show that we cannot achieve eqns. (9.14) and (9.17) for a static solution, i.e. one for which $\omega = 0$.

Solution: Eqns. (9.14) and (9.17) assume a solution of the form $\bar{h}^{\alpha\beta} = A^{\alpha\beta} \exp(ik_\alpha x^\alpha) = A^{\alpha\beta} \exp[i(-\omega t + k_j x^j)]$. This is a solution of the weak-field vacuum Einstein equations eqn. (9.3) when

$$0 = \eta^{\mu\nu} k_\mu k_\nu \bar{h}^{\alpha\beta} = k_j k^j \bar{h}^{\alpha\beta} = 0. \qquad \text{imposed } \omega = 0 \qquad (9.31)$$

With $\omega = 0$ in eqn. (9.31) only the trivial solutions $\bar{h}^{\alpha\beta} = 0$ remain. Note we require a non-zero $\vec{R}$ for a nontrivial gauge transformation.

9.7   Give a more rigorous proof that

$$\frac{\mathrm{d}}{\mathrm{d}\tau} U^\alpha + \Gamma^\alpha{}_{\mu\nu} U^\mu U^\nu = 0 \qquad\qquad \text{Schutz Eq. (9.22) (9.32)}$$

$$\left(\frac{\mathrm{d}U^\alpha}{\mathrm{d}\tau}\right)_0 = -\Gamma^\alpha{}_{00} = -\frac{1}{2}\eta^{\alpha\beta}(h_{\beta 0,0} + h_{0\beta,0} - h_{00,\beta}) \quad \text{Schutz Eq. (9.23) (9.33)}$$

imply that a free particle initially at rest in the TT gauge remains at rest.

Solution: As stated by Schutz $h^{TT}_{0\beta} = 0$, so it is clear from eqn. (9.33) that $\Gamma^\alpha{}_{00} = 0$. Thus $U^\alpha = \delta^\alpha{}_0$ is a solution of eqn. (9.32), see eqn. (9.33). To argue that it is the unique solution we note that eqn. (9.32) is a first-order ordinary differential equation for $U^\alpha(\tau)$ of the very general form $y' \equiv \mathrm{d}y/\mathrm{d}x = F(x, y)$, see (e.g. Hassani, 1999, Eq. (13.5)). Here $y = U^\alpha$, $x = \tau$ and

$$F(x, y) = -\Gamma^\alpha{}_{\mu\nu} U^\mu U^\nu = -\eta^{\alpha\beta}\left(h_{\beta\mu,\nu} + h_{\beta\nu,\mu} - h_{\mu\nu,\beta}\right) U^\mu U^\nu \frac{1}{2}. \qquad (9.34)$$

The existence of a solution on a given domain follows from the continuity of $F(x, y)$ on that domain (Hassani, 1999, Peano Existence Theorem 13.2.2). The uniqueness of that solution, for a given $U^\alpha(\tau = 0)$ requires that $F(x, y)$ satisfies the Lipschitz condition: $|F(x, y_1) - F(x, y_2)| \le L|y_1 - y_2|$ for finite real constant L. Here this requires

$$|\eta^{\alpha\beta}\left(h_{\beta\mu,\nu} + h_{\beta\nu,\mu} - h_{\mu\nu,\beta}\right) V^\mu V^\nu - \eta^{\alpha\beta}\left(h_{\beta\mu,\nu} + h_{\beta\nu,\mu} - h_{\mu\nu,\beta}\right) U^\mu U^\nu| \le L|V^\alpha - U^\alpha|,$$

which will be true for finite $A_{\alpha\beta}$ and k_α. In that case $U^\alpha = \delta^\alpha{}_0$ is the unique solution.

9.9   Does the free particle of the discussion following Schutz Eq. (9.23) *see* any acceleration? To answer this, consider the two particles whose relative proper distance is calculated in

$$\Delta l \approx \left[1 + \frac{1}{2}h^{TT}_{xx}(x = 0)\right]\varepsilon. \qquad\qquad \text{Schutz Eq. (9.24)} \qquad (9.35)$$

Let the one at the origin send a beam of light towards the other, and let it be reflected by the other and received back at the origin. Calculate the amount of proper time elapsed at the origin between the emission and reception of the light (you may assume that the particles' separation is much less than a wavelength of the gravitational wave). By monitoring changes in this time, the particle at the origin can "see" the relative acceleration of the two particles.

Solution: The light signal follows a null geodesic, $ds^2 = 0$, so coordinate intervals are related through the metric by $g_{\mu\nu}dx^\alpha dx^\beta = 0$. Here we can orient the pseudo-Cartesian x-axis along the null geodesic separating the two particles so that $dy = dz = 0$ leaving

$$
\begin{aligned}
0 &= g_{\mu\nu}dx^\alpha dx^\beta && \text{null path} \\
&= g_{00}dt^2 + 2g_{0x}dt\,dx + g_{xx}dx^2 && \text{convenient coordinates} \\
d\tau = \sqrt{-g_{00}}dt &= \sqrt{g_{xx}}dx. && \text{used } g_{0x} = 0 \quad (9.36)
\end{aligned}
$$

We are ultimately interested in the proper time interval $\Delta\tau$ (measured by an ideal clock stationary at the particle at the origin) so the details of g_{00} don't concern us. Two questions arise: why did we set $g_{0x} = 0$? and what is g_{xx}? We answer these together.

Recall we reconstruct the metric from eqn. (8.57), $g_{\alpha\beta} = \eta_{\alpha\beta} + h_{\alpha\beta}$, with the perturbations $h_{\alpha\beta}$ given by the plane waves of eqn. (9.26). In the TT gauge with $\vec{U} = \vec{e}_t$ the amplitudes of a plane gravitational wave traveling along the z-axis were found to be

$$
(A_{\alpha\beta}^{\text{TT}}) = \begin{pmatrix} 0 & 0 & 0 & 0 \\ 0 & A_{xx} & A_{xy} & 0 \\ 0 & A_{xy} & -A_{xx} & 0 \\ 0 & 0 & 0 & 0 \end{pmatrix}. \qquad \text{Schutz Eq. (9.21)} \qquad (9.37)
$$

Furthermore

$$
\bar{h}_{\alpha\beta}^{\text{TT}} = h_{\alpha\beta}^{\text{TT}} \qquad \text{Schutz Eq. (9.20)} \qquad (9.38)
$$

because of the traceless condition on the $A_{\alpha\beta}^{\text{TT}}$. Eqn. (9.37) with eqn. (9.26) and eqn. (9.38) give

$$
(h_{\alpha\beta}^{\text{TT}}) = \begin{pmatrix} 0 & 0 & 0 & 0 \\ 0 & h_{xx}^{\text{TT}} & h_{xy}^{\text{TT}} & 0 \\ 0 & h_{xy}^{\text{TT}} & -h_{xx}^{\text{TT}} & 0 \\ 0 & 0 & 0 & 0 \end{pmatrix}. \qquad (9.39)
$$

We see that $g_{0x} = \eta_{0x} + A_{0x} = 0$ and

$$
\begin{aligned}
g_{xx} &\underset{\text{TT}}{=} \eta_{xx} + A_{xx}^{\text{TT}} e^{ik_\sigma x^\sigma} \\
&= 1 + \text{Re}(A_{xx}^{\text{TT}})\cos(kx - \omega t) - \text{Im}(A_{xx}^{\text{TT}})\sin(kx - \omega t). \qquad (9.40)
\end{aligned}
$$

Now the point in time at which $t = 0$ is arbitrary in this problem and can be chosen such that $A_{xx}^{\text{TT}} \in \mathbb{R}$, and g_{xx} simplifies to

$$
g_{xx} \underset{\text{TT}}{=} 1 + A_{xx}^{\text{TT}}\cos(kx - \omega t). \qquad \text{shift time such that } A_{xx}^{\text{TT}} \in \mathbb{R} \qquad (9.41)
$$

Substituting eqn. (9.41) into eqn. (9.36) we have a messy integral to solve. But here the situation simplifies to an almost stationary spacetime during the round-trip time of the light signal; because the gravitational wave wavelength is much greater than ε we also have that ωt is approximately the same for the emitted and reflected signal and

$x \approx 0$. Then round-trip time of the light signal simplifies to twice the proper distance,

$$\Delta \tau \approx 2\Delta l \approx 2 \left(1 + \frac{1}{2} A_{xx} \cos(\omega t) \right) \varepsilon. \qquad \text{used eqn. (B.5)} \qquad (9.42)$$

The proper distance measured in this way is sometimes called the "radar distance" (Rindler, 2006, e.g. Eq. (11.24)). The particle acceleration is revealed by the radar distance oscillating with the period of the gravitational wave.

Radar distance and proper distance are not one and the same in general; please see SP9.4.

9.11   (a) Derive Schutz Eq. (9.27) for the components of the Riemann tensor in terms of the components of the metric in the TT gauge

$$R^x{}_{0x0} = R_{x0x0} = -\frac{1}{2} h^{\text{TT}}_{xx,00},$$

$$R^y{}_{0x0} = R_{y0x0} = -\frac{1}{2} h^{\text{TT}}_{xy,00},$$

$$R^y{}_{0y0} = R_{y0y0} = -\frac{1}{2} h^{\text{TT}}_{yy,00} = -R^x{}_{0x0}, \qquad \text{Schutz Eq. (9.27)} \qquad (9.43)$$

with all other "independent" components vanishing.

Solution: In the solution to Exercise 9.9 we explained how one reconstructs the metric perturbations in the TT gauge for the case of a gravitational wave in the z-direction. Straightforward substitution of eqn. (9.39) above into eqn. (8.14), $R_{\alpha\beta\mu\nu} = \frac{1}{2}(h_{\alpha\nu,\beta\mu} + h_{\beta\mu,\alpha\nu} - h_{\alpha\mu,\beta\nu} - h_{\beta\nu,\alpha\mu})$, gives eqn. (9.43).

We're told that the remaining *independent* components of $R_{\alpha\beta\mu\nu} = 0$. Here apparently "independent" does *not* refer to the 20 independent components of the Riemann tensor in 4D spacetime. First let's systematically find all 20 terms; refer to Exercise 6.18. Starting with the six Riemann symmetry independent terms of the form $R_{0x\mu\nu}$ we find:

$$R_{0x0x} = -\frac{1}{2} h^{\text{TT}}_{xx,00} \qquad R_{0x0y} = -\frac{1}{2} h^{\text{TT}}_{xy,00} \qquad R_{0x0z} = 0$$

$$R_{0xxy} = 0 \qquad R_{0xxz} = \frac{1}{2} h^{\text{TT}}_{xx,0z} \qquad R_{0xyz} = \frac{1}{2} h^{\text{TT}}_{xy,0z}.$$

$$(9.44)$$

Of the six (Riemann symmetry) independent terms of the form $R_{0y\mu\nu} = 0$, one ($R_{0y0x} = R_{0x0y}$) was explicitly counted in eqn. (9.44) above, leaving five new terms:

$$R_{0y0y} = -\frac{1}{2} h^{\text{TT}}_{yy,00} \qquad R_{0y0z} = 0 \qquad R_{0yxy} = 0$$

$$R_{0yxz} = \frac{1}{2} h^{\text{TT}}_{xy,0z} \qquad R_{0yyz} = \frac{1}{2} h^{\text{TT}}_{yy,0z}.$$

$$(9.45)$$

This brings our count to 11 terms accounted for. All six (Riemann symmetry) independent terms $R_{0z\mu\nu} = 0$, of which two $0 = R_{0z0x} = R_{0z0y}$ were explicitly counted above, leaving four new, i.e. $0 = R_{0z0z} = R_{0zxy} = R_{0zxz} = R_{0zyz}$. This brings our count to 15 terms accounted for. Increment the first pair of indices; we find all six terms $R_{xy\mu\nu} = 0$. But three of these are explicitly accounted for above ($R_{xy0i} = 0$). And another degree of freedom is lost through $R_{0xyz} + R_{y0xz} + R_{xy0z} = 0$, bringing our total to 17 (Riemann symmetry) independent terms accounted for. Increment the first pair of indices again considering R_{xzij} (terms with 0 have been accounted for already). These give two independent terms:

$$R_{xzxz} = -\frac{1}{2}h^{\text{TT}}_{xx,zz} \qquad\qquad R_{xzyz} = -\frac{1}{2}h^{\text{TT}}_{xy,zz}. \qquad (9.46)$$

19 terms accounted for. Considering R_{yzij} adds the final term:

$$R_{yzyz} = -\frac{1}{2}h^{\text{TT}}_{yy,zz}. \qquad (9.47)$$

All 20 terms accounted for, 10 of which were non-zero.

Why did only three appear in eqn. (9.43)? One interpretation is that we can use the vacuum Einstein equations, eqn. (9.80), $R_{\alpha\beta} = 0$ to relate $\partial^2/\partial t^2$ to $\partial^2/\partial z^2$:

$$0 = R_{xx} = R^{\mu}{}_{x\mu x} = R^0{}_{x0x} + R^z{}_{xzx} = -R_{0x0x} + R_{zxzx} \implies h^{\text{TT}}_{xx,00} = h^{\text{TT}}_{xx,zz}$$
$$0 = R_{xy} = R^{\mu}{}_{x\mu y} = R^0{}_{x0y} + R^z{}_{xzy} = -R_{0x0y} + R_{zxzy} \implies h^{\text{TT}}_{xy,00} = h^{\text{TT}}_{xy,zz}$$
$$(9.48)$$

consistent with eqn. (9.3) of course. The other Einstein equations can be seen to lead to no contradictions without adding new information, see SP9.3. Furthermore, for the plane wave solution traveling in the z-direction, one finds $\partial/\partial t = -\partial/\partial z$, and $h^{\text{TT}}_{xx,0z} = -h^{\text{TT}}_{xx,00}$ so the other non-vanishing $R^{\alpha}{}_{\beta\mu\nu}$ are known in terms of the quantities in eqn. (9.43).

9.11 (b) Solve the geodesic deviation equations

$$\frac{\partial^2}{\partial t^2}\xi^x = \frac{1}{2}\varepsilon\frac{\partial^2}{\partial t^2}h^{\text{TT}}_{xx} \qquad \frac{\partial^2}{\partial t^2}\xi^y = \frac{1}{2}\varepsilon\frac{\partial^2}{\partial t^2}h^{\text{TT}}_{xy} \qquad \text{Schutz Eq. (9.28a)}$$

and

$$\frac{\partial^2}{\partial t^2}\xi^y = -\frac{1}{2}\varepsilon\frac{\partial^2}{\partial t^2}h^{\text{TT}}_{xx} \qquad \frac{\partial^2}{\partial t^2}\xi^x = \frac{1}{2}\varepsilon\frac{\partial^2}{\partial t^2}h^{\text{TT}}_{xy} \qquad \text{Schutz Eq. (9.28b)} \quad (9.49)$$

for the motion of the test particles in the polarization rings shown in Schutz Fig. 9.1.

Solution: The term ε on the RHS of eqn. (9.49) is the magnitude of the initial separation vector, $\varepsilon = |\vec{\xi}(0)|$. The deviations are notoriously small, so to a very good approximation the ξ^β that contracts with $R^\alpha_{\mu\nu\beta}$ on the RHS of the geodesic deviation equation, Schutz Eq. (9.25), can be replaced by a constant, say $\xi^\beta(0)$ the value at $t = 0$. The solution for both Schutz Eqs. (9.28a) and (9.28b) is

$$\xi_i = \xi^j(0)\left(\delta_{ij} + \frac{1}{2}h^{TT}_{ij}\right). \tag{9.50}$$

9.13 One kind of background Lorentz transformation is a simple $45°$ rotation of the x- and y-axes in the $x-y$ plane. Show that under such a rotation from (x, y) to (x', y'), we have $h^{TT}_{x'y'} = h^{TT}_{xx}, h^{TT}_{x'x'} = -h^{TT}_{xy}$. This is consistent with Schutz Fig. 9.1.

Solution: For a rotation of the Cartesian coordinates $\pi/4$ about the z-axis, the new coordinates x' and y' are related to the old via $x = \cos\theta x' - \sin\theta y'$ and $y = \sin\theta x' + \cos\theta y'$. From these we find the appropriate derivatives and

$$h^{TT}_{x'x'} = \left(\frac{\partial x}{\partial x'}\right)^2 h^{TT}_{xx} + \left(\frac{\partial y}{\partial x'}\right)^2 h^{TT}_{yy} + 2\left(\frac{\partial x}{\partial x'}\frac{\partial y}{\partial x'}\right)h^{TT}_{xy} = 2\cos\left(\frac{\pi}{4}\right)\sin\left(\frac{\pi}{4}\right)h^{TT}_{xy} = h^{TT}_{xy},$$

$$h^{TT}_{x'y'} = \left(\frac{\partial x}{\partial x'}\frac{\partial x}{\partial y'}\right)h^{TT}_{xx} + \left(\frac{\partial x}{\partial x'}\frac{\partial y}{\partial y'}\right)h^{TT}_{xy} + \left(\frac{\partial y}{\partial x'}\frac{\partial x}{\partial y'}\right)h^{TT}_{xy} + \left(\frac{\partial y}{\partial x'}\frac{\partial y}{\partial y'}\right)h^{TT}_{yy} = -h^{TT}_{xx}. \tag{9.51}$$

The signs disagree with the relations given, but of course the agreement with Schutz Fig. 9.1 is the important point of the exercise; the "+" vs. "×" polarization depends upon the orientation of the coordinates.

For general reference, the sense of a rotation can be determined from the right-hand rule; with the thumb pointing in along the axis of rotation the fingers curl in the direction of positive rotation; e.g. a counterclockwise rotation in the $x-y$ plane.

9.23 Derive[2]

$$R = \frac{\frac{1}{2}l_0 A\Omega^2}{B} = \frac{\frac{1}{2}l_0 A\Omega^2}{\sqrt{(\Omega^2 - \omega_0^2)^2 + 4\gamma^2\Omega^2}} \qquad \text{Schutz Eq. (9.48)} \tag{9.52}$$

$$\tan\phi = \frac{2\Omega\gamma}{\Omega^2 - \omega_0^2} \qquad \text{Schutz Eq. (9.49)} \tag{9.53}$$

and derive the general solution of the ODE

$$\xi_{,00} + 2\gamma\xi_{,0} + \omega_0^2\xi = \frac{1}{2}l_0 h^{TT}_{xx,00} \qquad \text{Schutz Eq. (9.45)} \tag{9.54}$$

for arbitrary initial data at $t = 0$, given the forcing

$$h^{TT}_{xx} = A\cos(\Omega t). \qquad \text{Schutz Eq. (9.46)} \tag{9.55}$$

[2] We have corrected a sign error in eqn. (9.53).

Solution: It's easier to work with complex numbers. Let $h_{xx}^{TT} = A \exp(i\Omega t)$ and write the steady solution as $\xi = R \exp[i(\Omega t + \phi)]$ with both $A, R \in \mathbb{R}$, cf. eqn. (9.55) and Schutz Eq. (9.47). Substituting these into the ODE eqn. (9.54) above gives an algebraic equation:

$$[\Omega^2 - i2\gamma\Omega - \omega_0^2]R \exp(i\phi) = \frac{1}{2}l_0 A\Omega^2. \tag{9.56}$$

The RHS is real so the term in square parentheses on the LHS must be $B \exp(-i\phi)$, $B \in \mathbb{R}$, to make the LHS real. Euler's formula, see Boas (1983, Chapter 2 Eq. (4.1)) or Felder and Felder (2014, Chapter 3) then gives

$$\tan\phi = \frac{2\Omega\gamma}{\Omega^2 - \omega_0^2} \qquad R = \frac{\frac{1}{2}l_0 A\Omega^2}{B} = \frac{\frac{1}{2}l_0 A\Omega^2}{\sqrt{(\Omega^2 - \omega_0^2)^2 + 4\gamma^2\Omega^2}}, \tag{9.57}$$

in agreement with eqns. (9.52) and (9.53).

The general solution of a second-order linear ODE can be written as $\xi_g = \xi_h + \xi_p$, with ξ_h a linear combination of two linearly independent solutions of the corresponding homogeneous ODE (set the RHS = 0 in eqn. (9.54) above) and ξ_p a particular solution (Felder and Felder, 2014, §1.6 and §3.5) or (Boas, 1983, Chapter 8 Eq. (6.8)). Here the latter is the steady solution, Schutz Eq. (9.47). Trial and error reveals that ξ_h are proportional to $\exp(\lambda t)$. Substituting into eqn. (9.54) above with RHS = 0 confirms this and gives two solutions for λ:

$$\lambda = -\gamma \pm i\sqrt{\omega_0^2 - \gamma^2}. \tag{9.58}$$

Assuming the system is underdamped ($\gamma^2 < \omega_0^2$), it is convenient to choose complex constants with ratio proportional to $\exp(i\pi/2)$ for this gives the two independent solutions:

$$\xi_h = a \exp(-\gamma t) \cos\left(t\sqrt{\omega_0^2 - \gamma^2}\right) + b \exp(-\gamma t) \sin\left(t\sqrt{\omega_0^2 - \gamma^2}\right). \tag{9.59}$$

The two real constants a and b can be chosen to match the initial data at $t = 0$. For instance the initial data could be $\xi(t = 0)$ and $\xi_{,t}(t = 0)$. The constant a gives the initial departure from the steady solution $\xi_p(t = 0) = R\cos\phi$, i.e. $a = \xi(t = 0) - R\cos\phi$. And b can be set so that

$$\xi_{,t}(t = 0) = -R\Omega\sin\phi + b\sqrt{\omega_0^2 - \gamma^2 - a\gamma}.$$

9.25 Derive

$$Q = \frac{\omega_0}{2\gamma} \qquad\qquad \text{Schutz Eq. (9.56)} \tag{9.60}$$

from the given definition of Q.

Solution: The quality factor Q is defined such that $1/Q$ is the average fraction of the energy E of the undriven oscillator lost to friction in one radian of oscilation:

$$\frac{1}{Q} = \frac{1}{\omega}\frac{\dot{E}}{E}. \tag{9.61}$$

$\dot{E}$ is the sum of the kinetic and potential energy of the oscillator, and for a detector at rest at $t = 0$, is given by Schutz Eq. (9.51):

$$E = \frac{1}{4}m[(\xi_{,t})^2 + \omega_0^2\xi^2]. \tag{9.62}$$

For $\gamma \ll \omega$ the frequency of the free oscillations $\omega \approx \omega_0$ and general solution, eqn. (9.59) above, becomes

$$\xi \approx a\exp(-\gamma t)\cos(\omega_0 t) + b\exp(-\gamma t)\sin(\omega_0 t). \tag{9.63}$$

Substitute this into eqn. (9.62) above and take the average over one period, denoted by $\langle\ \rangle$. Use the fact that $\gamma \ll \omega_0$ so that $\exp(2\gamma t)$ is almost constant over a period, $\langle\sin(\omega_0 t)\cos(\omega_0 t)\rangle = 0$, $\langle\sin^2(\omega_0 t)\rangle = \langle\cos^2(\omega_0 t)\rangle = 1/2$. After some algebra one finds:

$$\frac{1}{\omega}\frac{\dot{E}}{E} = 2\frac{\gamma}{\omega_0}, \tag{9.64}$$

consistent with eqn. (9.60).

(See SP9.5 for a solution that directly estimates the rate of energy dissipation from the damping force.)

9.27 (a) Derive the full three-term return relation,

$$\frac{dt_{\text{return}}}{dt_{\text{start}}} = 1 + \frac{1}{2}\{(1 - \sin\theta)h_+(t_{\text{start}} + 2L) - (1 + \sin\theta)h_+(t_{\text{start}})$$
$$+ 2\sin\theta h_+[t_{\text{start}} + (1 - \sin\theta)L]\} \qquad \text{Schutz Eq. (9.63)} \tag{9.65}$$

for the rate of change of the return time for a beam traveling through a plane wave h_+ along the x-direction, when the wave is moving at an angle θ to the z-axis in the $x-z$ plane.

Solution: The light beam is in the x-direction so we only need the h_{xx} component. Our strategy will be to first work in a TT coordinate system $x^{\alpha'}$ in which the gravitational wave travels in the z'-direction with h_+ polarization, so that we can use our previous results in eqn. (9.39). Then we can perform a coordinate system rotation about the y'-axis to obtain the h_{xx} we need.

For a gravitational wave with h_+ polarization traveling in the z'-direction in the TT coordinate system (t', x', y', z') we have $h_{x'x'}^{\text{TT}} = A'\cos[(z' - t')\omega]$, $h_{y'y'}^{\text{TT}} = -h_{x'x'}^{\text{TT}}$, all other components vanishing, see eqn. (9.39) and recall the h_+ polarization implies the off-diagonal terms vanish. Let (t, x, y, z) be the pseudo-Cartesian coordinate system obtained by a rotation of the (t', x', y', z') coordinates about the y'-axis by an angle θ. Then

$$h_{xx} = \left(\frac{\partial x'}{\partial x}\right)^2 h_{x'x'}^{\text{TT}} + \left(\frac{\partial y'}{\partial x}\right)^2 h_{y'y'}^{\text{TT}} = \cos^2\theta \, h_{x'x'}^{\text{TT}} = \cos^2\theta \, A' \cos[\omega(\cos\theta z - \sin\theta x - t)]$$

$$= A \cos[-\omega(\sin\theta x + t)]. \tag{9.66}$$

We have defined $A \equiv \cos^2\theta A'$ and without loss of generality, set $z = 0$. Now substitute h_{xx} into the expression for the time for the light beam to travel from $x = 0$ to $x = L$:

$$t_{\text{far}} = t_{\text{start}} + \int_0^L \sqrt{1 + h_{xx}} \, \mathrm{d}x \qquad\qquad \text{Schutz Eq. (9.60)}$$

$$\tag{9.67}$$

$$\simeq t_{\text{start}} + L + \frac{1}{2}\int_0^L h_{xx} \, \mathrm{d}x \qquad\qquad \text{used eqn. (B.5)}$$

$$= t_{\text{start}} + L + \frac{1}{2}\int_0^L A \cos[-\omega(x(1 + \sin\theta) + t_{\text{start}})]\mathrm{d}x. \tag{9.68}$$

The third line used $t(x) \approx t_{\text{start}} + x$ in eqn. (9.66) above. On the return travel from $x = L$ (when $t = t_{\text{far}}$) to $x = 0$ (when $t = t_{\text{return}}$) we instead have $t(x) \approx t_{\text{start}} + 2L - x$.

$$\int_{t_{\text{far}}}^{t_{\text{return}}} \mathrm{d}t = -\int_L^0 \sqrt{1 + h_{xx}} \, \mathrm{d}x = \int_0^L \sqrt{1 + h_{xx}} \, \mathrm{d}x. \quad \text{negative root from Schutz Eq. (9.59)}$$

$$\tag{9.69}$$

Integrating both sides

$$t_{\text{return}} = t_{\text{far}} + L + \frac{1}{2}\int_0^L h_{xx}\mathrm{d}x$$

$$= t_{\text{far}} + L + \frac{1}{2}\int_0^L A \cos[-\omega(x(\sin\theta - 1) + t_{\text{start}} + 2L)]\mathrm{d}x$$

$$= t_{\text{start}} + 2L + \frac{1}{2}\int_0^L A \cos[-\omega(x(1 + \sin\theta) + t_{\text{start}})]\mathrm{d}x$$

$$+ \frac{1}{2}\int_0^L A \cos[-\omega[x(\sin\theta - 1) + t_{\text{start}} + 2L)]\mathrm{d}x, \qquad \text{used eqn. (9.68)}$$

$$\tag{9.70}$$

which generalizes Schutz Eq. (9.61). Differentiating with respect to t_{start} gives:

$$\frac{\mathrm{d}t_{\text{return}}}{\mathrm{d}t_{\text{start}}} = 1 + \frac{1}{2}\left(\frac{1}{1 - \sin\theta}h_{xx}(t_{\text{start}} + 2L) - \frac{1}{1 + \sin\theta}h_{xx}(t_{\text{start}})\right.$$

$$\left. + \left(\frac{1}{1 + \sin\theta} - \frac{1}{1 - \sin\theta}\right)h_{xx}(t_{\text{start}} + L(1 + \sin\theta))\right)$$

$$= 1 + \frac{1}{2} \big[(1 + \sin\theta) h'_{xx}(t_{\text{start}} + 2L) - (1 - \sin\theta) h'_{xx}(t_{\text{start}})$$
$$- 2\sin\theta h'_{xx}(t_{\text{start}} + L(1 + \sin\theta)) \big]. \qquad (9.71)$$

A factor of $1/\cos^2\theta$ was absorbed when h_{xx} was replaced with h'_{xx} (recall $A = \cos^2\theta A'$ above). Our expression differs from eqn. (9.65) because we defined the wave direction as θ *clockwise* from the z-axis.

9.27 (b) Show that, in the limit where L is small compared to a wavelength of the gravitational wave, the derivative of the return time is the derivative of $t + \delta L$, where $\delta L = L \cos^2\theta h(t)$ is the excess proper distance for small L. Explain where the factor of $\cos^2\theta$ comes from.

Solution: When the wavelength is much longer than L the integrand in eqn. (9.70) above becomes independent of x so the return time is simply

$$t_{\text{return}} = t_{\text{start}} + 2L + L\, h_{xx}(t). \qquad (9.72)$$

Recall we made use of a similar simplification in Exercise 9.9. Taking the derivative

$$\frac{dt_{\text{return}}}{dt_{\text{start}}} = 1 + L\frac{d}{dt_{\text{start}}} h_{xx}(t_{\text{start}}) = 1 + L\cos^2\theta \frac{d}{dt_{\text{start}}} h'_{xx}(t_{\text{start}}). \qquad (9.73)$$

The $\cos^2\theta$ is simply the $(\partial x'/\partial x)^2$ term that came from transformation of the metric under a rotation of the coordinates about the y-axis.

9.27 (c) Examine the limit of the three-term formula in (a) when the gravitational wave is traveling along the x-axis too ($\theta = \pm\pi/2$): what happens to light going parallel to a gravitational wave?

Solution: Setting $\theta = \pm\pi/2$ turns the z'-axis so that it lies on top of the x-axis (pointing in the opposite or same direction respectively), so the gravitational wave travels parallel to the direction of the light beam. Simple substitution $\theta = \pm\pi/2$ in eqn. (9.65) gives

$$\frac{dt_{\text{return}}}{dt_{\text{start}}} = 1. \qquad (9.74)$$

Light parallel to the gravitational wave is not redshifted.

9.2 Supplementary problems

SP 9.1 Show that the dispersion relation eqn. (9.5) implies the phase speed and group velocity are unity. What is wave dispersion?

Solution

The *phase speed* of a classical wave is given by the ratio of the frequency divided by the magnitude of the wavenumber vector, $C_p = \omega/|\mathbf{k}|$. Curves of constant phase are observed to propagate in the direction of the wavenumber vector $\mathbf{k}$ at a speed C_p. The term "speed" emphasizes that this quantity is not a traditional vector. Here

$$C_p = \frac{\omega}{|\mathbf{k}|} = 1. \tag{9.75}$$

The *group velocity* of a classical wave is given by the gradient of the frequency with respect to the components of the wavenumber vector:

$$\mathbf{C_g} = \frac{\partial \omega}{\partial k^x}\,\hat{\mathbf{x}} + \frac{\partial \omega}{\partial k^y}\,\hat{\mathbf{y}} + \frac{\partial \omega}{\partial k^z}\,\hat{\mathbf{z}}, \tag{9.76}$$

where $\hat{\mathbf{x}}, \hat{\mathbf{y}}, \hat{\mathbf{z}}$ are traditional unit three-vectors in the x-, y-, and z-directions. In contrast to the phase speed, the group velocity is a traditional three-vector quantity. Let's get started with just a single partial derivative

$$\frac{\partial \omega}{\partial k^x} = \frac{\partial |\mathbf{k}|}{\partial k^x} = \frac{\partial \sqrt{(k^x)^2 + (k^y)^2 + (k^z)^2}}{\partial k^x} = \frac{1}{2}\frac{1}{|\mathbf{k}|}\frac{\partial (k^x)^2}{\partial k^x} = \frac{k^x}{|\mathbf{k}|}. \tag{9.77}$$

So now it's clear

$$\mathbf{C_g} = \frac{k^x}{|\mathbf{k}|}\,\hat{\mathbf{x}} + \frac{k^y}{|\mathbf{k}|}\,\hat{\mathbf{y}} + \frac{k^z}{|\mathbf{k}|}\,\hat{\mathbf{z}}, \tag{9.78}$$

and clearly $|\mathbf{C_g}| = 1$.

Wave dispersion occurs when the phase speed differs from the magnitude of the group velocity. Roughly speaking, the wave spreads out.

SP 9.2 Use the Einstein field equations to show that in general the Ricci scalar R is related to the trace of the stress–energy tensor T by

$$R = -8\pi T. \tag{9.79}$$

Use this to show that the vacuum field equations can be written as

$$G_{\mu\nu} = R_{\mu\nu} = 0. \tag{9.80}$$

So the Einstein and Ricci tensors exactly vanish in a vacuum spacetime. However, the Riemann tensor only vanishes if the spacetime is flat.

SP 9.3   Write out the remaining vacuum Einstein field equations $R_{\mu\nu} = 0$, revealing that these simply do not contradict the information already presented up to eqn. (9.48) above in the solution to Exercise 9.11(a). They do not add any new information.

Solution

In Exercise 9.11(a) we found two components of the vacuum Einstein equations, $R_{xx} = 0$ and $R_{xy} = 0$; see eqn. (9.48). We "learned" that $h^{TT}_{xx,00} = h^{TT}_{xx,zz}$ and $h^{TT}_{xy,00} = h^{TT}_{xy,zz}$. Because of the symmetry $R_{\alpha\beta} = R_{\beta\alpha}$, there are only eight more equations to consider. They are

$$0 = R_{yy} = R^{\mu}{}_{y\mu y} = R^{0}{}_{y0y} + R^{z}{}_{yzy} = -\frac{1}{2}h^{TT}_{xx,00} + \frac{1}{2}h^{TT}_{xx,zz}$$

$$0 = R_{00} = R^{\mu}{}_{0\mu 0} = R^{x}{}_{0x0} + R^{y}{}_{0y0} = -\frac{1}{2}h^{TT}_{xx,00} - \frac{1}{2}h^{TT}_{yy,00}$$

$$0 = R_{zz} = R^{\mu}{}_{z\mu z} = R^{x}{}_{zxz} + R^{y}{}_{zyz} = -\frac{1}{2}h^{TT}_{xx,zz} - \frac{1}{2}h^{TT}_{yy,zz}$$

$$0 = R_{0x} = R^{y}{}_{0yx} = 0$$

$$0 = R_{0y} = R^{x}{}_{0xy} = 0$$

$$0 = R_{0z} = R^{\mu}{}_{0\mu z} = R^{x}{}_{0xz} + R^{y}{}_{0yz} = -\frac{1}{2}h^{TT}_{xx,0z} - \frac{1}{2}h^{TT}_{yy,0z}$$

$$0 = R_{xz} = R^{0}{}_{x0z} = 0$$

$$0 = R_{yz} = R^{0}{}_{y0z} = 0. \tag{9.81}$$

Taking account the traceless condition, $h^{TT}_{xx} = -h^{TT}_{yy}$, the equations here reduce to $h^{TT}_{xx,00} = h^{TT}_{xx,zz}$ or are trivial.

SP 9.4   Find the round-trip proper time to send a light signal radially in the Schwarzschild spacetime from an observer, stationary in the standard Schwarzschild coordinates at r_1, to a stationary mirror at r_2, where it is reflected back to the observer. This time, in geometric units, is twice the so-called *radar distance*. Compare this to the proper distance.

Hint

Exercise 9.9 should get you started with the radar distance calculation. See SP7.8 to verify your calculation of proper distance.

SP 9.5   Solve Exercise 9.25 by directly finding the rate of energy dissipation from the damping force.

Solution

Consider an oscillator initially at rest so that $x_{1,t} = -x_{2,t} = -\xi_{,t}/2$. In Schutz Eq. (9.41) the damping force has the form of a Rayleigh damping

$$mx_{1,tt} = +k\xi + v\xi_{,t} = +k\xi - 2vx_{1,t} \tag{9.82}$$

with damping coefficient $2v$. The rate of working of this force is $2v(x_{1,t})^2$ and since there are two masses we have a total dissipation rate of $\dot{E} = 4v(x_{1,t})^2 = 4\gamma m(x_{1,t})^2$. Substitution into eqn. (9.61) above, again with the approximation that $\gamma \ll \omega_0$ so that $\langle(x_{1,t})^2\rangle \approx \omega^2\langle\xi^2\rangle$, gives eqn. (9.60).

SP 9.6 In Exercise 9.3 we took the Fourier transform of the wave equation and integrated by parts. This led to boundary terms that we happily assumed were zero. Can we use the square integrability condition, $\int dx^{\alpha*}|\bar{h}^{\mu\nu}|^2 < \infty$ for any fixed index $\alpha*$, to defend the assumption that boundary terms like eqn. (9.7) vanished? Consider the function[3] $f(x) = e^{-x}\sin(e^{2x})$. Is it square integrable? What happens to its derivative at infinity? To make rigorous sense of these issues, it is necessary to extend the framework of the Fourier transform to the space of so-called tempered distributions. The curious reader will need to delve into more advanced mathematics (e.g. Friedlander and Joshi, 1999, Chapter 8).

---

[3] We thank Johan Huisman for this example, and Jean-Philippe Nicolas for additional mathematical advice.

10 Spherical solutions for stars

In fact, neutron star matter is the most complex and fascinating state of matter that astronomers have yet discovered.

Bernard Schutz §10.7

10.1 Exercises

10.1 Starting with $ds^2 = \eta_{\alpha\beta}\,dx^\alpha dx^\beta$, show that the coordinate transformation $r = (x^2 + y^2 + z^2)^{1/2}, \theta = \arccos(z/r), \phi = \arctan(y/x)$ leads to

$$ds^2 = -dt^2 + dr^2 + r^2(d\theta^2 + \sin^2\theta d\phi^2). \qquad \text{Schutz Eq. (10.1)} \qquad (10.1)$$

Hint: The problem is to transform the coordinates of the $\binom{0}{2}$ tensor $\eta_{\alpha\beta}$, the Minkowski metric tensor, between the two coordinate systems. This is straightforward, as we learned in Chapter 5, see for example Exercise 5.12(b). See also Schutz Eqs. (6.21, 6.22, 6.25). The transformation matrix we need has elements that are the derivatives of the original coordinates with respect to the new coordinates, terms like $\partial x/\partial r$ etc. It helps tremendously to recognize the new coordinates r, θ, ϕ as the spherical coordinates. So we can immediately write $x(t, r, \theta, \phi) = r\sin(\theta)\cos(\phi)$ etc., see eqn. (B.13) in Appendix B. Exercise 6.28 guides you through this exercise (solution not available herein) for the spatial part. Exercise 7.7(b) extends this to 4D spacetime. The required matrix is given in eqn. (B.14) of Appendix B.

10.3 The locally measured energy of a particle, given by

$$E^* = -\vec{U} \cdot \vec{p} = e^{-\Phi}\,E, \qquad \text{Schutz Eq. (10.11)} \qquad (10.2)$$

is the energy the same particle would have in SR if it passed the observer with the same speed. (See SP10.3(a) for its derivation.) It therefore contains no information about gravity, about the curvature of spacetime. By referring to

$$-p_0 \approx m(1 + \phi + \mathbf{p}^2/2m^2) = m + m\phi + \mathbf{p}^2/2m, \quad \text{Schutz Eq. (7.34)} \quad (10.3)$$

show that the difference between E^* and E in the weak-field limit is, for particles with small velocities, just the gravitational potential energy.

Solution: Here $E \equiv -p_0$, is the particle energy, cf. Schutz Eq. (10.10), which in the Newtonian limit was found in Chapter 7 to be well-approximated by the sum of three terms:

$$E \equiv -p_0 \approx m + m\phi + \frac{\mathbf{p}^2}{2m}.$$

This energy will be different from that found by the local observer, E^*. The local observer measures that found in SR where the particle energy is defined as the $E^* \equiv p^0$. In the small-velocity limit, one finds p^0 is well-approximated by the sum of rest-mass energy plus classical kinetic energy:

$$E^* \equiv p^0 = mU^0 = m\frac{1}{\sqrt{1-v^2}} \qquad \text{used eqn. (2.39)}$$

$$\simeq m\left(1 + \frac{1}{2}v^2\right) \qquad \text{used eqn. (B.8)}$$

$$= m + \frac{\mathbf{p}^2}{2m}. \qquad (10.4)$$

Taking the difference between E and E^* in this limit we find:

$$E - E^* = \left(m + m\phi + \frac{\mathbf{p}^2}{2m}\right) - \left(m + \frac{\mathbf{p}^2}{2m}\right) = m\phi.$$

Working in the flat spacetime of SR we had $E = -p_0 = p^0 = E^*$ globally. Now in the curved spacetime of GR these two energy quantities differ because of the metric term, and for metric eqn. (10.53) the factor $e^{-\Phi}$ in particular. In the Newtonian limit we have a nice interpretation of the difference; the curved spacetime version includes a gravitational potential.

10.5 Show that a static star must have $U^r = U^\theta = U^\phi = 0$ in our coordinates, by examining the result of the transformation $t \to -t$.

Solution: By analogy with the notion of a static spacetime, for which the geometry is unchanged by time reversal, a static fluid must have the motion unchanged by a time reversal. As we'll now argue, this implies that the spatial components of the fluid four-velocity must be nil. Let U^α be the four-velocity of a fluid element of a static star. The three-velocity of any fluid element must remained unchanged under time reversal. But of course if there is a non-zero spatial four-velocity component then reversing the direction of time causes the corresponding three-velocity component to reverse direction. We can see this more formally by letting U^α, v^i and $U^{\bar{\alpha}}$, $v^{\bar{i}}$ be the four-velocity and three-velocity of a given fluid element in the original and time-reversed coordinate systems. Then the coordinate transformation $t \to -t$ results in:

$$U^{\bar{\alpha}} = \frac{\partial x^{\bar{\alpha}}}{\partial x^{\alpha}} U^{\alpha} = \begin{pmatrix} -1 & 0 & 0 & 0 \\ 0 & 1 & 0 & 0 \\ 0 & 0 & 1 & 0 \\ 0 & 0 & 0 & 1 \end{pmatrix} \begin{pmatrix} U^0 \\ U^1 \\ U^2 \\ U^3 \end{pmatrix} = \begin{pmatrix} -U^0 \\ U^1 \\ U^2 \\ U^3 \end{pmatrix}. \qquad (10.5)$$

From $U^{\bar{\alpha}}$ we reconstruct the corresponding three-velocity from

$$v^{\bar{i}} = \frac{dx^{\bar{i}}}{d\bar{t}} = \frac{d\tau}{d\bar{t}} \frac{dx^{\bar{i}}}{d\tau} = \frac{U^{\bar{i}}}{U^{\bar{0}}} = -v^i,$$

as we expected from simply reversing the direction of time. Thus the star appears different when the "film runs backwards" if the $U^i \neq 0$; however, if the spatial four-velocity components vanish, i.e. $U^i = 0$, the star appears the same under a time reversal.

10.7 Describe how to construct a static stellar model in the case that the equation of state has the form

$$p = p(\rho, S). \qquad \text{Schutz Eq. (10.24)} \qquad (10.6)$$

Show that we must give an additional arbitrary function, such as $S(r)$ or $S(m(r))$.

Solution: We go through the steps in Schutz §10.3 verifying that each one is valid for the more general equation of state eqn. (10.6) above. The conservation of energy equation, eqn. (7.20) or Schutz Eq. (10.26), still holds and the stress tensor, Schutz Eqs. (10.20)–(10.23) is unchanged, so we still have:

$$(\rho + p)\frac{d\Phi}{dr} = -\frac{dp}{dr}, \qquad \text{Schutz Eq. (10.27)} \qquad (10.7)$$

because it followed from the r component of eqn. (7.20). The Einstein tensor, eqn. (B.21) or Schutz Eqs. (10.14)–(10.17), is unchanged, so the Einstein equations remain the same. In particular the $(0,0)$ and (r,r) components still give:

$$\frac{dm(r)}{dr} = 4\pi r^2 \rho, \qquad \text{Schutz Eq. (10.30)}$$

$$\frac{d\Phi}{dr} = \frac{m(r) + 4\pi r^3 p}{r[r - 2m(r)]}. \qquad \text{Schutz Eq. (10.31)} \qquad (10.8)$$

With equation of state in eqn. (10.6), we now have four equations but five unknowns: $\rho, p, S, \Phi(r), m(r)$. So we need another equation to give S as a function of the independent variable r, or equivalently some function of r, like $S(m(r))$. Once that relation is known then the equation of state reduces to $p = p(\rho)$, a function implicitly including a known quantity $S(r)$, and we are left with four equations and four unknowns.

10.9 (a) Define a new radial coordinate, $\bar{r}$, in terms of the Schwarzschild r by the implicit equation:

$$r = \bar{r}\left(1 + \frac{M}{2\bar{r}}\right)^2.$$ Schutz Eq. (10.88)   (10.9)

Notice that as $r \to \infty, \bar{r} \to r$, while at the horizon $r = 2M$, we have $\bar{r} = M/2$. Show that the metric for spherical symmetry takes the form

$$ds^2 = -\left[\frac{1 - M/2\bar{r}}{1 + M/2\bar{r}}\right]^2 dt^2 + \left[1 + \frac{M}{2\bar{r}}\right]^4 [d\bar{r}^2 + \bar{r}^2 d\Omega^2].$$ Schutz Eq. (10.89)

(10.10)

Solution: We are required to transform the Schwarzschild metric, given in

$$ds^2 = -\left(1 - \frac{2M}{r}\right) dt^2 + \left(1 - \frac{2M}{r}\right)^{-1} dr^2 + r^2 d\Omega^2,$$ Schutz Eq. (10.36)

(10.11)

with $d\Omega^2 = d\theta^2 + \sin^2\theta \, d\phi^2$ (same as metric (ii) of Exercise 7.7), under the transformation given:

$$x^0 \equiv t = x^{\bar{0}}, \quad x^1 \equiv r = \bar{r}(1 + M/2\bar{r})^2, \quad x^2 \equiv \theta = x^{\bar{2}}, \quad x^3 \equiv \phi = x^{\bar{3}}.$$

So formally

$$g_{\bar{\alpha}\bar{\beta}} = g_{\alpha\beta} \frac{\partial x^\alpha}{\partial x^{\bar{\alpha}}} \frac{\partial x^\beta}{\partial x^{\bar{\beta}}}.$$

But since the Schwarzschild metric is diagonal, and only the r component has been transformed, we immediately see that only the g_{rr} component is transformed. For the temporal component,

$$g_{\bar{0}\bar{0}} = g_{\alpha\beta} \frac{\partial x^\alpha}{\partial x^{\bar{0}}} \frac{\partial x^\beta}{\partial x^{\bar{0}}} = g_{\alpha\beta} \delta^\alpha_{\bar{0}} \delta^\beta_{\bar{0}} = -\left(1 - \frac{2M}{r}\right) = -\frac{\left(1 - \frac{M}{2\bar{r}}\right)^2}{\left(1 + \frac{M}{2\bar{r}}\right)^2}.$$ (10.12)

The final step involved simple algebra after substituting r from eqn. (10.9). For the θ and ϕ components,

$$g_{\bar{2}\bar{2}} = g_{\alpha\beta} \frac{\partial x^\alpha}{\partial x^{\bar{2}}} \frac{\partial x^\beta}{\partial x^{\bar{2}}} = g_{\alpha\beta} \delta^\alpha_{\bar{2}} \delta^\beta_{\bar{2}} = g_{22} = r^2 = \bar{r}^2\left(1 + \frac{M}{2\bar{r}}\right)^4,$$

$$g_{\bar{3}\bar{3}} = g_{\alpha\beta} \frac{\partial x^\alpha}{\partial x^{\bar{3}}} \frac{\partial x^\beta}{\partial x^{\bar{3}}} = g_{\alpha\beta} \delta^\alpha_{\bar{3}} \delta^\beta_{\bar{3}} = g_{33} = r^2 \sin^2\theta = \bar{r}^2\left(1 + \frac{M}{2\bar{r}}\right)^4 \sin^2\theta.$$

(10.13)

For the r component we require the derivative

$$\frac{\partial r}{\partial \bar{r}} = \left(1 + \frac{M}{2\bar{r}}\right)^2 - \frac{M}{\bar{r}}\left(1 + \frac{M}{2\bar{r}}\right) = \left(1 + \frac{M}{2\bar{r}}\right)\left(1 - \frac{M}{2\bar{r}}\right),$$

which we use in the transformation:

$$g_{\bar{1}\bar{1}} = g_{\alpha\beta}\frac{\partial x^\alpha}{\partial x^{\bar{1}}}\frac{\partial x^\beta}{\partial x^{\bar{1}}} = g_{rr}\left(\frac{\partial r}{\partial \bar{r}}\right)^2 = \left(1 + \frac{M}{2\bar{r}}\right)^4.$$

In simplifying the algebra in the last step above it is useful to observe that $g_{rr} = 1/g_{tt}$ so you can reuse the expression for g_{tt} in eqn. (10.12).

10.9 (b) Define quasi-Cartesian coordinates by the usual equations $x = \bar{r}\cos\phi\sin\theta$, $y = \bar{r}\sin\phi\sin\theta$, and $z = \bar{r}\cos\theta$ so that (as in Exercise. 10.1), $d\bar{r}^2 + \bar{r}^2 d\Omega^2 = dx^2 + dy^2 + dz^2$. Thus, the metric has been converted into coordinates (x, y, z), which are called *isotropic coordinates*. Now take the limit as $\bar{r} \to \infty$ and show

$$ds^2 = -\left[1 - \frac{2M}{\bar{r}} + O\left(\frac{1}{\bar{r}^2}\right)\right]dt^2 + \left[1 + \frac{2M}{\bar{r}} + O\left(\frac{1}{\bar{r}^2}\right)\right](dx^2 + dy^2 + dz^2).$$

(10.14)

This proves Schutz Eq. (10.38) [which is eqn. (10.14) with "=" replaced by "≈", applicable when $\bar{r} \gg 1$].

Solution: In the limit $\bar{r} \to \infty$ the term $M/2\bar{r} \to 0$ so we can use the binomial series approximation, see Appendix B eqn. (B.2). The coefficient of dt^2 simplifies,

$$-\frac{\left(1 - \frac{M}{2\bar{r}}\right)^2}{\left(1 + \frac{M}{2\bar{r}}\right)^2} = -\left(1 - \frac{M}{2\bar{r}}\right)^2\left(1 + \frac{M}{2\bar{r}}\right)^{-2}$$

$$= -\left(1 - \frac{M}{\bar{r}}\right)\left(1 - \frac{M}{\bar{r}}\right) + O\left(\frac{1}{\bar{r}^2}\right) \qquad \text{used eqn. (B.2)}$$

$$= -\left(1 - \frac{2M}{\bar{r}}\right) + O\left(\frac{1}{\bar{r}^2}\right). \qquad (10.15)$$

The coefficient of $d\bar{r}^2$ simplifies immediately with the binomial series approximation:

$$\left(1 + \frac{M}{2\bar{r}}\right)^4 = \left(1 + \frac{2M}{\bar{r}}\right) + O\left(\frac{1}{\bar{r}^2}\right). \qquad \text{used eqn. (B.2)} \qquad (10.16)$$

In Exercise 10.1 we found

$$r^2(dr^2 + d\Omega^2) = (dx^2 + dy^2 + dz^2).$$

So just replace r by $\bar{r}$ and substitute this into Schutz Eq. (10.89), along with eqn. (10.15) and eqn. (10.16) to find eqn. (10.14), which agrees with Schutz Eq. (10.38) for $\bar{r} \gg 1$.

10.11 Derive the restrictions in,

$$p < p_*, \qquad \rho < 7p_*. \qquad \text{Schutz Eq. (10.57)} \qquad (10.17)$$

Solution: The local speed of sound c_s must be less than the speed of light to respect causality:

$$c_s = \left(\frac{dp}{d\rho}\right)^{1/2} < 1. \tag{10.18}$$

Recall that the restrictions in eqn. (10.17) apply to the Buchdahl equation of state,

$$\rho = 12(p_* p)^{1/2} - 5p. \qquad\qquad \text{Schutz Eq. (10.55)} \qquad (10.19)$$

So here eqn. (10.18) implies that

$$1 < \frac{d\rho}{dp} = \frac{d}{dp}\left(12\sqrt{p_* p} - 5p\right) = \frac{6\sqrt{p_*}}{\sqrt{p}} - 5, \quad \text{used eqn. (10.19)}$$

$$p < p_*, \qquad\qquad\qquad\qquad\qquad\qquad\qquad \text{simplified} \quad (10.20)$$

in agreement with eqn. (10.17). Now ρ is monotonically increasing with p because $\partial\rho/\partial p > 1$ for all p; that's what we just imposed via eqn. (10.20)! So the upper bound on ρ occurs at the upper bound on p, and the latter is simply $p = p_*$ by eqn. (10.20), so

$$\rho < \rho(p_*) = 7p_*,$$

as stated in eqn. (10.17).

10.13 Derive Schutz Eqs. (10.66) and (10.67):

$$z_s = \frac{1}{\sqrt{1-2\beta}} - 1, \qquad\qquad\qquad \text{Schutz Eq. (10.66)} \quad (10.21)$$

$$M = \frac{\pi\beta(1-\beta)}{(1-2\beta)A} = \left[\frac{\pi}{288 p_*(1-2\beta)}\right]^{1/2} \beta(1-\beta). \quad \text{Schutz Eq. (10.67)} \quad (10.22)$$

for the surface redshift z_s of a star with the Buchdahl equation of state, eqn. (10.19), with β the value of M/R at the surface and M the mass of the star.

Solution: The gravitational redshift $z = \nu_{em}/\nu_{rec} - 1$ at infinity of light emitted at some radial coordinate value r_{em} in the static, spherically symetric spacetime eqn. (10.53) with $g_{00} = -\exp(2\Phi)$ is given by

$$z = \exp(-\Phi(r_{em})) - 1. \qquad\qquad \text{Schutz Eq. (10.13)} \qquad (10.23)$$

For light emitted from the surface of a (non-rotating) star the Schwarzschild metric, eqn. (10.11), applies so we substitute $\exp(-\Phi(R))$ from

$$\exp(2\Phi) = \exp(-2\Lambda) = 1 - 2\beta, \qquad\qquad \text{Schutz Eq. (10.64)} \qquad (10.24)$$

into eqn. (10.23),

$$z = \exp(-\Phi(R)) - 1 = \frac{1}{\sqrt{1 - 2\beta}} - 1,$$

confirming eqn. (10.21). To derive eqn. (10.22) we must evaluate the integral

$$M = \int_0^R 4\pi r^2 \rho \, dr, \qquad \text{Schutz Eq. (10.41)} \qquad (10.25)$$

which is not straightforward. Fortunately Schutz gives us a useful change of variables defined by

$$r(r') := r' \frac{1 - \beta + u(r')}{1 - 2\beta}, \qquad \text{Schutz Eq. (10.59)} \qquad (10.26)$$

with

$$u(r') := \beta \frac{\sin(Ar')}{Ar'}, \qquad A^2 := \frac{288\pi p_*}{1 - 2\beta}, \qquad \text{Schutz Eq. (10.58)} \qquad (10.27)$$

with respect to which the density is given by

$$\rho(u(r')) = \frac{2A^2(1 - 2\beta)(1 - \beta - 3u/2)u}{8\pi(1 - \beta + u)^2}. \qquad \text{Schutz Eq. (10.63)} \qquad (10.28)$$

In the accompanying Maple™ worksheet, we evaluate eqn. (10.25) by integrating over dr':

$$M = \int_0^{\pi/A} 4\pi \, (r(r'))^2 \, \rho(u(r')) \frac{dr}{dr'} \, dr' = \frac{\pi\beta(1 - \beta)}{A(1 - 2\beta)},$$

which is the first equality in eqn. (10.22). The second equality follows immediately from substituting for A from eqn. (10.27).

10.15 Calculations of stellar structure more realistic than Buchdahl's solution must be done numerically. But

$$\frac{dp}{dr} = -\frac{(\rho + p)(m + 4\pi r^3 p)}{r(r - 2m)} \qquad \text{Schutz Eq. (10.39)} \qquad (10.29)$$

has a zero denominator at $r = 0$, so the numerical calculation must avoid this point. One approach is to find a power-series solution to

$$\frac{dm}{dr} = 4\pi r^2 \rho \qquad \text{Schutz Eqs. (10.30)} \qquad (10.30)$$

and eqn. (10.29) valid near $r = 0$, of the form

$$m(r) = \sum_j m_j r^j, \quad p(r) = \sum_j p_j r^j, \quad \rho(r) = \sum_j \rho_j r^j. \quad \text{Schutz Eq. (10.91)}$$

$$(10.31)$$

Assume that the equation of state $p = p(\rho)$ has the expansion near the central density ρ_c

$$p = p(\rho_c) + \frac{p_c\Gamma_c}{\rho_c}(\rho - \rho_c) + \cdots, \qquad \text{Schutz Eq. (10.92)} \qquad (10.32)$$

where Γ_c is the adiabatic index $\mathrm{d}(\ln p)/\mathrm{d}(\ln \rho)$ evaluated at ρ_c. Find the first two non-vanishing terms in each power series in eqn. (10.31), and estimate the largest radius r at which these terms give an error no larger than 0.1% in any power series. Numerical integrations may be started at such a radius using the power series to provide the initial values.

Solution: Substituting the power series into eqn. (10.30) we find:

$$\frac{\mathrm{d}m}{\mathrm{d}r} = m_1 + 2m_2 r + 3m_3 r^2 + 4m_4 r^3 + 5m_5 r^4 + \text{HOT}$$

$$= 4\pi r^2 \rho = 4\pi \left(\rho_0 r^2 + \rho_1 r^3 + \rho_2 r^4 + \text{HOT}\right),$$

where HOT stands for higher order terms. Equating like powers in r we conclude that

$$
\begin{array}{ccccc}
r^0 & r^1 & r^2 & r^3 & r^4 \\
m_1 = 0 & m_2 = 0 & 3m_3 = 4\pi\rho_0 & 4m_4 = 4\pi\rho_1 & 5m_5 = 4\pi\rho_2.
\end{array}
\qquad (10.33)
$$

For instance, there is no constant term in the second line of the ODE above so nothing to balance the m_1 term in the first line, hence $m_1 = 0$. Similarly, there is no term linear in r in the second line, so the $2m_2 r = 0$, and implying $m_2 = 0$, etc. (If this is mysterious, it is easy to find a nice introduction to power series solution of ODEs (e.g. Boas, 1983, §12.1) or (e.g. Riley et al., 2006, Chapter 16).) Henceforth we drop m_1 and m_2.

Substituting the power series into eqn. (10.29) is a bit messy. Write eqn. (10.29) as

$$-(r^2 - 2mr)\frac{\mathrm{d}p}{\mathrm{d}r} = (\rho + p)(m + 4\pi r^3 p), \qquad (10.34)$$

and we'll deal with the LHS and RHS separately. First the LHS gives:

$$-(r^2 - 2mr)\frac{\mathrm{d}p}{\mathrm{d}r} = -(r^2 - 2m_0 r - 2m_3 r^4 + \cdots)$$

$$(p_1 + 2p_2 r + 3p_3 r^2 + 4p_4 r^3 + \cdots) + \text{HOT}$$

$$= 2m_0 p_1 r - (p_1 - 4m_0 p_2)r^2 - (2p_2 - 6m_0 p_3)r^3$$

$$- (-2m_3 p_1 + 3p_3 - 8m_0 p_4)r^4 + \text{HOT}, \quad (10.35)$$

The RHS of eqn. (10.34) gives

$$\text{RHS} = (\rho + p)(m + 4\pi r^3 p)$$

$$= \left(\rho_0 + p_0 + (\rho_1 + p_1)r + (\rho_2 + p_2)r^2 + \cdots\right)\left(m_0 + (m_3 + 4\pi p_0)r^3 + (m_4 + 4\pi p_1)r^4\right)$$

$$= (\rho_0 + p_0)m_0 + (\rho_1 + p_1)m_0 r + (\rho_2 + p_2)m_0 r^2 + [(\rho_0 + p_0)(m_3 + 4\pi p_0) + (\rho_3 + p_3)m_0]r^3$$

$$+ [(\rho_0 + p_0)(m_4 + 4\pi p_1) + (\rho_1 + p_1)(m_3 + 4\pi p_0) + (\rho_4 + p_4)m_0]r^4 + \text{HOT}. \quad (10.36)$$

Equating eqns. (10.35) and (10.36), we immediately conclude from terms proportional to r^0 that $(\rho_0 + p_0)m_0 = 0$. But since $\rho_0 = \rho_c$ and $p_0 = p_c$, the central density and pressure respectively (clearly non-zero quantities!), we conclude that $m_0 = 0$. Furthermore, from each power of r in turn we learn that:

$$\begin{matrix} r^1 & r^2 & r^3 & r^4 \end{matrix}$$
$$(p_1 - \rho_1) = 0, \quad p_1 = 0, \quad 2p_2 = -(\rho_0 + p_0)(m_3 + 4\pi p_0), \quad 3p_3 = -m_4(\rho_0 + p_0). \tag{10.37}$$

Combining results from eqns. (10.33) and (10.37) and using the EOS in eqn. (10.32) we find that the first two non-vanishing terms of all three series can be written in terms of the central density $\rho_c = \rho_0$. From the EOS we have $p_c = p(\rho_c) = p_0$. From eqn. (10.37) and m_3 from eqn. (10.33) we have

$$p_2 = -\frac{1}{2}(\rho_c + p_c)(m_3 + 4\pi p_c) = -\frac{2}{3}\pi(\rho_c + p_c)(\rho_c + 3p_c). \tag{10.38}$$

Differentiate ρ with respect to r twice to relate p_2 and ρ_2:

$$\begin{aligned} 2\rho_2 &= \left.\frac{\partial^2 \rho}{\partial r^2}\right|_{r=0} = \left.\frac{\partial}{\partial r}\left(\frac{\partial \rho}{\partial p}\frac{\partial p}{\partial r}\right)\right|_{r=0} \\ &= \left.\left(\frac{\partial \rho}{\partial p}\right)\left(\frac{\partial^2 p}{\partial r^2}\right)\right|_{r=0} && \text{because } p_1 = 0 \\ &= \frac{1}{c_s^2}2p_2 && c_s^2 = \frac{\partial p}{\partial \rho}, \text{ see eqn. (10.18)} \\ \rho_2 &= \frac{\rho_c}{\Gamma_c p_c}p_2. && \text{used eqn. (10.32)} \end{aligned} \tag{10.39}$$

Finally to construct the power series for m we substitute these results eqns. (10.33), (10.37), (10.38), (10.39) into the power series for $m(r)$ to find

$$\begin{aligned} m(r) &= \frac{4}{3}\pi\rho_c r^3 + \frac{4}{5}\pi\rho_2 r^5 + \text{HOT} \\ &= \frac{4}{3}\pi\rho_c r^3 - \frac{8}{15}\pi^2\frac{\rho_c}{\Gamma_c p_c}(\rho_c + p_c)(\rho_c + 3p_c)r^5 + \text{HOT}. \end{aligned} \tag{10.40}$$

And for the other two series:

$$p(r) = p_c - \frac{2}{3}\pi(\rho_c + p_c)(\rho_c + 3p_c)r^2 + \text{HOT},$$

$$\rho(r) = \rho_c - \frac{\rho_c}{\Gamma_c p_c}\frac{2}{3}\pi(\rho_c + p_c)(\rho_c + 3p_c)r^2 + \text{HOT}. \tag{10.41}$$

To find the maximum value of the radial coordinate r_{max} such that the error is less than about $1/10\%$ we estimate the error term as about the square of the second term in the series. We choose this because we note that $p_3 = 0$, again from eqns. (10.33) and (10.37). (But going to r^5 we find $p_4 \neq 0$.) Assuming a constant c_s near ρ_c one finds ρ_3 vanishes too. So our error term is $O(r^4)$. To express the error as a percentage we divide it by the first term. For instance the ρ power series suggests we want

$$\left|\frac{\rho_2 r^2}{\rho_c}\right| < \sqrt{0.001},$$

which gives

$$r_{max}^2 = \left| \sqrt{0.001} \frac{\rho_c}{\rho_2} \right|.$$

Considering the m power series gives a very slightly less restrictive range. The p power series gives

$$r_{max}^2 = \left| \sqrt{0.001} \frac{\rho_c}{\rho_2 \Gamma_c} \right|,$$

which, depending upon Γ_c might be more restrictive. Obviously the most restrictive criterion should be used.

10.17 (This problem requires access to a computer.) Numerically construct a sequence of stellar models using the equation of state:

$$p = \begin{cases} k\rho^{4/3}, & \rho \leq \rho_* \\ \frac{1}{3}\rho, & \rho \geq \rho_* \end{cases},$$ (10.42)

where $\rho_* = 1/(27k^3)$, where k was defined in Schutz Eq. (10.81):

$$k = \frac{2\pi}{3h^3} \left(\frac{3h^3}{8\pi \mu m_p} \right)^{4/3},$$

where h is Planck's constant, μ is the ratio of the number of nucleons to electrons (1 or 2), and m_p is the mass of the proton. This is a crude approximation to a realistic "stiff" neutron-star equation of state. Construct the sequence by using the following values for ρ_c : $\rho_c/\rho_* = 0.1, 0.8, 1.2, 2, 5, 10$. Use the power series developed in Exercise 10.15 to start the integration. Does the sequence seem to approach a limiting mass, a limiting value of M/R, or a limiting value of the central redshift?

Solution: The complete solution is in the Maple™ worksheet. You will learn about Maple's powerful differential equation solver, dsolve. We numerically integrated eqns. (10.29) and (10.30) using boundary conditions for small r from the power series eqns. (10.40) and (10.41).

The mass function $m(r)$ is plotted vs. r in Fig. 10.1. For the two small values of the central density $\rho_c = 0.1\rho_*, 0.8\rho_*$ the mass function $m(r)$ appears to be constant, however, close inspection reveals that it increases very slowly with r. But for the central density $\rho_c > \rho_*$, the shape of $m(r)$ changes qualitatively; it increases steadily with r at small r then develops a sharp kink with an apparently constant value of m at larger r. Call the apparently constant value $m(r) = M$. With increasing ρ_c/ρ_*, M at first decreases but then appears to increase to a "limiting mass" of about $M \approx 354$ (≈ 85) m for $\mu = 1(= 2)$. And increasing the central density above $\rho_c \approx 5\rho_*$

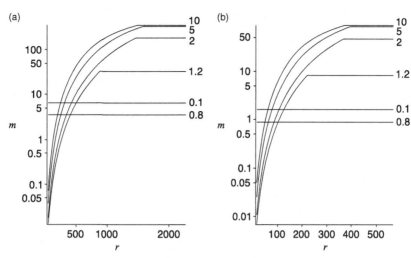

Figure 10.1   The mass function $m(r)$ in meters vs. radial coordinate r in meters. (a) $\mu = 1$. (b) $\mu = 2$. The six lines correspond to $\rho_c/\rho_* = (0.1, 0.8, 1.2, 2, 5, 10)$, with the value of ρ_c/ρ_* to the right of curve. Note that with increasing ρ_c/ρ_* the limiting mass increases except for the two lowest ρ_c/ρ_* which are reversed. This plot was made with the Maple^{tm} worksheet.

increases the limiting value of M by only a few percent, barely distinguishable in Fig. 10.1.

The radial coordinate r at the kink in $m(r)$ vs. r gives a convenient indicator of the boundary of the star, $r = R$. One can roughly estimate the r location of the kink in Fig. 10.1, but it is much easier if we take the second derivative $m'' = \mathrm{d}^2 m/\mathrm{d}r^2$. This is plotted in Fig. 10.2 for $\mu = 1$. Note we used only the four $\rho_c/\rho_* > 1$ and added two additional lines with even larger values of ρ_c/ρ_*.

In the Maple^{TM} worksheet of course we focused the horizontal axis near each value of R separately to obtain estimates of R to within about ± 2m. The corresponding sequence of M, R, and M/R are listed in Table 10.1. It's not clear from the sequence of ρ_c/ρ_* given in the exercise, or from the extended sequence in Table 10.1, if limiting values are reached. However, it is clear that "turnaround" values of M and M/R are reached indicating local maxima with respect to ρ_c/ρ_*.

The gravitational redshift of light emitted from the surface of the star, $r = R$, and received far from it, is a function only of M/R:

$$z = \exp(-\Phi(R)) - 1 \qquad\qquad \text{used eqn. (10.23)}$$

$$= \frac{1}{\sqrt{1 - \frac{2M}{R}}} - 1. \qquad\qquad \text{used Schutz Eq. (10.35)} \qquad (10.43)$$

10.19   Our Sun has an equatorial rotation velocity of about 2 km s^{-1}.

     (a) Estimate its angular momentum, on the assumption that the rotation is rigid (uniform angular velocity) and the Sun is of uniform density. As the true angular

Table 10.1 Star mass M, radial coordinate R, and ratio M/R for various ρ_c/ρ_* with R taken from Fig. 10.2 and M taken from Fig. 10.1

ρ_c/ρ_*	R	M	M/R
1.2	880	33.0	0.037
2	1465	184.7	0.16
5	1585	332.9	0.21
10	1490	353.8	0.24
20	1376	338.5	0.25
100	1198	275.8	0.23

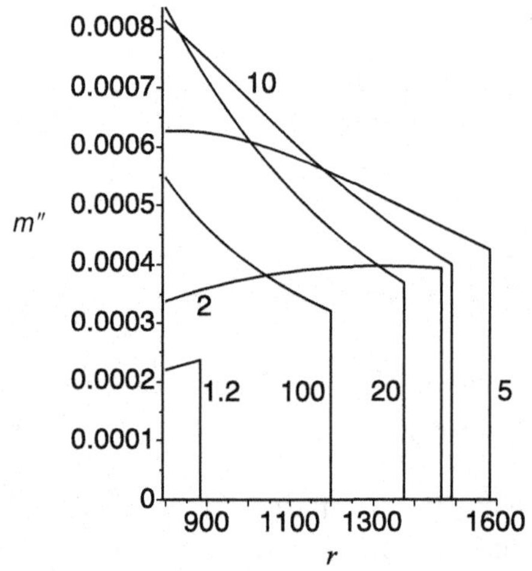

Figure 10.2 The second derivative of the mass function $m''(r)$ vs. radial coordinate r in meters with $\mu = 1$. Where the curve plunges toward zero helps identify the maximum radial coordinate of the star interior, $r = R$. The six lines correspond to $\rho_c/\rho_* = (1.2, 2, 5, 10, 20, 100)$, with the value of ρ_c/ρ_* next to the curve. Note the maximum R occurs for intermediate $\rho_c/\rho_* = 5$. This plot was made with the Maple^{tm} worksheet.

velocity is likely to increase inwards, this is a lower limit on the Sun's angular momentum.

Solution: The angular momentum of an element of mass dm with velocity v about the axis of rotation and at radius r from the axis of rotation is, by definition, $dL = rvdm = r^2\omega dm$. We simply integrate this over the sphere. Picture the sphere as composed of concentric pipes of wall thickness dr and height $2R\cos(\theta)$ so we can write:

$$dL = r^2\omega dm = r^2\omega\rho(2\pi r)(2R\cos(\theta))dr. \qquad (10.44)$$

Make the substitution $r = \sin\theta\, R$, so $dr = R\cos\theta\, d\theta$, and assuming constant density ρ we obtain an integral over polar angle:

$$L = \int dL = \int_0^{\pi/2} \rho 4\pi\omega R^5 \sin^3\theta\,\cos^2\theta\, d\theta$$

$$= \frac{8\pi}{15}\omega R^5 \rho \qquad\qquad \text{see Maple}^{\text{TM}} \text{ worksheet}$$

$$= 1.11 \times 10^{42} \text{ kg m}^2\,\text{s}^{-1}. \tag{10.45}$$

Note that we can define the moment of inertia I via $L = I\omega$, so the moment of inertia for the uniform solid ball of radius R is

$$I = \frac{8\pi}{15}R^5\rho = \frac{2}{15}R^2 M. \tag{10.46}$$

10.19 (b) If the Sun were to collapse to neutron-star size (say 10 km radius), conserving both mass and total angular momentum, what would its angular velocity of rigid rotation be? In non-relativistic language, would the corresponding centrifugal force exceed the Newtonian gravitational force on the equator?

Solution: Using conservation of mass, $M = $ constant, and of angular momentum, $L = $ constant, from eqn. (10.46) we have

$$\omega_{\text{new}} = \omega_{\text{sun}}\left(\frac{R_{\text{sun}}^2}{R_{\text{new}}^2}\right) \approx 1.4 \times 10^4 \text{rad s}^{-1}.$$

The centrifugal force, A_c per unit mass at the equator of the Sun would be:

$$A_c = \omega_{\text{new}}^2 R_{\text{new}} \approx 1.9 \times 10^{12}\text{m s}^{-2}.$$

This actually exceeds the gravitational constant at the surface,

$$g_{\text{new}} = \frac{GM}{R_{\text{new}}^2} \approx 1.3 \times 10^{12}\text{m s}^{-2}.$$

Computations were performed in the Maple^{TM} worksheet.

10.19 (c) A neutron star of 1 M$_\odot$ and radius 10 km rotates 30 times per second (typical of young pulsars). Again in Newtonian language, what is the ratio of centrifugal to gravitational force on the equator? In this sense the star is slowly rotating.

Solution: The ratio of centrigual force to gravitational constant at the surface, A_c/g at the equator of the your pulsar is:

$$\frac{A_c}{g} = \frac{\omega^2 R}{GM/R^2} = 2.7 \times 10^{-4}.$$

Computations were performed in the Maple™ worksheet.

10.19 (d) Suppose a main-sequence star of 1 $M_\odot$ has a dipole magnetic field with typical strength 1 gauss in the equatorial plane. Assuming flux conservation in this plane, what field strength should we expect if the star collapses to radius of 10 km? (The Crab pulsar's field is of the order of 10^{11} gauss.)

Solution: The Sun is a main-sequence star, so we can assume properties similar to the Sun. In particular, assume a solar radius of $R \approx 1.392 \times 10^9$ m. Conservation of flux in the equatorial plane gives,

$$B = \frac{R_{\text{sun}}^2}{(1 \times 10^4)^2} 1 \text{ gauss} \approx 5 \times 10^9 \text{ gauss}.$$

10.2 Supplementary problems

SP 10.1   Derive the equation for the $-g_{00}$ component of the metric in the interior region of a spherically symmetric star of constant density,

$$\exp(\Phi) = \frac{3}{2}(1 - 2M/R)^{1/2} - \frac{1}{2}(1 - 2Mr^2/R^3)^{1/2}, \quad r \le R, \quad \text{Schutz Eq. (10.54)}$$

(10.47)

from eqn. (10.7), i.e. the r component of the energy conservation law $T^{\alpha\beta}{}_{;\beta}$.

Hint

Use the T-O-V equation, Schutz Eq. (10.48), to eliminate dp/dr, and use the integral of the T-O-V equation from a central pressure of p_c, Schutz Eq. (10.49), to write p as a function of known quantities. Then eqn. (10.7) has the form

$$d\Phi = f(r)dr,$$

where $f(r)$ is a function of r and known parameters and can be integrated from the known (Schwarzschild) boundary condition at $r = R$.

SP 10.2   From the coordinate transformation $(t, r, \theta, \phi) \to (-t, r, \theta, \phi)$, derive the coordinate transformation of the metric tensor:

$$g_{\bar{0}\bar{0}} = \left(\Lambda^0{}_{\bar{0}}\right)^2 g_{00}$$

$$g_{\bar{0}\bar{r}} = \left(\Lambda^0{}_{\bar{0}}\, \Lambda^r{}_{\bar{r}}\right) g_{0r}$$

$$g_{\bar{r}\bar{r}} = \left(\Lambda^r{}_{\bar{r}}\right)^2 g_{rr}. \qquad\qquad \text{Schutz Eq. (10.6)} \qquad\qquad (10.48)$$

SP 10.3 (a) Fill in the missing steps in arriving at eqn. (10.2)

$$E^* = -\vec{U} \cdot \vec{p} = e^{-\Phi} E.$$

for the energy measured by a local inertial observer in the static spherically symmetric spacetime with metric eqn. (10.53), with $E = -p_0$ the energy at infinity.

Solution

A local observer can always construct a LIF in which she has four-velocity $\vec{U}$. From the principle of equivalence the local laws of physics are indistinguishable from SR and eqn. (2.27) for the observed energy of a particle with four-momentum $\vec{p}$ applies, giving

$$
\begin{aligned}
E^* &= -\vec{U} \cdot \vec{p} & \text{eqn. (2.27)} \\
&= -U^\alpha p^\beta g_{\alpha\beta} & \text{transform to curved spacetime coordinates} \\
&= -U^0 p^\beta g_{0\beta} = -U^0 p_0 & \text{observer stationary in coordinates of eqn. (10.53)} \\
&= -e^{-\Phi}(-E) = e^{-\Phi} E. & \text{used eqn. (7.67) and eqn. (10.53)}
\end{aligned}
$$

$$(10.49)$$

SP 10.3 (b) The above solution used,

$$E = -\vec{p} \cdot \vec{U}_{\text{obs}} \qquad\qquad \text{Schutz Eq. (2.35)} \qquad\qquad (10.50)$$

for the energy of a particle measured in a reference frame moving with four-velocity U_{obs}. This frame-invariant expression was derived in Schutz §2.6 assuming a particle of mass m, but for eqn. (12.25) we will be using it for a photon, with $m = 0$ and undefined four-velocity. Is eqn. (10.50) still valid for a photon?

Solution

Going through the derivation of eqn. (10.50) assuming a photon, we find the derivation still holds. All that is needed for the result eqn. (10.50) is that the particle have a four-momentum $\vec{p}$ with time component equal to the energy, $p^0 = E$. As argued by Schutz in §2.7, this is indeed the case. Thus

$$-\vec{p} \cdot \vec{U}_{\text{obs}} = -p^\alpha U_\alpha \underset{\text{MCLIRF}}{=} p^0, \qquad\qquad (10.51)$$

where we have coined the acronym MCLIRF for Momentarily Co-moving Local Inertial Reference Frame. Note that it is important that the frame be inertial for otherwise we do not necessarily have $U_0 = -1$ and $U_i = 0$ even when $U^0 = 1$ and $U^i = 0$ nor is p^0 necessarily the energy in a non-inertial frame.

SP 10.3   (c) Explain why in the paragraph after Schutz Eq. (10.11), when considering the photon emitted at radial coordinate r_1, the conserved constant was

$$E = h\nu_{\text{em}} e^{\Phi(r_1)},$$

where h was Planck's constant, ν_{em} was the emitted frequency. Why is there no factor of 2 in front of $\Phi(r_1)$ in the exponential term? In particular, what is wrong with arguing that the conserved constant is p_0, because g_{00} is time independent, and therefore

$$p_0 = p^\alpha g_{\alpha 0} = p^0 g_{00} = -h\nu_{\text{em}} e^{2\Phi(r_1)}? \tag{10.52}$$

Solution

The solution in eqn. (10.52) above makes the mistake of confusing fixed points in the standard Schwarzschild spacetime coordinates as an inertial frame, and inappropriately used the results of SR. For p^0 was the energy of a particle in SR (Minkowski spacetime in inertial coordinate systems) but this is not true in the general GR setting of curved spacetime with arbitrary coordinate systems.

Instead, the photon emitted by gas that is stationary at $r = r_1$ in standard Schwarzschild spacetime coordinates with frequency ν_{em} has *locally measured energy* $E^* = h\nu_{\text{em}}$. But E^* is related to energy at infinity, $E \equiv -p_0$, via eqn. (10.2). In Schwarzschild coordinates E^* is not simply p^0 nor p_0. One must transform the four-momentum vector to a MCLIRF before equating p^0 with the locally measured energy. Eqn. (10.2) offers a shortcut to this result.

SP 10.4   Generalize eqn. (10.23) for the gravitational redshift, $z = \exp(-\Phi(r_1)) - 1$, for a photon emitted at radial coordinate r_1 and received far away in a static spherically symmetric spacetime with line element eqn. (10.53) to the case where the photon is received by a stationary observer at r_2 where the spacetime is not flat.

SP 10.5   Use Maple™ (or any other symbolic mathematics software you prefer to use) to find the Einstein tensor for the metric given by

$$ds^2 = -e^{2\Phi}\,dt^2 + e^{2\Lambda}\,dr^2 + r^2\,d\Omega^2. \qquad \text{Schutz Eq. (10.7)} \tag{10.53}$$

Maple's tensor package uses a different convention resulting in a sign difference. Note that this is an exercise in using Maple™. One can easily find the Einstein tensor analytically, see Schutz Exercise 10.4.

Solution

See Maple™ worksheet. You should multiply your answer from Maple™ by (-1).
You can confirm your answer with eqn. (B.21), Schutz Eqs. (10.14)–(10.17).

SP 10.6   Recall that in deriving the boundary condition on $m(r)$ at $r = 0$ it was argued that
$m(r)$ goes to zero faster than r as $r \to 0$. Use the T-O-V equation, see eqn. (10.29), to
argue for a stricter requirement on the tendency of $m(r)$ so that the pressure gradient is not
singular at the origin $r = 0$.

SP 10.7   Where does

$$|g|^{1/2} d^3 x = \exp(\Lambda) r^2 \sin(\theta)\, dr\, d\theta\, d\phi \qquad \text{Schutz Eq. (10.42)} \qquad (10.54)$$

come from?

Solution

This is the expression for proper volume. It was introduced in Schutz §6.2; see
eqn. (6.102) and discussion in SP7.7. The expression $|g|$ is the determinant of the
matrix of the spatial metric,

$$|g| = \det \begin{pmatrix} \exp(2\Lambda) & 0 & 0 \\ 0 & r^2 & 0 \\ 0 & 0 & r^2 \sin^2\theta \end{pmatrix} = \exp(2\Lambda)\, r^4\, \sin^2\theta.$$

SP 10.8   How does one conclude from eqn. (10.24),

$$\exp(2\Phi) = \exp(-2\Lambda) = 1 - 2\beta,$$

that $\beta = M/R$ on the surface?

SP 10.9   Reconsider Exercise 10.15. The general method for a power series solution is
to substitute the power series into the ODEs and match terms of equal power. But we
can immediately eliminate some of the terms from the summation with some preliminary
analysis based on physical arguments. In particular, argue that $m_0 = m_1 = m_2 = 0$.
Furthermore, argue that $\rho_1 = p_1 = 0$.

Solution

Consider

$$m(r) = \sum_j m_j r^j = m_0 + m_1 r + m_2 r^2 + m_3 r^3 + m_4 r^4 + \cdots \qquad (10.55)$$

Substitute $r = 0$ to obtain $m(r = 0) = m_0$. But from eqn. (10.25) we can immediately say $m(r = 0) = 0$ so $m_0 = 0$. Next we recognize that this power series is identical to the Taylor series expansion of $m(r)$ about $r = 0$, so

$$m_1 = \frac{dm}{dr}\bigg|_{r=0}, \qquad m_2 = \frac{1}{2}\frac{d^2m}{dr^2}\bigg|_{r=0}, \qquad \text{etc.}$$

(Alternatively, just differentiate $m(r)$ with respect to r, and set $r = 0$ to isolate a single term.) Using eqn. (10.30) we immediately find

$$m_1 = \frac{dm}{dr}\bigg|_{r=0} = 0.$$

Differentiate eqn. (10.30) with respect to r to conclude

$$m_2 = \frac{1}{2}\frac{d^2m}{dr^2}\bigg|_{r=0} = \frac{1}{2}\left(8\pi r\rho + 4\pi r^2\frac{d\rho}{dr}\right)\bigg|_{r=0} = 0.$$

Differentiate eqn. (10.30) with respect to r twice to conclude

$$m_3 = \frac{1}{6}\frac{d^3m}{dr^3}\bigg|_{r=0} = \frac{1}{6}\left(8\pi\rho + 16\pi r\frac{d\rho}{dr} + 4\pi r^2\frac{d^2\rho}{dr^2}\right)\bigg|_{r=0} = \frac{4}{3}\pi\rho_c.$$

We can go even further to conclude that $m_4 = 0$ by differentiating eqn. (10.30) with respect to r three times:

$$m_4 = \frac{1}{24}\frac{d^4m}{dr^4}\bigg|_{r=0} = \frac{1}{24}\left(24\pi\frac{d\rho}{dr} + 24\pi r\frac{d^2\rho}{dr^2} + 4\pi r^2\frac{d^3\rho}{dr^3}\right)\bigg|_{r=0} = 0.$$

Here we assumed that $\rho(r)$ was a smooth function so that its second derivative at the origin $r = 0$ must be finite. This implies that $d\rho/dr = 0$ at $r = 0$ (otherwise there is a kink at the origin). Differentiating once more we find

$$m_5 = \frac{1}{120}\frac{d^5m}{dr^5}\bigg|_{r=0} = \frac{1}{120}\left(48\pi\frac{d^2\rho}{dr^2} + 32\pi r\frac{d^3\rho}{dr^3} + 4\pi r^2\frac{d^4\rho}{dr^4}\right)\bigg|_{r=0}$$
$$= \frac{6}{15}\pi\frac{d^2\rho}{dr^2}\bigg|_{r=0}.$$

For the same reason that $dm/dr\,|_{r=0} = 0$, we can argue that $dp/dr\,|_{r=0} = 0 = d\rho/dr\,|_{r=0}$.

SP 10.10 Convert two of the equations crucial for constructing the stellar model, the $(0,0)$ component of the Einstein equations, eqn. (10.30), and the T-O-V equation (10.29), from geometrized units to SI units.

SP 10.11 Recall the EOS given in eqn. (10.42). What are the dimensions of k? Express k in SI and geometric units. Use this EOS to find Γ, the adiabatic index defined in Exercise 10.15.

Solution

The dimensions of k are $[energy \times time \times mass^{-4/3}]$. We can obtain k in SI or geometric units by starting with h and m_p in the appropriate units. But suppose we had k in SI units. We multiply by $G^{-1/3}c^{-1/3}$, all in SI units, to obtain k in geometric units. To see this consider:

$$k[\text{J s kg}^{-4/3}] = k[\text{J s kg}^{-1}\text{kg}^{-1/3}].$$

Multiply on top by c and on bottom by c^2:

$$k[\text{J s kg}^{-1}\text{kg}^{-1/3}]\frac{c[\text{m/s}]}{c^2[\text{m}^2\text{s}^{-2}]} = \frac{k}{c}\frac{[\text{m}]}{[\text{kg}^{1/3}]}.$$

Now use eqn. (8.8) to convert [kg] to [m]:

$$\frac{k}{c}\frac{[\text{m}]}{[\text{kg}^{1/3}]}\left(\frac{c^2}{G}\right)^{1/3}\frac{[\text{kg}^{1/3}]}{[\text{m}^{1/3}]} = \frac{k}{c^{1/3}G^{1/3}}[\text{m}^{2/3}].$$

The adiabatic index was defined to be

$$\Gamma = \frac{d\ln(p)}{d\ln(\rho)} = \frac{\rho dp}{p d\rho} = \begin{cases} \frac{1}{3}\frac{\rho}{\rho/3} = 1, & \text{if } \rho > \rho_*, \\ \frac{\rho}{k\rho^{4/3}}k(4/3)\rho^{1/3} = 4/3, & \text{if } \rho < \rho_*. \end{cases} \tag{10.56}$$

SP 10.12 An experimentalist wishes to perform an experiment to test the role of the Ricci scalar on particle flux divergence; see Exercise 7.1(iii). Frustrated by the inability of a centrifuge or an electromagnetic field to produce a non-zero Ricci scalar, she reasons that she needs a strong matter density. She turns to her office mate, an astrophysicist, and asks: "What's the highest density object you know of?" The astrophysicist replies that a neutron star has densities around 10^{16} kg m^{-3}. However, her excitement wanes immediately when she looks up

$$p = \frac{1}{3}\rho \qquad\qquad \text{Schutz Eq. (10.87)} \qquad (10.57)$$

for the equation of state of a neutron star. Explain her disappointment.

Hint

Treat the neutron star as a perfect fluid. See also SP7.12, SP8.6, and SP12.8.

SP 10.13 In Newtonian gravity, a well-known result is that matter in the form of a spherical shell exerts no gravitational pull on objects inside the shell. Show that an analogous result hold in GR. That is, show that the spacetime metric for a vacuum within a static spherical shell of matter is that of Minkowski spacetime.

Solution

Because of the spherical symmetry and static matter field, the metric must be static and spherically symmetric and can be written with Schwarzschild coordinates in the form eqn. (10.53). The Einstein tensor has only four non-zero components, eqn. (B.21), the first two of which are:

$$G_{tt} = \frac{1}{r^2} \exp(2\Phi) \frac{\mathrm{d}}{\mathrm{d}r} [r(1 - \exp(-2\Lambda)], \qquad \text{Schutz Eq. (10.14)} \qquad (10.58)$$

$$G_{rr} = -\frac{1}{r^2} \exp(2\Lambda)(1 - \exp(-2\Lambda) + \frac{2}{r}\Phi'. \qquad \text{Schutz Eq. (10.15)} \qquad (10.59)$$

The stress–energy tensor $T_{\mu\nu}$ inside the vacuum vanishes. So the Einstein equations become $G_{\mu\nu} = 0$. Setting eqn. (10.58) to zero we find:

$$G_{tt} = 0 = \frac{\mathrm{d}}{\mathrm{d}r} [r(1 - \exp(-2\Lambda)],$$

which integrates to

$$r(1 - \exp(-2\Lambda)) = -C_1,$$

$$\exp(2\Lambda) = \frac{r}{r + C_1}. \qquad (10.60)$$

Setting eqn. (10.59) to zero we find:

$$G_{rr} = 0 = -\frac{1}{r^2} \exp(2\Lambda)(1 - \exp(-2\Lambda) + \frac{2}{r}\Phi'$$

$$\Phi' = \frac{1}{2r^2} \exp(2\Lambda)C_1 \qquad \text{used eqn. (10.60), rearranged}$$

$$\Phi(r) = \frac{C_1}{2} \int \frac{1}{r^2 + rC_1} \mathrm{d}r$$

$$= \frac{1}{2} \ln \left(\frac{r}{r + C_1} \right) + C_2$$

$$\exp(2\Phi(r)) = C_3 \left(\frac{r}{C_1 + r} \right). \qquad (10.61)$$

Now we apply the boundary condition at $r = 0$ that we seek a non-degenerate metric. Eqn. (10.60) demands that $C_1 = 0$. This implies that $g_{tt} = -C_3$ in eqn. (10.60). Furthermore, we can absorb the C_3 into the definition of t, and obtain the Minkowski metric in spherical coordinates:

$$\mathrm{d}s^2 = -\mathrm{d}t^2 + \mathrm{d}r^2 + r^2(\mathrm{d}\theta^2 + \sin^2\theta \mathrm{d}\phi^2). \qquad (10.62)$$

A careful study of the [Schwarzschild geometry's] timelike and null geodesics – the paths of freely moving particles and photons – is the key to understanding the physical importance of this metric.

Bernard Schutz, §11.1

Everything inside $r = 2M$ is trapped and, moreover, doomed to encounter the singularity at $r = 0$, since $r = 0$ is in the future of every timelike and null world line inside $r = 2M$. Once a particle crosses the surface $r = 2M$, it cannot be seen by an external observer, since to be seen means to send out a photon [that] reaches the external observer. This surface is therefore called a *horizon* ...

Bernard Schutz, §11.2

11.1 Exercises

11.1 Consider a particle or photon in an orbit in the Schwarzschild metric with a certain E and L, at a radius $r \gg M$. Show that if spacetime were really *flat*, the particle would travel on a straight line which would pass a distance $b := L/[E^2 - m^2]^{1/2}$ from the center of coordinates $r = 0$. This ratio b is called the *impact parameter*. Show also that photon orbits that follow from

$$\left(\frac{dr}{d\lambda}\right)^2 = E^2 - \left(1 - \frac{2M}{r}\right)\frac{L^2}{r^2} \qquad \text{Schutz Eq. (11.12)} \qquad (11.1)$$

depend *only* on b.

Solution: If we rotate the coordinate system such that $\theta = \pi/2$ then $p^\theta = 0$ always. For a particle or a photon $p^\alpha p_\alpha = -m^2$ where m is the rest mass, which is nil for a photon (recall Schutz Eqs. (2.33) and (2.40)). Imagine the particle has mass. (The analysis is the same for a photon, but instead of proper time τ and four-velocity $\vec{U}$ we should use a general affine parameter and four-momentum $\vec{p}$.) At the point where r is a minimum

$$p^r = mU^r = m\frac{dr}{d\tau} = 0$$

because it's an extremum of the particle path. The Schwarzschild metric is diagonal so that these results also imply $p_\theta = 0$ everywhere and $p_r = 0$ at $r = b$. So we have only two non-zero components of the momentum, p_t and p_ϕ:

$$-m^2 = p^\alpha p_\alpha = p^t p_t + p^\phi p_\phi \qquad\qquad\qquad p^r = 0, p^\theta = 0$$
$$= g^{t\alpha} p_\alpha p_t + g^{\phi\alpha} p_\alpha p_\phi$$
$$= g^{tt} p_t p_t + g^{\phi\phi} p_\phi p_\phi \qquad\qquad\qquad \text{diagonal metric}$$
$$= -\left(1 - \frac{2M}{r}\right)^{-1} E^2 + r^{-2}L^2 \quad \text{used defs eqns. (11.51) and (11.52)}$$
$$= -E^2 + r^{-2}L^2 \qquad\qquad\qquad \text{flat space approximation}$$
$$= -E^2 + b^{-2}L^2. \qquad\qquad\qquad r = b \quad (11.2)$$

Solve for b and choose the positive root since b is the spatial distance:

$$b = \frac{L}{\sqrt{E^2 - m^2}}, \qquad\qquad\qquad (11.3)$$

as we were required to show.

For a photon $m = 0$ so the expression for impact parameter eqn. (11.3) reduces to $b = L/E$. Dividing the basic equation for the photon orbit eqn. (11.1) by E^2 the RHS becomes a function of $L/E = b$. And we can absorb the E on the LHS into the affine parameter:

$$\frac{1}{E^2}\left(\frac{dr}{d\lambda}\right)^2 = 1 - \left(1 - \frac{2M}{r}\right)\frac{b^2}{r^2} \qquad\qquad \text{substituted } L = bE$$

$$\left(\frac{dr}{d\gamma}\right)^2 = 1 - \left(1 - \frac{2M}{r}\right)\frac{b^2}{r^2}, \qquad\qquad \text{new affine parameter } \gamma \quad (11.4)$$

so besides M, the only parameter is b. Note that we defined the new parameter $\gamma = E\lambda$ that is still affine, consistent with eqn. (6.42), so the paths are still geodesics.

11.2 Prove

$$\text{particles:} \quad r = \frac{\tilde{L}^2}{2M}\left[1 \pm \left(1 - \frac{12M^2}{\tilde{L}^2}\right)^{1/2}\right]; \quad \text{Schutz Eq. (11.17)} \quad (11.5)$$

$$\text{photons:} \quad r = 3M. \qquad\qquad\qquad\qquad \text{Schutz Eq. (11.18)} \quad (11.6)$$

Hint: Differentiate, don't integrate!

11.3 Plot $\tilde{V}^2$ against r/M for the three cases $\tilde{L}^2 = 25M^2, \tilde{L}^2 = 12M^2, \tilde{L}^2 = 9M^2$ and verify the qualitative correctness of Schutz Figs. 11.1 and 11.3.

Solution: One limitation of plotting $\tilde{V}^2$ vs. r is that the asymptotic behavior for large r is not revealed. (Of course one can infer this, see Hint in SP11.6.) If one is not

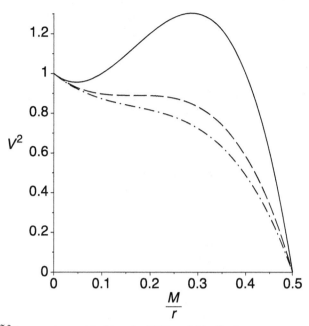

Figure 11.1 Effective potential $\tilde{V}^2$ for a massive particle, Schutz Eq. (11.13), vs. M/r. The three cases $\tilde{L}^2 = 25M^2, \tilde{L}^2 = 12M^2, \tilde{L}^2 = 9M^2$ correspond to line styles: solid, dashed, and dash-dot. The qualitative difference from Schutz Fig. 11.1 results from the different choice in horizontal coordinate. Plot made with Maple^{tm} worksheet.

> interested in the behavior near $r = 0$, a simple solution is to plot $\tilde{V}^2$ vs $1/r$, as we have done in fig. 11.1. Then it is immediately clear that with $\tilde{L}^2 > 12M^2$ we have two extrema, with $\tilde{L}^2 = 12M^2$ an inflection point, and $\tilde{L}^2 < 12M^2$ there are no extrema or inflection points on the interval $2M < r < \infty$.

11.5 (a) Find the radius $R_{0.01}$ at which $-g_{00}$ differs from the "Newtonian" value $1 - 2M/R$ by only 1%. (b) How many normal [Sun-like] stars can fit in the region between $R_{0.01}$ and the radius $2M$?

[In Schutz's online solutions for instructors he suggests that: *The question is badly worded, since the value $1 - 2M/R$ is the relativistic value of $-g_{00}$, not its Newtonian approximation.*]

Please see the supplementary problem SP11.1 below for an alternative question that is meant to be in the same spirit.

11.7 A clock is in a circular orbit at $r = 10M$ in a Schwarzschild metric.

(a) How much time elapses on the clock during one orbit? (Integrate the proper time $d\tau = |ds^2|^{1/2}$ over an orbit.)

Solution:

(a) The clock will read proper time; recall item (II) of Schutz §7.1. The equation for the period T follows from the definition of the azimuthal four-velocity component:

$$U^\phi \equiv \frac{\mathrm{d}\phi}{\mathrm{d}\tau}$$

$$\int_0^{\Delta\tau} \mathrm{d}\tau = \left| \int_0^{2\pi} \frac{1}{U^\phi} \mathrm{d}\phi \right| \qquad \text{rearranged, integrated over period}$$

$$\Delta\tau = \frac{2\pi}{|U^\phi|} = \frac{2\pi m}{|p^\phi|}. \qquad \text{circular orbit } U^\phi \text{ constant} \qquad (11.7)$$

The absolute value was introduced to ensure a positive period even when U^ϕ happens to be negative. Express p^ϕ in terms of the specific angular momentum $\tilde{L}$, which in turn is known for a circular orbit of fixed r and given M. Because the Schwarzschild metric is diagonal,

$$\frac{m}{|p^\phi|} = \frac{m}{|g^{\phi\alpha}p_\alpha|} = \frac{m}{|g^{\phi\phi}p_\phi|} = \frac{r^2}{|\tilde{L}|}. \qquad \text{cf. Schutz Eq. (11.22)} \qquad (11.8)$$

Schutz Eq. (11.20) gives $\tilde{L}$ for a stable circular orbit of given M and r,

$$\tilde{L}^2 = \frac{Mr}{1 - 3M/r} \qquad \text{Schutz Eq. (11.20)} \qquad (11.9)$$

$$\frac{1}{|\tilde{L}|} = \sqrt{\frac{1 - 3M/r}{Mr}}. \qquad \text{rearranged} \qquad (11.10)$$

Combining eqns. (11.7), (11.8), and (11.10) we find the period

$$\Delta\tau = 2\pi \sqrt{\frac{r^3(1 - 3M/r)}{M}}. \qquad (11.11)$$

When $r \gg 3M$ this reduces to Kepler's third law, cf. eqn. (8.68) in the solution to Exer. 8.19(c). Here $r = 10M$, so eqn. (11.11) gives a period of $\Delta\tau = 20\sqrt{7}\pi M$.

11.7 (b) The clock sends out a signal to a distant observer once each orbit. What time interval does the distant observer measure between receiving any two signals?

Solution: The distant observer's clock agrees with the Schwarzschild coordinate time. Why? Because the Schwarzschild spacetime is *asymptotically flat*, see Schutz Eqs. (10.7) and (10.8). Assume that the distant observer is stationary with respect to the origin of the Schwarzschild coordinates so that there is no Doppler shift in the rate of signals sent by the clock due to motion of the receiver. (SP11.8 considers the role of Doppler shift in this exercise.)

The coordinate time period Δt of a circular orbit can be found from eqn. (8.61), see solution to Exercise 8.19(c),

$$\Delta t = 2\pi \left| \frac{\partial t}{\partial \phi} \right| = 2\pi \left| \frac{p^t}{p^\phi} \right|. \tag{11.12}$$

For the Schwarzschild metric eqn. (11.12) gives (see Schutz Eq. 11.25):

$$\Delta t = 2\pi \frac{\tilde{E}}{1 - 2M/r} \frac{r^2}{|\tilde{L}|} = 2\pi \sqrt{\frac{r^3}{M}}. \tag{11.13}$$

With $r = 10M$, eqn. (11.13) gives a period of $\Delta t = 20\sqrt{10\pi}\,M$.

Note $\Delta t > \Delta \tau$ found in Exercise 11.7(a). The distant observer concludes that the orbiting clock is running slowly. The effect is partly due to its motion relative to the observer and partly due to its proximity to the mass forming the Schwarzschild spacetime. The former is the time dilation we encountered in SR. The latter is the gravitational redshift of signals sent, classically speaking, from "low gravitational potential," and received by distant "higher potential" observers. Exercise 11.7(c) helps you to separate these two effects.

11.7   (c) A second clock is located at rest at $r = 10M$ next to the orbit of the first clock. (Rockets keep it there.) How much time elapses on it between successive passes of the orbiting clock?

Solution: Feel free to cogitate long hours on the rates of clocks doing various things in curved spacetime, but the safe thing to do to answer this question is to start with the line element. In this case the calculation is very simple for this clock that is stationary in the Schwarzschild coordinates: $dr = d\theta = d\phi = 0$, so

$$ds^2 = -(d\tau)^2 = g_{tt}(dt)^2$$
$$d\tau = \sqrt{-g_{tt}}\,dt$$
$$\Delta \tau = \int_0^\tau d\tau = \int_0^{\Delta t} \sqrt{-g_{tt}}\,dt = \sqrt{-g_{tt}}\Delta t. \tag{11.14}$$

Of course $\Delta \tau$ is the proper time interval, i.e. that measured by the clock in question. How much coordinate time, Δt, passes between orbital passes? We just found that in Exercise 11.7(b); $\Delta t = 20\sqrt{10\pi}\,M$. So

$$\Delta \tau = \sqrt{1 - \frac{2M}{r}}\, 2\pi \sqrt{\frac{r^3}{M}}. \qquad \text{used eqn. (11.13)} \tag{11.15}$$

With $r = 10M$, eqn. (11.15) gives $\Delta \tau = 20\sqrt{8}\,\pi\,M$.

The shorter period measured by this stationary clock compared to the stationary clock at infinity leads the distant observer to conclude that the clock at "lower potential" is running slowly, but not quite as slowly as the moving clock at the same proximity to the mass. The distant observer interprets the slowing down of the stationary clock as purely an effect of gravity; it is an example of what some call *gravitational time dilation* Weinberg (1972) and is of course related to the gravitational redshift that would be observed in the wavelength of radiation used to send the signals.

On the other hand the stationary clock held at $r = 10M$ sees the orbiting clock as running slowly. This is purely due to the latter's motion relative to the observer as is related to the time dilation we saw in SR. This connection is explored in SP11.9.

11.7   (d) Calculate (b) again in seconds for an orbit at $r = 6M$ where $M = 14M_\odot$. This is the minimum fluctuation time we expect in the X-ray spectrum of Cyg X-1: why?

Hint: Be careful about the units!

Solution: Plugging $r = 6M$ into eqn. (11.13) gives a period of $\Delta t = 6.4$ ms. The X-ray spectrum emitted from the hot gas orbiting Cyg X-1 will fluctuate with the orbital period. We, as observers, are far from it so eqn. (11.13) applies. The minimum stable orbit for massive particles is $r = 6M$, see Schutz Eq. (11.19). The orbital period increases monotonically with r so this corresponds to the minimum period and fluctuation time.

11.7   (e) If the orbiting "clock" is the twin Artemis, in the orbit in (d), how much does she age during the time her twin Diana lives 40 years far from the black hole and at rest with respect to it?

Solution: The far-away twin, residing in approximately flat spacetime, ages at the rate of Schwarzschild coordinate time t. The twin in the minimum stable circular orbit, $r = 6M$, ages with the clocks at rest in her orbit. The two are related via

$$\frac{d\tau}{dt} = \frac{1}{U^\phi} = \frac{1 - 2M/r}{\tilde{E}}. \qquad \text{used Schutz Eq. (11.23)} \qquad (11.16)$$

For the stable circular orbit we can use Schutz Eq. (11.21) for the orbiting particle's specific energy $\tilde{E}$:

$$\tilde{E} = \frac{1 - 2M/r}{\sqrt{1 - 3M/r}}. \qquad (11.17)$$

Substituting eqn. (11.17) into eqn. (11.16), gives

$$\frac{d\tau}{dt} = \sqrt{1 - 3M/r}.$$ (11.18)

Note we could have obtained this from the results of Exercises 11.7(a) and (b) by dividing eqn. (11.11) by eqn. (11.13). Setting $r = r_{min} = 6M$, and integrating we find

$$\Delta\tau = \Delta t/\sqrt{2} \approx 28.28 \text{ yr.}$$ (11.19)

11.9 (This problem requires access to a computer.)

(a) Integrate numerically either

$$\left(\frac{dr}{d\phi}\right)^2 = \frac{\tilde{E}^2 - (1 - 2M/r)(1 + \tilde{L}^2/r^2)}{\tilde{L}^2 r^{-4}} \qquad \text{Schutz Eq. (11.26) or} \quad (11.20)$$

$$\left(\frac{du}{d\phi}\right)^2 = \frac{\tilde{E}^2}{\tilde{L}^2} - (1 - 2Mu)\left(\frac{1}{\tilde{L}^2} + u^2\right) \qquad \text{Schutz Eq. (11.28)} \quad (11.21)$$

for the orbit of a particle (i.e. for r/M as a function of ϕ) when $\tilde{E}^2 = 0.91$ and $(\tilde{L}/M)^2 = 13.0$. Compare the perihelion shift from one orbit to the next with

$$\Delta\phi = \frac{2\pi}{k} = 2\pi\left(1 - \frac{6M^2}{\tilde{L}^2}\right)^{-1/2}. \qquad \text{Schutz Eq. (11.37)} \quad (11.22)$$

Solution: First rewrite eqn. (11.21) with the definition $u' = Mu = M/r$ so that:

$$\left(\frac{du'}{d\phi}\right) = \pm\left[\frac{\tilde{E}^2}{\tilde{L}^2/M^2} - (1 - 2u')\left(\frac{1}{\tilde{L}^2/M^2} + u'^2\right)\right]^{1/2}$$

$$= \pm\left[\frac{0.91}{13} - (1 - 2u')\left(\frac{1}{13} + u'^2\right)\right]^{1/2}. \qquad (11.23)$$

Solving for $du'/d\phi$ was necessary for a numerical solution but the square root introduced the ambiguity of the sign on the RHS. The positive root applies when the orbiting body approaches $r = 0$ (u' growing), and vice versa. Fortunately it's rather straightforward to find the periastron and apastron. Then we can initiate the integration from the apastron (smallest u') with the positive RHS in eqn. (11.23). Next we initiate the solution with the largest u' (periastron) and integrate with the negative RHS in eqn. (11.23).

To find the periastron and apastron we simply set the LHS of eqn. (11.21) equal to zero and solve (numerically) this cubic. This gives three solutions but when we plot the RHS of eqn. (11.21) vs. u' we immediately find that it is only defined between the two smaller u' (larger two radial coordinate). So clearly this smaller pair of u' correspond to the apastron and periastron.

For the example given the numerical integration found roughly $\Delta\phi = 9\pi/4$ between apastron and periastron, or $\Delta\phi = 9\pi/2$ to return to the same orbital position.

So the periastron shift was about 2.5π. One the other hand Schutz Eq. (11.37) gives only $\Delta\phi = 2.7\pi$ implying a periastron shift of about 0.7π. The orbit was highly non-circular so the approximations leading to Schutz Eq. (11.37) do not apply.

Numerical computations can be found in the accompanying Maple™ worksheet.

11.9   (b) Integrate again when $\tilde{E}^2 = 0.95$ and $(\tilde{L}/M)^2 = 13.0$. How much proper time does this particle require to reach the horizon from $r = 10M$ if its initial radial velocity is negative?

Solution: Now we want the proper time so start with the basic orbit equation for a massive particle,

$$\left(\frac{dr}{d\tau}\right)^2 = \tilde{E}^2 - \left(1 - \frac{2M}{r}\right)\left(1 + \frac{\tilde{L}^2}{r^2}\right), \qquad \text{Schutz Eq. (11.11)} \qquad (11.24)$$

and we arrange this so we can integrate numerically, which means the ODE must involve only real numbers and the independent and dependent variables. A convenient choice is: $r' \equiv r/M$, $L' \equiv \tilde{L}/M$, and $\tau' \equiv \tau/M$. Substituting these into eqn. (11.24) above and rearranging

$$\frac{dr'}{d\tau'} = -\sqrt{\tilde{E}^2 - \left(1 - \frac{2}{r'}\right)\left(1 + \frac{L'^2}{r'^2}\right)}. \quad \text{--ve root for --ve radial velocity (11.25)}$$

In fact the problem separates and reduces to a challenging integral in r'. Maple™ has done the work for us and found $\tau' \approx 34.2$, so $\tau \approx 34.2M$ is the proper time for the particle to reach the horizon.

11.11   The right-hand side of eqn. (11.21) is a polynomial in u. Trace the u^3 term back through the derivation and show that it would not be present if we had started with the Newtonian version of Schutz Eq. (11.9). Interpret this term as a redshift effect on the orbital kinetic energy. Show that it is responsible for the maximum in the curve in Schutz Fig. 11.1.

Solution: Eqn. (11.21) above, came from eqn. (11.20) via the substitution $u \equiv 1/r$. This substitution was useful because $du = -dr/r^2$, which, upon squaring, eliminated the r^4 in the numerator of eqn. (11.21). So the u^3 term came from the product of the two underlined terms below, which appeared in eqn. (11.20)

$$\left(1 - \frac{2M}{r}\right)\left(1 + \frac{\tilde{L}^2}{r^2}\right) = \tilde{V}^2. \qquad \text{from numerator of eqn. (11.20)} \qquad (11.26)$$

And eqn. (11.20) came from the square of $dr/d\tau$ divided by $d\phi/d\tau$. The key product in eqn. (11.26) came exclusively from $dr/d\tau$. And eqn. (11.24) for $dr/d\tau$ was simply a rearrangement of Schutz Eq. (11.9), $\vec{p} \cdot \vec{p} = -m^2$ for a massive particle. In the solution to SP11.3 we show that ignoring the key product in eqn. (11.26), and using a binomial series to keep only terms of order $O(v^2)$ and $O(\phi = -M/r)$, eqn. (11.24) reduces to a simple mechanical energy equation of a particle in a Newtonian gravitational potential $\phi = -M/r$, plus the rest mass.

For a circular orbit, the only kinetic energy comes from $\tilde{L}^2/r^2$ (see solution to SP11.3). To find its redshift factor, compare the periods of a body in circular orbit observed by two different observers, both stationary with respect to the spatial coordinates of the standard Schwarzschild coordinate system. The "near observer" at r equal to that of the orbiting body (like the clock in Exercise 11.7(c)) measures $\Delta\tau_{\text{near}}$ given by eqn. (11.15). And the other observer, very far $r \gg M$ (like the observer in Exercise 11.7(b)), measures $\Delta\tau_{\text{far}} = \Delta t$ given by eqn. (11.13). The ratio gives the inverse ratio of the corresponding frequencies:

$$\frac{\Delta\tau_{\text{near}}}{\Delta\tau_{\text{far}}} = \sqrt{1 - \frac{2M}{r}} = \frac{\nu_{\text{far}}}{\nu_{\text{near}}}. \tag{11.27}$$

Consider again eqn. (11.24) but for a circular orbit:

$$\tilde{E} = \sqrt{1 - \frac{2M}{r}}\left(1 + \frac{\tilde{L}^2}{r^2}\right)^{1/2}$$

$$= \sqrt{1 - \frac{2M}{r}}\left(1 + \frac{1}{2}\frac{\tilde{L}^2}{r^2} + O(\tilde{L}^4/r^4)\right), \tag{11.28}$$

so the redshift factor we deduced from eqn. (11.27) is applied to the kinetic energy of the orbital velocity, and the u^3 term arises from this.

Schutz Fig. 11.1 or fig. 11.1 herein is the plot of $\tilde{V}^2$ versus r or u respectively. The maximum comes from the u^3 term. To see this, recall that we noted in eqn. (11.26) above that the key product arises in the expression for $\tilde{V}^2$. Writing $\tilde{V}^2$ in terms of u and completing the square we obtain

$$\tilde{V}^2 = (1 - 2Mu)\left(1 + \tilde{L}^2 u^2\right)$$

$$= \left(u\tilde{L} - \frac{M}{\tilde{L}}\right)^2 + 1 - \frac{M^2}{\tilde{L}^2} - 2M\tilde{L}^2 u^3. \tag{11.29}$$

Ignoring temporarily the u^3 term on the RHS of eqn. (11.29) we have a concave upward quadratic (a parabola centered on $u = M/\tilde{L}^2$ where there is a minimum of $1 - M^2/\tilde{L}^2$). This parabola has no maximum; there is no maximum without the u^3 term.

11.13 (a) Derive[1]

$$\left(\frac{dy}{d\phi}\right)^2 = \frac{\tilde{E}^2 + M^2/\tilde{L}^2 - 1}{\tilde{L}^2} + \frac{2M^4}{\tilde{L}^6} + \frac{6M^3}{\tilde{L}^4} y$$
$$- \left(1 - \frac{6M^2}{\tilde{L}^2}\right) y^2 \qquad \text{Schutz Eq. (11.34)} \quad (11.30)$$

in the approximation that y is small. What must it be small compared to?

Solution: The derivation is simply a matter of substituting $y \equiv u - M/\tilde{L}^2$ into eqn. (11.21) for $(du/d\phi)^2$. One finds two terms linear in y cancel: $-2yM/\tilde{L}^2$ and $+2yM/\tilde{L}^2$. The two other terms in y combine: $4yM^3/\tilde{L}^4 + 2yM^3/\tilde{L}^4 = 6yM^3/\tilde{L}^4$. Similarly two terms in y^2 combine: $2y^2M^2/\tilde{L}^2 + 4y^2M^2/\tilde{L}^2 = 6y^2M^2/\tilde{L}^2$. The result is

$$\left(\frac{dy}{d\phi}\right)^2 = \frac{\tilde{E}^2 + M^2/\tilde{L}^2 - 1}{\tilde{L}^2} + \frac{2M^4}{\tilde{L}^6} + \frac{6M^3}{\tilde{L}^4} y - \left(1 - \frac{6M^2}{\tilde{L}^2}\right) y^2 + 2My^3.$$
$$(11.31)$$

Dropping the final y^3 term in eqn. (11.31) gives eqn. (11.30) above. For $\tilde{L}^2 \gg M^2$ the leading order contribution from the $O(y^2)$ term in eqn. (11.31) is just $-y^2$. Comparing this with the $O(y^3)$ term in eqn. (11.31) we conclude that we require $2My^3 \ll y^2$ or $y \ll 1/M$ to legitimately ignore the y^3 term.

11.13 (b) Derive

$$y = y_0 + A\cos(k\phi + B) \qquad \text{Schutz Eq. (11.35)} \qquad (11.32)$$

and[2]

$$k = \left(1 - \frac{6M^2}{\tilde{L}^2}\right)^{1/2} \qquad (11.33)$$

$$y_0 = \frac{3M^3}{k^2\tilde{L}^4} \qquad (11.34)$$

$$A = \frac{1}{k}\left[\frac{\tilde{E}^2 + M^2/\tilde{L}^2 - 1}{\tilde{L}^2} + \frac{2M^4}{\tilde{L}^6} + k^2 y_0^2\right]^{1/2} \qquad \text{Schutz Eq. (11.36)} \quad (11.35)$$

from eqn. (11.30).

[1] We have corrected a typo in the original Schutz Eq. (11.34): $\frac{6M^3}{\tilde{L}^2} y \to \frac{6M^3}{\tilde{L}^4} y$.
[2] We have corrected a typo in eqn. (11.34): $\tilde{L}^2 \to \tilde{L}^4$ and in eqn. (11.35): $-y_0^2 \to +k^2 y_0^2$. It propagated from the error in corrected eqn. (11.30).

Solution: Well one way to proceed is to simply substitute eqns. (11.32) and (11.35) into eqn. (11.30) and show that it is consistent. More instructive is to follow Schutz's suggestion, just before his Eq. (11.35), to complete the square in eqn. (11.30). It's an equation of the form:

$$\left(\frac{dy}{d\phi}\right)^2 = -a\left(y^2 + by + c\right) \qquad \text{absorb } \tilde{E}, M \text{ and } \tilde{L} \text{ into } a, b, c$$

$$= -a\left[\left(y + \frac{b}{2}\right)^2 + c - \frac{b^2}{4}\right] \qquad \text{completed square}$$

$$\left(\frac{dy'}{d\phi}\right)^2 = -a\left[(y')^2 + d\right]. \qquad \text{change of variables} \quad (11.36)$$

We introduced $y' \equiv y + b/2$ and $d \equiv c - b^2/4$. For completeness we note that

$$a = 1 - \frac{6M^2}{\tilde{L}^2}, \quad (-ab) = \frac{6M^3}{\tilde{L}^4}, \quad (-ac) = \frac{\tilde{E}^2 + M^2/\tilde{L}^2 - 1}{\tilde{L}^2} + \frac{2M^4}{\tilde{L}^6}. \quad (11.37)$$

Now the analogy with the ODE Schutz Eq. (11.31) is close enough that one might guess a solution of the form $y' = A\cos(k\phi + B)$ or

$$y = y_0 + A\cos(k\phi + B), \qquad \text{with } y_0 = -\frac{b}{2}. \qquad (11.38)$$

Let's verify this guess and in the process try to find A and k. Substituting $y' = A\cos(k\phi + B)$ into eqn. (11.36) gives:

$$\left(\frac{dy'}{d\phi}\right)^2 = k^2 A^2 \sin^2(k\phi + B) = -a\left[A^2 \cos^2(k\phi + B) + d\right]$$

$$= -k^2\left[A^2 \cos^2(k\phi + B) + d\right]. \quad \text{chose } k^2 = a$$

$$(11.39)$$

Things simplify because we set $k^2 = a$, allowing the two trignometric terms to combine to give unity, so

$$A = \sqrt{-d} = \sqrt{-c + b^2/4} = \frac{1}{\sqrt{a}}\sqrt{-ac + ab^2/4} \quad \text{divided eqn. (11.39) by } k^2$$

$$= \frac{1}{k}\left(\frac{\tilde{E}^2 + M^2/\tilde{L}^2 - 1}{\tilde{L}^2} + \frac{2M^4}{\tilde{L}^6} + k^2 y_0^2\right)^{1/2}. \quad \text{used eqns. (11.37), (11.38)}$$

$$(11.40)$$

This agrees with eqns. (11.33)–(11.35). In summary we found eqn. (11.38) is indeed a solution of eqn. (11.30) with A as in eqn. (11.40) and

$$k = \sqrt{a} = \sqrt{1 - \frac{6M^2}{\tilde{L}^2}} \qquad\qquad y_0 = -\frac{b}{2} = \frac{3M^3}{k^2 \tilde{L}^4}. \qquad (11.41)$$

11.13 (c) Verify the remark after Schutz Eq. (11.36) that $y = 0$ is not the correct circular orbit for the given $\tilde{E}$ and $\tilde{L}$ by using eqn. (11.9) and eqn. (11.17) to find the correct value of y and comparing it to y_0 in eqn. (11.34).

Solution: Eqn. (11.9) gives $\tilde{L}$ for a massive particle in a stable circular orbit in the Schwarzschild geometry without approximation. It can of course be rearranged to give a quadratic in r; in fact eqn. (11.9) was derived from the *solution* to this quadratic, eqn. (11.5). The latter gives for r for the stable circular orbit:

$$
\begin{aligned}
r &= \frac{\tilde{L}^2}{2M}\left[1 + \left(1 - \frac{12M^2}{\tilde{L}^2}\right)^{1/2}\right] \\
&= \frac{\tilde{L}^2}{2M}\left[1 + \left(1 - \frac{1}{2}\frac{12M^2}{\tilde{L}^2} - \frac{1}{8}\left(\frac{12M^2}{\tilde{L}^2}\right)^2 + O\left(\left(\frac{12M^2}{\tilde{L}^2}\right)^3\right)\right)\right] \quad \text{used eqn. (B.6)} \\
&= \frac{\tilde{L}^2}{M}\left[1 - 3\frac{M^2}{\tilde{L}^2} - 9\frac{M^4}{\tilde{L}^4} + O\left(\frac{M^6}{\tilde{L}^6}\right)\right]. \quad\quad (11.42)
\end{aligned}
$$

(The stable orbit corresponds to the +ve root; point B, not A, in Schutz Fig. 11.1.) Recall that eqn. (11.35) dropped the term in y^3, which corresponds to dropping terms $O(M^6/\tilde{L}^6)$. For consistency with this degree of approximation we work here to the same order.

Now recall the solution for the general orbit in eqns. (11.32) and (11.35) was centered on radial coordinate $r = \tilde{L}/M$ and with y defined as the perturbation to this circular orbit through $1/r \equiv u = M/\tilde{L}^2 + y$, cf. Schutz Eq. (11.30). If A were small then eqn. (11.32) would approach a circular orbit with radial coordinate:

$$
\begin{aligned}
r &= \frac{1}{u} = \frac{1}{\frac{M}{\tilde{L}^2} + y_0} = \frac{\tilde{L}^2}{M}\left(\frac{1}{1 + \frac{\tilde{L}^2}{M}y_0}\right) = \frac{\tilde{L}^2}{M}\left(\frac{1}{1 + \frac{3M^2/\tilde{L}^2}{1 - 6M^2/\tilde{L}^2}}\right) \quad \text{used eqn. (11.41)} \\
&= \frac{\tilde{L}^2}{M}\left(\frac{1}{\frac{\tilde{L}^2 - 3M^2}{\tilde{L}^2 - 6M^2}}\right) = \frac{\tilde{L}^2}{M}\left(\frac{1 - 6M^2/\tilde{L}^2}{1 - 3M^2/\tilde{L}^2}\right) \quad \text{rearranged} \\
&= \frac{\tilde{L}^2}{M}\left(1 - 6\frac{M^2}{\tilde{L}^2}\right)\left[1 + 3\frac{M^2}{\tilde{L}^2} + 9\frac{M^4}{\tilde{L}^4} + O\left(\frac{M^6}{\tilde{L}^6}\right)\right] \quad \text{used eqn. (B.4)} \\
&= \frac{\tilde{L}^2}{M}\left[1 - 3\frac{M^2}{\tilde{L}^2} - 9\frac{M^4}{\tilde{L}^4} + O\left(\frac{M^6}{\tilde{L}^6}\right)\right] \quad \text{expanded}
\end{aligned}
$$

$$(11.43)$$

in agreement to $O(M^6/\tilde{L}^6)$ with eqn. (11.42) above.

11.13 (d) Show from the expression for the effective potential,

$$
\tilde{V}^2(r) = \left(1 - \frac{2M}{r}\right)\left(1 + \frac{\tilde{L}^2}{r^2}\right) \quad\quad \text{Schutz Eq. (11.13)} \quad\quad (11.44)
$$

that a particle that has an inner turning point in the "Newtonian" regime, i.e. for $r \gg M$, has a value $\tilde{L} \gg M$. Use this to justify the step from $\Delta\phi$ (the change in ϕ from one perihelion to the next),

$$\Delta\phi = \frac{2\pi}{k} = 2\pi \left(1 - \frac{6M^2}{\tilde{L}^2}\right)^{-1/2}, \qquad \text{Schutz Eq. (11.37)} \qquad (11.45)$$

to its approximation for the Newtonian orbits

$$\Delta\phi \simeq 2\pi \left(1 + \frac{3M^2}{\tilde{L}^2}\right), \qquad \text{Schutz Eq. (11.38)} \qquad (11.46)$$

Solution: The turning point will occur where $\tilde{E}^2 = \tilde{V}^2$, i.e. somewhere between points A (maximum $\tilde{V}^2$) and B (minimum $\tilde{V}^2$) in Schutz Fig. 11.1 We now argue that the possibility of having a turning point in the Newtonian regime is dictated by having r_B sufficiently large. The radial coordinate of these extrema are found from setting $d\tilde{V}^2/dr = 0$, see eqn. (11.5). The inner one is point A with

$$r_A = \frac{\tilde{L}^2}{2M} \left[1 - \left(1 - \frac{12M^2}{\tilde{L}^2}\right)^{1/2}\right], \qquad (11.47)$$

from which it follows that $3M \leq r_A \leq 6M$. Clearly we never have $r_A \gg M$. The maximum r_{turning} is r_B. How big can r_B get? For $\tilde{L}^2 = 12M^2$ it is small, $r_B = 6M$, that of the minimum stable circular orbit. But when $\tilde{L}^2 \gg M^2$ then

$$\frac{r_B}{M} = \frac{\tilde{L}^2}{2M^2} \left[1 + \left(1 - \frac{12M^2}{\tilde{L}^2}\right)^{1/2}\right]$$

$$= \frac{\tilde{L}^2}{2M^2} [2 - O(M^2/\tilde{L})^2] \to \frac{\tilde{L}^2}{M^2}. \qquad (11.48)$$

So to ensure that a turning point in the Newtonian regime is possible we require $r_B/M \gg 1$ which implies $\tilde{L}^2/M^2 \gg 1$.

The inequality $\tilde{L}^2/M^2 \gg 1$ allows us to simplify the expression for $\Delta\phi$, eqn. (11.45) above, because we can use a binomial series expansion keeping just the first term:

$$\Delta\phi = 2\pi \left(1 - \frac{6M^2}{\tilde{L}^2}\right)^{-1/2} = 2\pi \left(1 + \frac{3M^2}{\tilde{L}^2}\right) + O\left(\frac{M^4}{\tilde{L}^4}\right). \quad \text{Schutz Eq. (11.37)}$$

$$(11.49)$$

11.38 Consider equatorial motion of particles with $m \neq 0$ in the Kerr metric. Find the analogue of the following equation

$$\left(\frac{dr}{d\lambda}\right)^2 = g^{rr}[(-g^{tt})E^2 + 2g^{t\phi}EL - g^{\phi\phi}L^2]$$

$$= g^{rr}(-g^{tt})\left[E^2 - 2\omega EL + \frac{g^{\phi\phi}}{g^{tt}}L^2\right], \quad \text{Schutz Eq. (11.91)} \quad (11.50)$$

where $\tilde{E}$ and $\tilde{L}$ are the constants of motion related to energy and angular momentum:

$$\tilde{E} \equiv -\frac{p_t}{m}, \qquad \text{Schutz Eq. (11.5)} \qquad (11.51)$$

$$\tilde{L} \equiv \frac{p_\phi}{m}. \qquad \text{Schutz Eq. (11.6)} \qquad (11.52)$$

Also find the analogues of

$$\left(\frac{dr}{d\lambda}\right)^2 = \frac{(r^2+a^2)^2 - a^2\Delta}{r^4}\left[E^2 - \frac{4Mra}{(r^2+a^2)^2 - a^2\Delta}EL\right.$$
$$\left. - \frac{r^2 - 2Mr}{(r^2+a^2)^2 - a^2\Delta}L^2\right] \qquad \text{Schutz Eq. (11.92)}$$
$$(11.53)$$

$$\left(\frac{dr}{d\lambda}\right)^2 = \frac{(r^2+a^2)^2 - a^2\Delta}{r^4}\,(E - V_+)\,(E - V_-), \qquad \text{Schutz Eq. (11.93)}$$
$$(11.54)$$

where

$$V_\pm(r) = [\omega \pm (\omega^2 - g^{\phi\phi}/g^{tt})^{1/2}]L \qquad \text{Schutz Eq. (11.94)} \qquad (11.55)$$

$$= \frac{4Mra \pm r^2\Delta^{1/2}}{(r^2+a^2)^2 - a^2\Delta}L. \qquad \text{Schutz Eq. (11.95)} \qquad (11.56)$$

Plot $\tilde{V}_\pm$ for $a = 0.5M$ and $\tilde{L}/M = 20, 12,$ and 6. Discuss the qualitative features of the trajectories. For arbitrary a determine the relations among $\tilde{E}, \tilde{L},$ and r for circular orbits with either sense of rotation. What is the minimum radius of a stable circular orbit? What happens to circular orbits in the ergosphere?

Solution: For massive particles we start with $p^\alpha p_\alpha = -m^2$. Because we want an expression involving the covariant components of four-momentum, p_t and p_ϕ, through the $\tilde{E}$ and $\tilde{L}$ terms, we write this as

$$g^{\alpha\beta}p_\alpha p_\beta = -m^2$$
$$g^{rr}(p_r)^2 = -\left(m^2 + g^{tt}(p_t)^2 + g^{\phi\phi}(p_\phi)^2 + 2g^{\phi t}p_\phi p_t\right). \quad \text{equatorial motion}$$
$$(11.57)$$

But we need $dr/d\tau$ on the LHS so we substitute $p_r = p^\alpha g_{\alpha r} = p^r g_{rr}$:

$$g_{rr}(p^r)^2 = -\left(m^2 + g^{tt}(p_t)^2 + g^{\phi\phi}(p_\phi)^2 + 2g^{\phi t}p_\phi p_t\right)$$
$$g_{rr}m^2\left(\frac{dr}{d\tau}\right)^2 = -\left(m^2 + g^{tt}(m\tilde{E})^2 + g^{\phi\phi}(m\tilde{L})^2 - 2g^{\phi t}(m\tilde{L})(m\tilde{E})\right)$$

used eqns. (11.51,11.52)

$$\left(\frac{dr}{d\tau}\right)^2 = -g^{rr}\left(1 + g^{tt}\tilde{E}^2 + g^{\phi\phi}\tilde{L}^2 - 2g^{\phi t}\tilde{L}\tilde{E}\right)$$

$$= -g^{rr}g^{tt}\left(\frac{1}{g^{tt}} + \tilde{E}^2 + \frac{g^{\phi\phi}}{g^{tt}}\tilde{L}^2 - 2\omega\tilde{L}\tilde{E}\right). \tag{11.58}$$

In the last step we used the expression for the angular velocity of the so-called "dragging of inertial frames,"

$$\omega = \frac{g^{\phi t}}{g^{tt}}. \qquad \text{from Schutz Eq. (11.77)} \tag{11.59}$$

Eqn. (11.58) for massive particles is the analogue of eqn. (11.50) for photons.

Next we simply substitute the definitions of the inverse metric terms using

$$\Delta := r^2 - 2rM + a^2$$
$$\rho^2 := r^2 + a^2\cos^2\theta \qquad \text{Schutz Eq. (11.72)} \tag{11.60}$$

and Schutz Eqs. (11.86) and (11.89), with $\theta = \pi/2$ because we're in the equatorial plane, and using Schutz Eq. (11.90) for ω. This leads directly to the analogue of eqn. (11.53):

$$\left(\frac{dr}{d\tau}\right)^2 = \frac{(r^2+a^2)^2 - a^2\Delta}{r^4}\left[-\frac{r^2\Delta}{(r^2+a^2)^2 - a^2\Delta} + \tilde{E}^2 - \frac{4Mra}{(r^2+a^2)^2 - a^2\Delta}\tilde{L}\tilde{E}\right.$$
$$\left. - \frac{r^2 - 2Mr}{(r^2+a^2)^2 - a^2\Delta}\tilde{L}^2\right]. \tag{11.61}$$

Note that for the $\tilde{L}^2$ term we kept Δ in the denominator, but in the numerator we simplified $\Delta - a^2 = r^2 - 2Mr + a^2 - a^2 = r^2 - 2Mr$.

To determine the excluded region we need to find the energy levels $\tilde{E}$ such that $dr/d\tau = 0$. So we search for the conditions under which the RHS of eqn. (11.61) is zero. That is we want the quantity in square brackets to be zero, ignoring or assuming the factor outside $g^{rr}g^{tt} \neq 0$. Write the square bracket term as a quadratic in $\tilde{E}$:

$$\tilde{E}^2 + b\tilde{E} + c = 0, \quad \text{where } b \equiv \frac{-4Mra}{(r^2+a^2)^2 - a^2\Delta}\tilde{L}, \ c \equiv \frac{-r^2\Delta - (r^2 - 2Mr)\tilde{L}^2}{(r^2+a^2)^2 - a^2\Delta}. \tag{11.62}$$

This was the situation with photon orbits, accept of course the c term didn't include the factor of $1/g^{tt}$. Solving for the two roots of $\tilde{E}$, and calling them $\tilde{V}_+$ and $\tilde{V}_-$ gives

$$\tilde{V}_\pm = \frac{-b \pm \sqrt{b^2 - 4c}}{2}$$

$$= \frac{2Mra}{(r^2+a^2)^2 - a^2\Delta}\tilde{L} \pm \frac{1}{2}\left(\left(\frac{-4Mra}{(r^2+a^2)^2 - a^2\Delta}\tilde{L}\right)^2 + 4\frac{r^2\Delta + (r^2 - 2Mr)\tilde{L}^2}{(r^2+a^2)^2 - a^2\Delta}\right)^{1/2}. \tag{11.63}$$

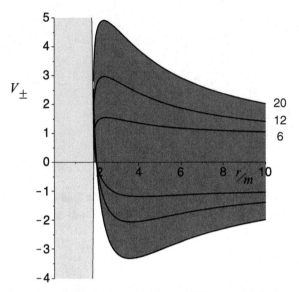

Effective potential $V_\pm$ for massive particles, from eqn. (11.65), vs. r/M for $a = M/2$. The three thick lines correspond to $\tilde{L}/M = 6, 12, 20$, as indicated to the right of the line. The region inside the horizon $r < r_+$, is shaded with lighter gray.

This is a bit messy, but any quadratic, say in x, can be written as $(x - x_-)(x - x_+)$ where $x_\pm$ are the two roots. So just as for photons we can write eqn. (11.61) as

$$\left(\frac{dr}{d\tau}\right)^2 = \frac{(r^2 + a^2)^2 - a^2\Delta}{r^4}\left[(\tilde{E} - \tilde{V}_+)(\tilde{E} - \tilde{V}_-)\right], \qquad (11.64)$$

which is similar to eqn. (11.54). And now $\tilde{V}_\pm$ are the roots found in eqn. (11.63), which are to be contrasted with eqns. (11.55) and (11.56) for the corresponding effective potentials for photon equatorial orbits in the Kerr spacetime.

For plotting purposes, define the non-dimensional variables $r' \equiv r/M$, $\Delta' \equiv \Delta/M^2$, $\tilde{L}' \equiv \tilde{L}/M$, $a' \equiv a/M$ so setting $a = M/2$ gives $a' = 1/2$. Then divide eqn. (11.63) top and bottom by M^4 giving

$$\tilde{V}_\pm = \frac{2r'a'}{(r'^2 + a'^2)^2 - a'^2\Delta'}\tilde{L}'$$

$$\pm \frac{1}{2}\left(\left(\frac{-4r'a'}{(r'^2 + a'^2)^2 - a'^2\Delta'}\tilde{L}'\right)^2 + 4\frac{r'^2\Delta' + (r'^2 - 2r')\tilde{L}'^2}{(r'^2 + a'^2)^2 - a'^2\Delta'}\right)^{1/2}. \qquad (11.65)$$

Then the horizon, for which $g_{rr} = \rho^2/\Delta = \infty$, corresponds to $r'_+ = 1 + \sqrt{3}/2$; set $a = M/2$ and $\Delta = 0$ in eqn. (11.60). The region inside the horizon is shaded light gray in fig. 11.2. The effective potential for three values of $\tilde{L}/M$ are plotted as thick lines that outline the corresponding forbidden regions shaded in darker gray.

11.2 Supplementary problems

SP 11.1 Recall from Exercise 7.5(d) that g_{00} is "closely related to" $-\exp(2\phi)$, where ϕ is the Newtonian potential for a similar [non-relativistic] situation. This was derived for a hydrostatic fluid in Exercise 7.5(d). Use this fact here to find at what distance from a black hole of mass $10^6 \, M_\odot$ the Schwarzschild g_{00} differs by 1% from $-\exp(2\phi)$, where ϕ is the Newtonian potential $\phi = -M/r$.

Solution

The Schwarzschild metric has, precisely,

$$-g_{00} = 1 - \frac{2M}{r},$$

with $G = 1$ of course. When the weak gravity limit applies, the Newtonian potential $\phi = -\frac{M}{r}$ for a similar [non-relativistic] situation has $|\phi| \ll 1$ and we have from Exercise 7.5(d),

$$-g_{00} \approx \exp(2\phi). \qquad\qquad \text{see eqn. (7.43)} \qquad\qquad (11.66)$$

Use a Taylor series about $\phi = 0$ to approximate the exponential function,

$$-g_{00} \simeq \exp(2\phi) = 1 + 2\phi + 2\phi^2 + \frac{4}{3}\phi^3 \cdots \qquad \text{used eqn. (B.10)}$$

$$1 - \frac{2M}{r} \simeq 1 - 2\frac{M}{r} + 2\frac{M^2}{r^2} + O(M^3/r^3). \qquad\qquad (11.67)$$

For $|\phi| \ll 1$ the LHS of eqn. (11.67) differs from the RHS by approximately $2M^2/r^2$. Setting this difference to 1% of $-g_{00} \approx 1$ we assign the corresponding radius to $r = R_{0.01}$:

$$2\phi^2 = 2\left(-\frac{M}{R_{0.01}}\right)^2 = \frac{1}{100} \qquad\qquad \text{for 1\% error}$$

$$R_{0.01} = 10\sqrt{2}\, M. \qquad\qquad (11.68)$$

In the case when $M = 10^6 \, M_\odot$, this corresponds to

$$R_{0.01} = 10\sqrt{2} \times 10^6 \times 1.5 \, \text{km} \approx 2 \times 10^{10} \, \text{m}, \qquad\qquad (11.69)$$

where we used $M_\odot \approx 1.5$ km, see Schutz Table 8.1.

SP 11.2 Recall that when deriving the orbits of massive particles and photons in the Schwarzschild metric, see Schutz §11.1, it was argued that motion is always confined to a single plane because of the spherical symmetry. Spell out this argument in more detail.

SP 11.3 The equation $\vec{p} \cdot \vec{p} = -m^2$ in the Schwarzschild geometry became the basic equation for the orbit, eqn. (11.24). Consider the case that $r \gg 2M$ and $r \gg |\tilde{L}|$. Argue that this looks like a classical energy equation for the total energy $m\tilde{E}$, matching each term with its Newtonian mechanics counterpart. Which term doesn't have a Newtonian mechanics counterpart?

Solution

Multiply eqn. (11.24) by m^2 so that it is an equation in energy rather than specific energy. Isolating $m^2\tilde{E}^2$ in eqn. (11.24) and ignoring the product of two small terms we find

$$m^2\tilde{E}^2 = m^2\left(1 - \frac{2M}{r} + \frac{\tilde{L}^2}{r^2}\right) + m^2\left(\frac{\mathrm{d}r}{\mathrm{d}\tau}\right)^2 \qquad \text{for } r \gg 2M, |\tilde{L}|$$

$$m\tilde{E} = m\left[1 - \frac{M}{r} + \frac{1}{2}\frac{\tilde{L}^2}{r^2} + \frac{1}{2}\left(\frac{\mathrm{d}r}{\mathrm{d}\tau}\right)^2\right] \qquad \text{used eqn. (B.5)}$$

$$= m - m\frac{M}{r} + \frac{1}{2}m\left[(v^\phi)^2 + (v^r)^2\right] \qquad (11.70)$$

where we've interpreted the specific angular momentum in classical terms using $\tilde{L} = rv^\phi$, v^ϕ being the three-velocity component in the azimuthal direction (see also eqn. (8.102) of SP8.7). Similarly we've interpreted the radial component of the four-velocity as the corresponding component of the three-velocity, v^r. Together they correspond to the classical kinetic energy term, the final term in eqn. (11.70). The middle term corresponds to the classical Newtonian gravitational potential energy (recall $G = 1$ in geometric units). Of course the key, non-classical term in this total energy equation is the term associated with the rest mass (the initial m on the RHS of eqn. (11.70)).

SP 11.4 Recall that when deriving the orbits of photons in the Kerr metric, see Schutz §11.3, it was argued that if initially $p^\theta = 0$ it would remain so because of the symmetry of the metric about the equatorial plane. Spell out this argument in more detail.

SP 11.5 (a) Two twins are in the stable circular orbit about a Schwarzschild black hole with the minimum radial coordinate, $r_A = r_{\mathrm{MIN}} = 6M$. One of the twins jumps into a rocket and quickly accelerates away from his brother, then turns off his engines and coasts in a circular orbit with radial coordinate $r_B = ar_{\mathrm{MIN}}$, with $a > 1$. After a long and prosperous life coasting in the outer orbit the adventurous twin uses his rocket to quickly return to visit his brother who has remained in the r_A orbit. Show that the twin who spent most of this life in the r_B orbit is no older than $\sqrt{2}$ times his less adventurous sibling.
(b) The adventerous brother fathered a pair of twin girls born in the outer stable orbit r_B. The more curious of these two girls decides to visit her uncle whom, she has heard, lives on the inner orbit. She uses a rocket to quickly reach the inner orbit. After a very long visit with her uncle, she returns to see her sister. How much *younger* can the curious sister be?

Solution

The twins age with the proper time, i.e. that measured by a standard clock that follows their world line. We can write the proper time in terms of coordinate time using the definition of the four-velocity of the twin, as in Schutz Eq. (11.23):

$$\frac{dt}{d\tau(r)} \equiv U^t = \frac{p^t}{m} = \frac{g^{tt} p_t}{m} = -g^{tt} \tilde{E} = \frac{\tilde{E}}{1 - \frac{2M}{r}}. \qquad (11.71)$$

The derivative of proper time at r_A with respect to that at r_B follows by dividing these two expressions, giving:

$$\frac{d\tau(r_A)}{d\tau(r_B)} = \frac{1 - \frac{2M}{r_A}}{1 - \frac{2M}{r_B}} \frac{\tilde{E}(r_B)}{\tilde{E}(r_A)}. \qquad (11.72)$$

The energy per unit mass $\tilde{E}$ must equal the effective potential, for otherwise the orbit would not be circular since:

$$\left(\frac{dr}{d\tau}\right)^2 = \tilde{E}^2 - \tilde{V}^2, \qquad \text{eqn. (11.24)} \qquad (11.73)$$

and hence $\tilde{E}$ is given by eqn. (11.17). Writing $r_B = ar_A$ and substituting these results into Eq. (11.72) above, we find after a bit of algebra:

$$\frac{d\tau(r_A)}{d\tau(r_B)} = \sqrt{\frac{1 - \frac{3M}{r_A}}{1 - \frac{3M}{r_B}}} = \sqrt{\frac{1 - \frac{1}{2}}{1 - \frac{1}{2a}}} = \sqrt{\frac{a}{2a - 1}}. \qquad (11.74)$$

The journey between orbits is assumed to be quick and has negligible influence on their ages. During his stay on the r_B orbit the adventurous brother ages at most $\sqrt{2}$ times that of his brother at r_A, as is clear from taking the limit $a \to \infty$ in eqn. (11.74) above. Similarly the sister who stays at r_B ages up to $\sqrt{2}$ times that of her sister who accelerates to and from the inner orbit. It's not simply the one who accelerates who always ends up younger, as we saw in Exercise 11.7.[3] The philosopher Tim Maudlin explores this point as well (Maudlin, 2012).

SP 11.6   In §11.1 Schutz notes that a particle at point G in the potential diagram $\tilde{V}^2(r)$ vs. r, his Fig. 11.1, has outward radial acceleration and therefore returns to infinity. What aspect of Fig. 11.1 indicates that the acceleration is outward. How do we know that such a particle actually reaches infinity (why does it not turn around and approach $r = 0$)?

Hint

The solution should include a discussion of the number of possible extrema in $\tilde{V}^2$ vs. r and the asymptotic value of $\tilde{V}^2$ as $r \to \infty$.

---

[3] We thank Jean-Philippe Nicolas and Alexander Afrait for bringing a similar example to our attention.

SP 11.7 Recall from the discussion of photon orbits in the equatorial plane of the Kerr metric, propagating in the direction of the rotation ($aL > 0$), it was concluded from eqn. (11.54) that the photon can move only in regions where $E > V_+$ or $E < V_-$. Argue that this is the case when $L > 0$, and be sure to prove that the first factor

$$\frac{(r^2 + a^2)^2 - a^2\Delta}{r^4}$$

is necessarily positive for all r, M and a. Show that for $L < 0$ the correct statement is $E < V_+$ or $E > V_-$.

Solution

From the definition eqn. (11.60) it is clear that

$$\Delta := r^2 - 2Mr + a^2 < r^2 + a^2,$$

since $M > 0$ and $r > 0$. Also $a^2 < r^2 + a^2$. So clearly $(r^2 + a^2)^2 > a^2\Delta$ which implies that the first factor is positive. The sign of eqn. (11.54) is determined by $(E - V_+)(E - V_-)$. From the definition of $V_\pm$ in eqn. (11.56) it is clear that $V_+ > V_-$ and $V_+ > 0$ when $L > 0$. So $E > V_+$ implies $E > V_-$ and the RHS of eqn. (11.54) is positive. When $a, L < 0$ and $a < 0$ (clearly the same physical situation but with different coordinate axes orientation) we can write

$$V_\pm = \frac{-2Mr|a| \pm r^2\Delta^{1/2}}{(r^2 + a^2)^2 - a^2\Delta}(-|L|)$$

$$= \frac{2Mr|a| \mp r^2\Delta^{1/2}}{(r^2 + a^2)^2 - a^2\Delta}|L|, \qquad (11.75)$$

hence the roles of V_+ and V_- are reversed so that the region of positive RHS corresponds to $E < V_+$ or $E > V_-$.

SP 11.8 Consider the clock in a stable circular orbit about a non-rotating star, as in Exercise 11.7. The clock sends out a radio signal of wavelength λ to a distant observer once per orbit at the moment that the clock directly approaches the observer. What is the role of Doppler shift regarding the rate at which the signals are received and the wavelength of the signal received both for an observer stationary with respect to the star and one moving away from the star at constant velocity $v \ll c$.

SP 11.9 Consider a non-rotating black hole with a clock A in a stable circular orbit at $r = 10M$ and another clock B held by rockets at fixed spatial coordinates (in the standard Schwarzschild coordinate system) at $r = 10M$ as in Exercise 11.7(c). Show that the clock B is related to the clock A via a time dilation factor as in SR. But why can we not say the same for an observer stationary with clock A regarding his observations of clock B?

Solution

An observer stationary with clock B sees the A clock pass by with three-velocity

$$v^\phi = r \frac{d\phi}{d\tau} = r \frac{d\phi}{dt} \frac{dt}{d\tau} = r \sqrt{\frac{M}{r^3}} \frac{1}{\sqrt{-g_{tt}}} = \sqrt{\frac{M}{r-2M}} \quad \text{used Schutz Eq. (11.24)}$$

$$(11.76)$$

This is the three-velocity of A measured by B. Of course B uses *his* clocks, not say Schwarzschild coordinate time, to measure time and velocities. And furthermore, his clock time τ is related to Schwarzschild time, $dt/d\tau = 1/\sqrt{-g_{tt}}$ as in Exercise 11.7(c) and not say that found by Schutz Eq. (11.23). For Schutz Eq. (11.23) is applicable to the clock in a circular orbit – different line element implies different proper time! Observer B will apply a Lorentz factor

$$\gamma = \frac{1}{\sqrt{1 - (v^\phi)^2}} = \frac{1}{\sqrt{1 - \frac{M}{r-2M}}} = \sqrt{\frac{1 - 2M/r}{1 - 3M/r}} \quad (11.77)$$

to obtain the time dilation of the moving clock A. So observer B anticipates

$$\Delta\tau_A = \frac{\Delta\tau_B}{\gamma} = 2\pi \sqrt{\frac{r^3}{M}\left(1 - \frac{2M}{r}\right)} \sqrt{\frac{1 - 3M/r}{1 - 2M/r}} = 2\pi \sqrt{\frac{r^3}{M}\left(1 - \frac{3M}{r}\right)},$$

$$(11.78)$$

in agreement with eqn. (11.11) found in Exercise 11.7(a).

SP 11.10 Analyze the situation in SP11.9 from the perspective of a local inertial frame instantenously at rest with clock B.

Cosmology

In recent years, with the increasing power of ground- and space-based astronomical observatories, cosmology has become a precision science, one which physicists look to for answers to some of their most fundamental questions.

Bernard Schutz, §12.1

... the correct relativistic description of the expanding universe is that, in our neighborhood, there exists a *preferred choice of time*, whose hypersurfaces are homogeneous and isotropic, and with respect to which [Hubble's law] Eq. (12.1) is valid in the local inertial frame of *any* observer who is at rest with respect to these hypersurfaces at *any* location.

Bernard Schutz, §12.2

12.1 Exercises

12.1 Use the metric of the two-sphere to prove the statement associated with Schutz Fig. 12.1 that the rate of increase of the distance between any two points as the sphere expands (as measured on the sphere) is proportional to the distance between them.

Solution: The metric of the two-sphere is apparent from the line element

$$dl^2 = r^2(d\theta^2 + \sin^2\theta \; d\phi^2).$$

See Exercise 6.28 for a derivation of this. Here we assume that $r = r(t)$ and pick two points on the sphere, p and q. We can always rotate the reference system such that the two points are on the equator, $\theta = \pi/2$, so the distance is

$$\int dl = l = \int r(t) \; d\phi = r(t)(\phi_p - \phi_q).$$

Differentiating with respect to time one obtains the rate of increase of their separation

$$\dot{l} = \dot{r}(\phi_p - \phi_q) = l\frac{\dot{r}}{r}. \tag{12.1}$$

So for a given r and rate of change of r, the rate of change of the distance l is proportional to l. It is for this reason that the easily visualized expanding two-sphere is often used as an analogue of the expanding RW universe in which the recessional speed of galaxies is proportional to their distance from us. This is the balloon model of the expanding universe discussed around Schutz Fig. 12.1.

12.4 Show that if $h_{ij}(t_1) \neq f(t_1, t_0) h_{ij}(t_0)$ for all i and j in

$$dl^2(t_1) = f(t_1, t_0) h_{ij}(t_0) \, dx^i \, dx^j$$
$$= h_{ij} dx^i \, dx^j, \qquad \text{Schutz Eq. (12.3)} \qquad (12.2)$$

then distances between galaxies would increase anisotropically: the Hubble law would have to be written as

$$v^i = H^i{}_j x^j, \qquad \text{Schutz Eq. (12.66)} \qquad (12.3)$$

for a matrix $H^i{}_j$ not proportional to the identity.

Solution: We replace

$$h_{ij}(t_1) = f(t_1, t_0) h_{ij}(t_0)$$

by a more general expression

$$h_{ij}(t) = a_{ij}(t) h_{ij}(t_0),$$

where $a_{ij}(t)$ represents six independent smooth functions of time t (six, not nine, because we demand $a_{ij} = a_{ji}$ to ensure that $h_{ij} = h_{ji}$), which of course reduce to unity at $t = t_0$ but are otherwise arbitrary. We wish to find the proper distance from our Galaxy to an arbitrary galaxy at rest in these coordinates. The line element at time t for a displacement in the radial direction dr is

$$dl^2 = h_{ij} \, dx^i \, dx^j = h_{ij} \frac{\partial x^i}{\partial r} \frac{\partial x^j}{\partial r} \, dr^2 = h_{rr}(\theta, \phi, t) \, dr^2. \qquad (12.4)$$

The factors $\partial x^i / \partial r$ and $\partial x^j / \partial r$ form the coordinate transformation of the metric from Cartesian to spherical coordinates, so if you like you can think of this as a coordinate transformation of the metric to spherical coordinates to facilitate the discussion of an arbitrary direction. From eqn. (B.14) we note that $\partial x^i / \partial r$ involves trigonometric functions of the polar and azimuthal angles θ and ϕ but, most importantly, not r. The proper distance to the galaxy at arbitrary time is obtained by integrating the square root of this line element,

$$l^r \equiv \int dl = \int_0^R \sqrt{h_{rr}} \, dr = \sqrt{h_{rr}} \, R. \qquad (12.5)$$

The recessional velocity of this galaxy is given by the time derivative of the proper distance:

$$\frac{dl^r}{dt} \equiv v^r = \frac{d\sqrt{h_{rr}}}{dt} R = \frac{\dot{h}_{rr}}{2\sqrt{h_{rr}}} R. \qquad (12.6)$$

Why is there no dR/dt term? Because we are considering a galaxy that has no random motion (zero peculiar velocity) and thus has fixed coordinates in our co-moving coordinate system. So the recessional velocity is that due to the expansion of the universe, the so-called Hubble flow. We can eliminate R in eqn. (12.6) using eqn. (12.5) to give

$$v^r = \frac{\dot{h}_{rr}(\theta, \phi, t)}{2h_{rr}(\theta, \phi, t)} l^r. \qquad (12.7)$$

This demonstrates the desired result that the Hubble law now depends upon direction (recall the trigonometric terms in θ and ϕ buried in h_{rr}). To express this in the form eqn. (12.3) multiple eqn. (12.7) by $\partial x^i / \partial r$, replace x^i on the RHS by $\delta^i{}_j x^j$ and call

$$\frac{\dot{h}_{rr}}{2h_{rr}} \delta^i{}_j \equiv H^i{}_j.$$

Note that we have written $x^i = l^r \partial x^i / \partial r$, so in eqn. (12.3) it represents the i-component of the proper distance to the galaxy, as we expected for a generalization of Hubble's law; the d on the RHS of Schutz Eq. (12.1) is the proper distance to the galaxy. We've also discovered that $H^i{}_j$ is diagonal but, as stipulated in the question, is not proportional to the identity matrix because of the polar and azimuthal angular dependence.

12.5 Show that if galaxies are assumed to move along the lines $x_i = $ const., and if we see the local universe as homogeneous, then g_{0i} in

$$ds^2 = -dt^2 + g_{0i}\, dt\, dx^j + R^2(t)h_{ij}\, dx^i\, dx^j \qquad \text{Schutz Eq. (12.5)} \qquad (12.8)$$

must vanish.

Hint: Add the condition that the universe is also isotropic. See SP12.13 and SP12.12 for alternative problems.

12.6 (a) Prove the statement leading to

$$G_{rr} = -\frac{1}{r^2}e^{2\Lambda}(1 - e^{-2\Lambda}), \qquad G_{\theta\theta} = -re^{-2\Lambda}\Lambda',$$
$$G_{\phi\phi} = \sin^2\theta\, G_{\theta\theta} \qquad \text{Schutz Eq. (12.8)} \qquad (12.9)$$

that we can deduce G_{ij} of our three-spaces by setting $\Phi = 0$ in eqn. (B.21).

Solution: In eqn. (B.21) the function $\Phi(r)$ appears in several terms through its first and second derivative. Simply setting $\Phi = 0$ we immediately reproduce eqn. (12.9). To verify this is the Einstein tensor we wanted of course we can simply calculate the Einstein tensor for the three-spaces with metric

$$dl^2 = e^{2\Lambda(r)}\, dr^2 + r^2 d\Omega^2. \qquad \text{Schutz Eq. (12.7)} \qquad (12.10)$$

Indeed we reproduce eqn. (12.9), see Maple™ worksheet.

Why does it work? There are a few subtleties buried here. The Einstein tensor is ultimately determined completely from the metric tensor, and the G_{ij} components in eqn. (B.21) were found for the static and spherically symmetric spacetime metric eqn. (10.53). Setting $\Phi = 0$ in eqn. (10.53) we obtain:

$$ds^2 = -e^{2\Phi}\, dt^2 + e^{2\Lambda}\, dr^2 + r^2\, d\Omega^2, \qquad \text{Schutz Eq. (10.7)}$$

$$= -dt^2 + e^{2\Lambda}\, dr^2 + r^2\, d\Omega^2. \qquad (12.11)$$

This certainly looks promising because the spatial part of eqn. (12.11) matches that of the 3D space we're interested in, eqn. (12.10). You might recall the warning in SP7.7 about inferring induced metrics on hypersurfaces, but fortunately these complications don't worry us here because eqn. (12.11) is diagonal, cf. eqn. (7.78).

You might also ask yourself the following question. We wanted the Einstein tensor for a 3D subspace. Are we allowed to simply extract the spatial part of the Einstein tensor for the 4D spacetime? Clearly in general the answer is *no*! To see this recall the metric eqn. (10.53) with $g_{tt} = -e^{2\Phi}$. As we just saw terms in Φ appeared in the *spatial* components of the Einstein tensor, e.g. in eqn. (10.59) we had $G_{11} = -\frac{e^{2\Lambda}}{r^2}\left[(1 - e^{-2\Lambda}\right] + 2\frac{\Phi'}{r}$. And this is to be anticipated more generally, for G_{ij} in a 4D spacetime is composed of the Ricci tensor $R_{ij} = R^{\alpha}_{\ i\alpha j}$, and its contraction $R^{\mu}_{\ \mu}$. So for instance R_{ij} contains the "extra" term $R^{t}_{\ itj}$ that is not present for the Einstein tensor of the 3D space. In short, the effect of the time dimension can sneak into G_{ij} in subtle ways. So in our particular case where we extracted the spatial part of a 4D Einstein tensor we were quite fortunate to get the correct answer!

12.6 (b) Derive

$$G = G_{ij}\, g^{ij}$$

$$= -\frac{1}{r^2}\left[1 - (re^{-2\Lambda})'\right]. \qquad \text{Schutz Eq. (12.9)} \qquad (12.12)$$

Hint: To obtain eqn. (12.12) we substitute eqn. (12.9) for the diagonal components of the Einstein tensor into the equation for the trace, $G = G_{ij}\, g^{ij}$, and the rest is just algebra.

12.7 Show the metric in eqn. (12.10) is only flat at $r = 0$ if $A = 0$ in

$$g_{rr} = e^{2\Lambda} = \frac{1}{1 + \frac{1}{3}\kappa r^2 - \frac{A}{r}}. \qquad \text{Schutz Eq. (12.11)} \qquad (12.13)$$

Solution: Recall this was used in the derivation of the RW metric in Schutz §12.2. For the geometry to be flat we require the Riemann tensor to be zero,

$$R^\alpha{}_{\beta\mu\nu} = 0. \qquad\qquad \text{Schutz Eq. (6.71)} \qquad (12.14)$$

Lowering and then raising the first index in eqn. (12.14) we easily confirm that

$$R_{\alpha\beta\mu\nu} = 0 \iff R^\alpha{}_{\beta\mu\nu} = 0, \qquad (12.15)$$

which will simplify our calculations. Our metric eqn. (12.10) has the same spatial part as the spacetime metric in Exercise 6.35. Casting aside the warning given at the end of Exercise 12.6(a), we consider the Riemann tensor found there, see eqn. (6.101),

$$R_{r\theta r\theta} = r\Lambda', \quad R_{r\phi r\phi} = r\Lambda' \sin^2\theta, \quad R_{\theta\phi\theta\phi} = r^2 \sin^2\theta \left(1 - e^{-2\Lambda}\right), \quad (12.16)$$

where $\Lambda' = \partial\Lambda/\partial r$. All other components are obtainable from these three by symmetry operations or are zero. (We confirm that these correspond to the Riemann tensor of the 3D space in the Maple^{TM} worksheet.) Consider g_{rr} from eqn. (12.13) with $A \neq 0$. Then

$$\Lambda = \frac{1}{2}\log(e^{2\Lambda}) = \frac{1}{2}\log\left(\left[1 + \frac{1}{3}\kappa r^2 - \frac{A}{r}\right]^{-1}\right) = -\frac{1}{2}\log\left(1 + \frac{1}{3}\kappa r^2 - \frac{A}{r}\right).$$

$$(12.17)$$

Substituting this into eqn. (12.16) for the $R_{r\theta r\theta}$ component we find a non-zero value at $r = 0$:

$$R_{r\theta r\theta} = r\Lambda' = -\frac{1}{2}\frac{r\left(\frac{2}{3}\kappa r + \frac{A}{r^2}\right)}{\left(1 + \frac{1}{3}\kappa r^2 - \frac{A}{r}\right)} \to \frac{1}{2} \quad \text{as} \quad r \to 0. \quad \text{used eqn. (12.17)}$$

$$(12.18)$$

On the other hand, if $A = 0$ then $R_{r\theta r\theta} = R^r{}_{\theta r\theta} = 0$ at $r = 0$:

$$R_{r\theta r\theta} = r\Lambda' = -\frac{1}{2}\frac{r\left(\frac{2}{3}\kappa r\right)}{\left(1 + \frac{1}{3}\kappa r^2\right)} \to 0 \quad \text{as} \quad r \to 0. \qquad (12.19)$$

We still need to check the other two non-zero components of the Riemann tensor when $A = 0$ to confirm that space is flat there. The work is already done for the second one since

$$R^r{}_{\phi r\phi} = R^r{}_{\theta r\theta} \sin^2\theta = 0 \quad \text{at} \quad r = 0 \quad \text{when} \quad A = 0.$$

For the third Riemann tensor component

$$R_{\theta\phi\theta\phi} = r^2 \sin^2\theta \left(1 - \exp(-2\Lambda)\right) \qquad\qquad \text{used eqn. (12.16)}$$

$$= r^2 \sin^2\theta \left[1 - \left(1 + \frac{1}{3}\kappa r^2 - \frac{A}{r}\right)\right] \qquad \text{used eqn. (12.17)}$$

$$= \sin^2\theta \left(-\frac{1}{3}\kappa r^4 + rA\right) \qquad\qquad \text{rearranged}$$

$$= 0 \quad \text{at} \quad r = 0, \qquad (12.20)$$

with or without $A = 0$.

In summary space is flat at $r = 0$ iff $A = 0$ in eqn. (12.13).

12.9 (a) Show that a photon that propagates on a radial null geodesic of the RW metric,

$$ds^2 = -dt^2 + R^2(t)\left[\frac{dr^2}{1 - kr^2} + r^2 d\Omega^2\right], \qquad \text{Schutz Eq. (12.13)} \qquad (12.21)$$

has energy $-p_0$ inversely proportional to $R(t)$.

Solution: First we must change variables so that all components of the RW metric are independent of spatial coordinates. The coordinate transformation depends upon k:

$$r = \begin{cases} \sin\chi, & k = +1, \\ r, & k = 0, \\ \sinh\chi, & k = -1. \end{cases}$$

In all cases the radial line element for the RW metric becomes $ds^2 = -dt^2 + R^2(t)\,d\chi^2$, which gives the null geodesic

$$0 = -dt^2 + R^2(t)\,d\chi^2. \qquad \text{Schutz Eq. (12.20)} \qquad (12.22)$$

As stated in the text, p_χ is constant along a radial geodesic. (If you are concerned that the angular components of the metric are dependent on χ, see SP12.3.) And because the photon follows a null geodesic,

$$0 = p_\alpha p^\alpha = (p_0)^2 g^{00} + (p_\chi)^2 g^{\chi\chi}$$

$$p_0 = \pm\sqrt{\frac{-p_\chi^2\, g^{\chi\chi}}{g^{00}}} = \pm\sqrt{\frac{-p_\chi^2/R^2(t)}{(-1)}} = \pm\frac{p_\chi}{R(t)}, \qquad (12.23)$$

where the sign is determined by the direction of the photon (toward or away from the origin).

Why do we call $-p_0$ the "energy"? That was discussed in SP10.3(c). In fact $-p_0$ was called the "energy at infinity" in the Schwarzschild metric but clearly that term is not appropriate here in the RW metric; for instance, in the $k = 1$ universe there is no spatial infinity.

12.9 (b) Show that a photon emitted at time t_e and received at time t_r by observers at rest in the cosmological reference frame is redshifted by

$$1 + z = \frac{R(t_r)}{R(t_e)}. \qquad \text{Schutz Eq. (12.67)} \qquad (12.24)$$

Solution: The redshift parameter z is standard in cosmology and astrophysics, and is defined as

$$1 + z = \frac{\nu_e}{\nu_r}, \qquad \text{Schutz Eq. (10.12)} \qquad (12.25)$$

where ν_e and ν_r are the emitted and received photon frequencies. Multiplying top and bottom by Planck's constant h we see the relation with photon energy:

$$1 + z = \frac{h\nu_e}{h\nu_r} = \frac{E(t_e)}{E(t_r)} \qquad \text{used Einstein–de Broglie relation}$$

$$= \frac{\pm p_\chi / R(t_e)}{\pm p_\chi / R(t_r)} \qquad \text{used eqn. (12.23)}$$

$$= \frac{R(t_r)}{R(t_e)}. \tag{12.26}$$

12.11 (a) Prove that the redshift z of a galaxy that emitted radiation at cosmic time t is

$$z(t) = H_0(t_0 - t) + \frac{1}{2}(H_0^2 - \dot{H}_0)(t_0 - t)^2 + \cdots, \qquad \text{Schutz Eq. (12.29)}$$
$$\tag{12.27}$$

where t_0 is the time of observation, the present. Deduce from it the expression for look-back time as a function of redshift:

$$t_0 - t(z) = \frac{1}{H_0}\left[z - \frac{1}{2}(1 - \dot{H}_0/H_0^2)z^2 + \cdots \right]. \qquad \text{Schutz Eq. (12.30)} \quad (12.28)$$

Solution: The Taylor series in general for $z(t)$ evaluated about $t = t_0$ is

$$z(t) = z(t_0) + \dot{z}(t_0)(t - t_0) + \frac{1}{2}\ddot{z}(t_0)(t - t_0)^2 + \cdots \tag{12.29}$$

We find $z(t)$ and its derivatives from Schutz Eq. (12.28):

$$z(t) = \exp\left[-\int_{t_0}^{t} H(t')\, dt' \right] - 1 \qquad \text{Schutz Eq. (12.28)} \quad (12.30)$$

$$z(t_0) = \exp\left[-\int_{t_0}^{t_0} H(t')\, dt' \right] - 1 = \exp(0) - 1 = 0 \tag{12.31}$$

$$\dot{z}(t) = \exp\left[-\int_{t_0}^{t} H(t')\, dt' \right] \frac{d}{dt}\left[-\int_{t_0}^{t} H(t')\, dt' \right]$$

$$= (z(t) + 1)[-H(t)] \qquad \text{used eqn. (12.30)} \quad (12.32)$$

$$\dot{z}(t_0) = (z(t_0) + 1)[-H(t_0)] = -H_0 \qquad \text{used eqn. (12.31)} \quad (12.33)$$

$$\ddot{z}(t) = \frac{d}{dt}\Big((z(t) + 1)[-H(t)] \Big) = -\dot{z}H(t) - (1 + z)\dot{H} \qquad \text{used eqn. (12.32)}$$

$$\ddot{z}(t_0) = -(-H_0)H_0 - \dot{H}_0 = H_0^2 - \dot{H}_0. \qquad \text{used eqns. (12.33, 12.31)} \quad (12.34)$$

Using these results in the general Taylor series eqn. (12.29) gives eqn. (12.27).

The power series for $t(z)$ is also obtained from a Taylor series about t_0:

$$t(z) = t_0 + z\frac{dt}{dz} + \frac{1}{2}z^2\frac{d^2t}{dz^2} + \cdots$$

$$t_0 - t(z) = -\left(z\frac{1}{\dot{z}} + \frac{1}{2}z^2\frac{d}{dz}\left(\frac{1}{\dot{z}}\right) + \cdots\right)_{t=t_0}$$

$$t_0 - t(z) = \frac{z}{H_0} - \left(\frac{1}{2}z^2\frac{1}{\dot{z}^2}\frac{d}{dz}[H(1+z)]\right)_{t=t_0} + \cdots \quad \text{used eqns. (12.32), (12.33)}$$

$$t_0 - t(z) = \frac{1}{H_0}\left(z - \frac{1}{2}z^2\left(1 - \frac{\dot{H}_0}{H_0^2}\right) + \cdots\right), \qquad \text{used eqns. (12.32),}$$

$$(12.35)$$

in agreement with eqn. (12.28). The last step involved writing $dH/dz = \dot{H}dt/dz = \dot{H}/\dot{z}$.

12.11 (b) Fill in the indicated steps leading to the Hubble parameter as a function of redshift:

$$H(z) = H_0\left(1 - \frac{\dot{H}_0}{H_0^2}z + \cdots\right) \qquad \text{Schutz Eq. (12.31)}$$

Solution: From the Taylor series of $H(t)$ about t_0 we have

$$H(t) = H_0 + \dot{H}_0(t - t_0) + \cdots$$

$$H(z) = H_0 + \dot{H}_0\frac{-1}{H_0}\left[z - \frac{1}{2}\left(1 - \frac{\dot{H}_0}{H_0^2}\right)z^2 + \cdots\right] \qquad \text{used eqn. (12.28)}$$

$$= H_0\left[1 - \left(\frac{\dot{H}_0}{H_0^2}\right)z + \cdots\right]. \qquad \text{dropped terms } O(z^2) \text{ and higher}$$

$$(12.36)$$

12.13 Astronomers usually do not speak in terms of intrinsic luminosity and flux. Rather, they use absolute and apparent magnitude. The (bolometric) apparent magnitude of a star is defined by its flux F relative to a standard flux F_s:

$$m = -2.5\log_{10}(F/F_s), \qquad \text{Schutz Eq. (12.68)} \qquad (12.37)$$

where $F_s = 3 \times 10^{-8}$ J m^{-2} s^{-1} is roughly the flux of visible light at Earth from the brightest stars in the night sky. The absolute magnitude is defined as the apparent magnitude the object would have at a distance of 10 pc:

$$M = -2.5\log_{10}\left[L/4\pi(10\text{pc})^2 F_s\right]. \qquad \text{Schutz Eq. (12.69)} \qquad (12.38)$$

Using the series expansion for the luminosity distance,

$$d_L = R_0 \, r (1 + z) = \frac{z}{H_0} \left[1 + \left(1 + \frac{1}{2} \frac{\dot{H}_0}{H_0^2} \right) z \right] + \cdots , \quad \text{Schutz Eq. (12.42)}$$

(12.39)

with the deceleration parameter,

$$q_0 = -\frac{R_0 \ddot{R}_0}{\dot{R}_0^2} = -\left(1 + \frac{\dot{H}_0}{H_0^2} \right), \quad \text{Schutz Eq. (12.27)} \qquad (12.40)$$

rewrite the luminosity distance and brightness relation,

$$d_L = \left(\frac{L}{4 \pi F} \right)^{1/2} \qquad \text{Schutz Eq. (12.34)} \qquad (12.41)$$

in astronomers' language as:

$$m - M = 5 \log_{10} \left(\frac{z}{10 \, \text{pc} \, H_0} \right) + 1.09 (1 - q_0) z. \quad \text{Schutz Eq. (12.70)} \quad (12.42)$$

Astronomers call this the *redshift-magnitude relation*.

Solution: Subtracting the absolute magnitude from the apparent magnitude we have

$$m - M = -2.5 \log_{10} \left[\frac{F}{F_s} \right] - 2.5 \log_{10} \left[\frac{L}{4 \pi (10 \text{pc})^2 F_s} \right] \qquad \text{used eqns. (12.37, 12.38)}$$

$$= -2.5 \log_{10} \left[\frac{F 4 \pi (10 \text{pc})^2}{L} \right] \qquad \text{used properties of log}$$

$$= -2.5 \log_{10} \left[\frac{(10 \text{pc})^2}{d_L^2} \right] \qquad \text{used eqn. (12.41)}$$

$$= 5 \log_{10} \left[\frac{d_L}{10 \text{pc}} \right] \qquad \text{used properties of log}$$

$$\simeq 5 \log_{10} \left[\frac{\frac{z}{H_0} \left[1 + \left(1 + \frac{1}{2} \frac{\dot{H}_0}{H_0^2} \right) z \right]}{10 \text{pc}} \right] \qquad \text{used eqn. (12.39)}$$

$$= 5 \log_{10} \left[\frac{z}{H_0 10 \text{pc}} \right] + 5 \log_{10} \left[1 + \left(1 + \frac{1}{2} \frac{\dot{H}_0}{H_0^2} \right) z \right] \qquad \text{used properties of log}$$

$$= 5 \log_{10} \left[\frac{z}{H_0 10 \text{pc}} \right] + \frac{5}{\ln(10)} \ln \left[1 + \left(1 + \frac{1}{2} \frac{\dot{H}_0}{H_0^2} \right) z \right] \qquad \text{used properties of log}$$

$$= 5 \log_{10} \left[\frac{z}{H_0 10 \text{pc}} \right] + \frac{5}{\ln(10)} \ln \left[1 + \frac{1}{2} (1 - q_0) z \right] \qquad \text{used eqn. (12.40)}$$

$$\simeq 5 \log_{10} \left[\frac{z}{H_0 10 \text{pc}} \right] + \frac{5}{\ln(10)} \frac{1}{2} (1 - q_0) z. \qquad \text{used eqn. (B.9)}$$

12.15 Show from the matter equation for a radiation dominated universe,

$$\frac{d}{dt}\left(\rho R^4\right) = 0 \qquad\qquad \text{Schutz Eq. (12.49)} \qquad (12.43)$$

that if the radiation has a black-body spectrum of temperature T, then T is inversely proportional to R.

Solution: The integral over the black-body spectrum times the energy per photon $h\nu$ (h is Planck's constant and ν is the photon frequency) gives the total radiation energy density

$$\rho c^2 = a_B T^4, \qquad (12.44)$$

where a_B is the radiation constant. The spectrum is given by eqn. (12.82) below and the integral can be found in the accompanying Maple™ worksheet. Furthermore, for a radiation dominated universe, integrating the corresponding matter equation, eqn. (12.43), gives

$$\rho R^4 = \text{constant}. \qquad (12.45)$$

Equating the density given by eqns. (12.44, 12.45) we find

$$T^4 \propto R^{-4} \implies T \propto \frac{1}{R}.$$

12.17 Use the matter equation

$$\frac{d}{dt}\left(\rho R^3\right) = -p\frac{d}{dt}\left(R^3\right) \qquad\qquad \text{Schutz Eq. (12.46)} \qquad (12.46)$$

and the time-derivative of the temporal component of the Einstein equation

$$\frac{1}{2}\dot{R}^2 = -\frac{1}{2}k + \frac{4}{3}\pi R^2(\rho_m + \rho_\Lambda) \qquad\qquad \text{Schutz Eq. (12.54)} \qquad (12.47)$$

to derive the "equation of motion" for the scale factor R

$$\frac{\ddot{R}}{R} = -\frac{4}{3}\pi(\rho + 3p). \qquad\qquad \text{Schutz Eq. (12.55)} \qquad (12.48)$$

Make sure you use the fact that $p_\Lambda = -\rho_\Lambda$.

Solution: In cosmology the temporal component of the Einstein equation is called the Friedmann equation (e.g. Liddle, 2003, Eq. (3.10)). Let $\rho = (\rho_m + \rho_\Lambda)$ be the total density and rewrite the Friedmann equation by multiplying the last term by R/R, giving

$$\frac{1}{2}\dot{R}^2 = -\frac{1}{2}k + \frac{4\pi}{3}\frac{R^3\rho}{R}. \qquad (12.49)$$

Differentiating this equation with respect to time gives

$$\frac{d}{dt}\left(\frac{1}{2}\dot{R}^2\right) = \frac{d}{dt}\left(-\frac{1}{2}k + \frac{4\pi}{3}\frac{R^3\rho}{R}\right)$$

$$\dot{R}\ddot{R} = \frac{4\pi}{3}\left(\frac{1}{R}\frac{d}{dt}(\rho R^3) + \rho R^3\frac{d}{dt}\frac{1}{R}\right) \qquad \text{recall } k \text{ is constant}$$

$$= \frac{4\pi}{3}\left(\frac{1}{R}(-p3R^2\dot{R}) - \dot{R}\rho R\right) \qquad \text{used eqn. (12.46)}$$

$$\frac{\ddot{R}}{R} = -\frac{4\pi}{3}(3p + \rho). \qquad\qquad\qquad \text{simplified} \qquad (12.50)$$

Thus we've reproduced eqn. (12.48). Note that our derivation did not explicitly use the EOS for dark energy, $p_\Lambda = -\rho_\Lambda$.

12.19 Assuming the universe to be matter-dominated and to have zero cosmological constant, show that at times early enough for one to be able to neglect k in the Friedmann equation eqn. (12.47), the scale factor in the early matter-dominated era evolves with time as $R(t) \propto t^{2/3}$.

Solution: For a matter-dominated universe, the matter equation simplifies because the pressure vanishes,

$$\frac{d}{dt}\left(\rho R^3\right) = -p\frac{d}{dt}\left(R^3\right) \qquad\qquad \text{eqn. (12.46)}$$

$$\frac{d}{dt}\left(\rho_m R^3\right) = 0 \qquad\qquad\qquad \text{for matter dominated}$$

$$\rho_m(t) = \rho_{0,m}\frac{1}{R(t)^3}, \qquad\qquad \text{integrated} \qquad (12.51)$$

where $\rho_{0,m}$ is the matter density when $R(t_0) = R_0 = 1$. We can substitute this into the temporal component of the Einstein equation (Friedmann equation) with $k = 0$ and $\rho_\Lambda = 0$:

$$\frac{1}{2}\dot{R}^2 = \frac{4}{3}\pi R^2\rho_{0,m}\frac{1}{R^3}$$

$$\dot{R} = \pm A\frac{1}{\sqrt{R(t)}} \qquad\qquad \text{with } A^2 = \frac{8\pi}{3}\rho_{0,m}$$

$$\int_{R(t)}^{R_0}\sqrt{R'}dR' = A\int_t^{t_0}dt \qquad\qquad \text{chose growing solution}$$

$$\frac{2}{3}(R^{3/2} - 1) = A(t - t_0) \qquad\qquad \text{recall } R_0 = 1$$

$$R = \left(1 + \frac{3A}{2}(t - t_0)\right)^{2/3}$$

$$= \left(\frac{3A}{2}\right)^{2/3}t^{2/3}. \qquad\qquad \text{chose } t_0 = \frac{2}{3A} \qquad (12.52)$$

The final line exploited the freedom to shift the origin of the time axis.

12.21 Explore the possible futures and histories of an expanding cosmology with *negative* cosmological constant. You may wish to do this graphically, by drawing figures analogous to Schutz Fig. 11.1. See also Schutz Fig. 12.4.

Solution: We can interpret the Friedmann equation eqn. (12.47) like an energy equation, as was done by Schutz in §12.3, subsection "Dynamics of Robertson–Walker universes." Starting from an expanding universe, $\dot{R} > 0$, with negative cosmological constant, $\rho_\Lambda = \Lambda/8\pi < 0$,

$$\underbrace{\tfrac{1}{2}\dot{R}^2}_{\text{"kinetic energy"}} \quad \underbrace{-\tfrac{4}{3}\pi R^2(\rho_m + \rho_\Lambda)}_{\text{"potential energy"}} \quad \underbrace{= -\tfrac{1}{2}k.}_{\text{"total energy"}} \quad \text{eqn. (12.47)}$$

(12.53)

It is clear that the "kinetic energy" must always attain zero, regardless of curvature, $k = -1, 0, +1$, as we now explain. Consider $k = -1$ so the "total energy" $(-k/2)$ is positive. The matter contribution to "potential energy" decreases with increasing R, since it falls as $1/R$ for non-relativistic matter and $1/R^2$ for radiation and relativistic matter, while the cosmological constant contribution increases with R^2. Eventually for sufficiently large R the "potential energy" becomes dominated by the cosmological constant, which with $\rho_\Lambda < 0$ gives a positive "potential energy" that balances the "total energy." Expansion stops and the universe contracts. With $k = 0, +1$ the expansion stops at smaller R. As R decreases eventually the radiation term dominates, the "potential energy" term approaches

$$-\frac{4}{3}\pi R^2(\rho_m + \rho_\Lambda) \rightarrow -\frac{4}{3}\pi \frac{\rho_{0R}}{R^2},$$

where ρ_{0R} is a constant (the density of radiation energy at $R = 1$). This overwhelms the total energy, the "kinetic energy" becomes larger as $\dot{R}$ becomes more negative, and the universe contracts in a Big Crunch.

We can also say something regarding the history of the universe for this negative cosmological constant case. As we have just argued for sufficiently small R the radiation term dominates the "potential energy" term, which overwhelms the "total energy" term regardless of curvature $k = -1, 0, +1$. In this model the presently expanding universe must have come from a singularity at the Big Bang.

12.23 Calculate the redshift of decoupling by assuming that the cosmic microwave radiation has temperature 2.7 K today and had the temperature $E_i/20k_B$ at decoupling, where $E_i = 13.6$ eV is the energy needed to ionize hydrogen, see Exercise 12.22(c), and $k_B = 8.617 \times 10^{-5}$ eV/K is the Boltzmann constant.

Solution: The cosmic microwave radiation has been out of equilibrium with matter since the time of decoupling, and has cooled in proportion to the expansion of the universe,

$$T(t) = T_0 \frac{R_0}{R(t)},$$

where T_0 and R_0 are the radiation temperature and expansion factor today, see Exercise 12.15. Using the redshift formula,

$$1 + z(t) = \frac{R_0}{R(t)} \qquad\qquad \text{eqn. (12.24)}$$

$$z(t) = \frac{T(t)}{T_0} - 1 = \frac{E_i/20k_B}{2.7\text{K}} - 1 = \frac{E_i/20k_B}{2.7\text{K}} - 1 = 2922. \qquad (12.54)$$

The calculation can be found in the accompanying Maple™ worksheet.

12.25 Estimate the times earlier than which our uncertainty about the laws of physics prevents us drawing firm conclusions about cosmology as follows.

(a) Deduce that, in the radiation-dominated early universe, where the curvature term depending on the curvature constant $k \in \{0, 1, -1\}$ is negligible, the temperature T behaves as

$$T = \beta t^{-1/2}, \qquad\qquad \text{with } \beta = (45\hbar^3/32\pi^3)^{1/4} k_B^{-1}, \qquad (12.55)$$

where k_B is Boltzmann's constant.

Solution: During the radiation-dominated era, $R(t) \propto t^{1/2}$, cf. Schutz Eq. (12.57), when we choose the integration constant such that $R(0) = 0$. This implies that

$$H(t) = \frac{\dot{R}}{R} = \frac{1}{2t}. \qquad \text{proportionality constant cancels} \qquad (12.56)$$

Although the Hubble parameter usually is expressed in rather peculiar units, its dimensions are the inverse of time. Substitute eqn. (12.56) into the Friedmann equation eqn. (12.47) with $k = 0$

$$\frac{1}{4t^2} = \frac{8\pi G}{3} \rho_r, \qquad (12.57)$$

where we have included Newton's constant G for clarity, and the subscript in ρ_r reminds us that this applies during the radiation-dominated era. Recall from the quantum theory of radiation in a cavity, the theory that initially led Planck to discover quantum mechanics, that the energy density of thermal radiation is given by

$$\rho_r c^2 = a_B T^4, \qquad\qquad \text{with } a_B = \frac{8\pi^5 k_B^4}{15h^3 c^3}, \qquad (12.58)$$

where a_B is the *radiation density constant*, h is Planck's constant, and k_B is Boltzmann's constant, see (Weinberg, 2013, Eq. (1.1.6)), or (Eisberg and Resnick, 1985). Substituting eqn. (12.58) into eqn. (12.57)

$$\frac{1}{4t^2} = \frac{8\pi G}{3} \frac{8\pi^5 k_B^4}{15h^3 c^5} T^4 = \frac{8\pi G}{3} \frac{8\pi^5 k_B^4}{15\hbar^3 (2\pi)^3 c^5} T^4 = G\frac{8\pi^3 k_B^4}{45\hbar^3 c^5} T^4. \qquad (12.59)$$

Now solve for T:

$$T^4 = \frac{45\hbar^3 c^5}{32\pi^3 Gk_B^4}\frac{1}{t^2}$$

$$T = \beta\frac{1}{\sqrt{t}}, \qquad\qquad \text{with } \beta = \left(\frac{45\hbar^3 c^5}{32\pi^3 Gk_B^4}\right)^{1/4}. \qquad (12.60)$$

A slightly more general version of this (including the effects of neutrinos) was derived by Liddle in his very accessible textbook (Liddle, 2003, Eq. (11.12)).

12.25 (b) Assuming that our knowledge of particle physics is uncertain for $k_B T > 10^3$ GeV, find the earliest time t at which we can have confidence in the physics.

Solution: First evaluate β in eqn. (12.60). Using SI units we find (see Maple™ worksheet):

$$\beta \approx 1.518 \times 10^{10} \text{ K s}^{1/2}. \qquad (12.61)$$

(Liddle (2003) found $\beta = 1.3 \times 10^{10}$ K s$^{1/2}$, which is slightly smaller than ours because he included a factor 1.68 multiplying the radiation energy density to account for neutrinos.) Now we simply set $k_B T = 10^3$ GeV in eqn. (12.60) and solve for the time,

$$t = \left(\frac{k_B\beta}{10^{12}}\right)^2 \approx 1.7 \times 10^{-12} \text{ s}, \qquad (12.62)$$

where we used $k_B = 8.61733238 \times 10^{-5}$ eV/K. A thorough discussion of the early universe can be´found in the textbook by Mukhanov (2005); he trusts the laws of physics probed by particle accelerators to 10 TeV (about 10^{-14} s). Thereafter he quotes the Russian astrophysicist Yakov Zel'dovich as saying that the very early universe becomes "an accelerator for poor people"!

12.25 (c) Quantum gravity is probably important when a photon has enough energy $k_B T$ to form a black hole within one wavelength ($\lambda = h/k_B T$). Show that this gives $k_B T \sim h^{1/2}$. This is the *Planck temperature*. At what time t is this an important worry?

Solution: Recall we were able to form a Schwarzschild black hole when we could pack enough mass/energy within the Schwarzschild radius, $2M$. Now we're imagining that the energy comes from that of a single photon, $h\nu = 2M$. And we want this to occur within one wavelength, $\lambda = c/\nu$, so set

$$2M = \lambda \qquad\qquad \text{Schwarzschild radius} = \text{photon wavelength}$$

$$h\nu = \frac{1}{\nu} \qquad\qquad \text{mass/energy of a single photon}$$

$$h\nu = \sqrt{h} = k_B T. \qquad\qquad\qquad \text{rearranged} \qquad (12.63)$$

The second equality in the last line simply converted the photon energy to a temperature using the Boltzmann constant.

Remember that eqn. (12.63) is in geometric units, so we need to introduce factors of c and G to balance the units:

$$k_B T = \sqrt{h}\sqrt{\frac{c^5}{G}} \approx 5 \times 10^9 \text{ J} \qquad\qquad \text{convert from geometric to SI units}$$

$$\approx 3 \times 10^{28} \text{ eV} \approx 3 \times 10^{19} \text{ GeV}. \qquad\qquad (12.64)$$

To find the time corresponding to this estimate of the Planck temperature, we replace 100 GeV in eqn. (12.62) with 3×10^{19} GeV, giving $t \sim 2 \times 10^{-45}$ s. Mukhanov (2005) writes: "Near the Planckian scale, nonperturbative quantum gravity dominates and general relativity can no longer be trusted. ..., the question of cosmic singularities remains. It is expected that these problems will be properly addressed in an as yet unknown nonperturbative string/quantum gravity theory."

12.2 Supplementary problems

SP 12.1 Recall from SP1.11 that the procedure used to construct the $\bar{x}$-axis in Fig. 1.4 described by Schutz in §1.5 was essentially that used to synchronize clocks as described by Einstein in his 1905 article "On the electrodynamics of moving bodies" (Einstein, 1905).
(a) What assumption did Einstein make in 1905 that is *not* consistent with the Robertson–Walker universe and that explains why the clock-synchronization procedure would not work on large spatial scales?

Solution

The Robertson–Walker universe is spatially homogeneous, but Einstein assumed that space *and* time were homogeneous so that in the frame wherein the mirror was at rest, the time for the reflected light ray to return to the origin must be the same as the time it took for the light ray to travel from the origin to the mirror. In a Robertson–Walker universe the return time will be different because $R(t)$ changes with time.

(b) In the thought experiment of (a) let t_1 be the time taken for the light ray to leave the origin and reach the mirror at a radial coordinate r_1 from the origin, and t_2 the time when the light reflected from the mirror returns to the origin. Show that if the universe is expanding $t_2 > 2t_1$.

Solution

Using the Robertson–Walker metric, eqn. (12.21), we set $ds^2 = 0$ because it is a light ray and $d\Omega^2 = 0$ because the light moves radially from the origin. This gives,

$$\frac{dr}{\sqrt{1 - kr^2}} = \pm\frac{dt}{R(t)},$$

where the $\pm$ sign is chosen depending upon direction of the light ray. For the light ray leaving the origin at $t_0 = 0$ and arriving at r_1 at t_1 we have

$$\int_0^{r_1} \frac{dr}{\sqrt{1 - kr^2}} = \int_0^{t_1} \frac{dt}{R(t)}.$$

For the return path

$$-\int_{r_1}^0 \frac{dr}{\sqrt{1 - kr^2}} = \int_0^{r_1} \frac{dr}{\sqrt{1 - kr^2}} = \int_{t_1}^{t_2} \frac{dt}{R(t)}.$$

And by inspection, without doing the integrals, we can immediately see that $t_2 > 2t_1$ because the integrand in the later integral is smaller as $R(t)$ grows with the expanding universe.

SP 12.2   Let us get a feel for the order of magnitude of the terms in the expanding universe by calculating the following rates. (Bear in mind these calculations are hypothetical. The expansion applies to the space between the galaxies, not within a galaxy because a galaxy is held together by gravity. Furthermore Earth and rulers are held together by gravity and other forces.)

(a) At the current rate of expansion of the universe, how long will it take for the universe to double in size? By size we mean a linear dimension, not its volume. If the expansion of the universe were to apply locally, how long for a meter stick to double in length?

(b) If Hubble's law were to apply locally, how much longer would a meter stick be in 50 years?

(c) Plate tectonics is causing the Atlantic Ocean to spread at a rate of about 25 mm/year. So stationed in Brest, France one observes New York City to recede at $v \approx 25$ mm/year. If Hubble's law were to apply locally, compare this to the rate at which New York would recede from Brest due to expansion of the universe.

SP 12.3   In Schutz §7.4 we learned that "if all the components of $g_{\mu\nu}$ are independent of $x^{\beta*}$ for some fixed index $\beta*$, then $p_{\beta*}$ is a constant along any particle's trajectory." In deriving the cosmological redshift eqn. (12.81) in §12.2 Schutz used this conservation property for the RW metric written in terms of χ,

$$ds^2 = -dt^2 + R^2(t)\,d\chi^2 + \begin{cases} R^2(t)\sin^2\chi\,d\Omega^2 & k = +1, \\ R^2(t)\chi^2\,d\Omega^2 & k = 0, \\ R^2(t)\sinh^2\chi\,d\Omega^2 & k = -1. \end{cases} \qquad (12.65)$$

But clearly not *all* components of $g_{\alpha\beta}$ are independent of χ. How then could we argue that p_χ was conserved for the photon?

Solution

Recall the general geodesic equation could be written

$$m\frac{\mathrm{d}p_\beta}{\mathrm{d}\tau} = \frac{1}{2}g_{\mu\nu,\beta}p^\mu p^\nu. \qquad \text{eqn. (7.54)}$$

We were considering a photon propagating in the $\chi \equiv x^1$-direction (without loss of generality) so its four-momentum was $\vec{p} \underset{\text{RW}}{\to} (p^0, p^1, 0, 0)$. In this case only the derivatives of $g_{0\nu}$ and $g_{1\nu}$ are relevant in eqn. (7.54):

$$m\frac{\mathrm{d}p_\chi}{\mathrm{d}\tau} = \frac{1}{2}\left(g_{00,\chi}p^0 p^0 + 2g_{01,\chi}p^0 p^1 + g_{11,\chi}p^1 p^1\right) \quad \text{radially moving photon}$$

$$= \frac{1}{2}g_{00,\chi}p^0 p^0 + \frac{1}{2}g_{11,\chi}p^\chi p^\chi \qquad\qquad \text{diagonal metric}$$

$$= 0. \qquad\qquad\qquad\qquad g_{00} = -1, \ g_{11} = R^2(t)$$

$$\text{(12.66)}$$

SP 12.4   (a) Find the proper distance from the Milky Way for which Hubble's law suggests that galaxies should be receding from us by speeds greater than the speed of light.

(b) Recall from Schutz Fig. 5.2 described in §5.1 that in the presence of gravity it was not possible to construct global Lorentz frames. Use the result from (a) to construct another argument against global Lorentz frames by considering frame $\mathcal{O}$ at rest at the center of the Milky Way Galaxy, and another $\overline{\mathcal{O}}$ at rest at the center of a galaxy further away than the proper distance found in (a).

SP 12.5   Derive the *fluid equation*

$$0 = \dot{\rho} + 3(\rho + p)\frac{\dot{R}}{R} \qquad\qquad \text{Schutz Eq. (12.72)} \qquad \text{(12.67)}$$

by applying the (classical) first law of thermodynamics,

$$\mathrm{d}E + p\,\mathrm{d}V = T\,\mathrm{d}S,$$

to a fluid of volume V, energy $E = \rho V$, pressure p, and temperature T, undergoing reversible adiabatic expansion $\mathrm{d}S = 0$. Consider the case of a small spherical ball with $r \ll 1$ in the general RW spacetime.

Solution

For time interval $\mathrm{d}t$ we have

$$\frac{\mathrm{d}E}{\mathrm{d}t} + p\frac{\mathrm{d}V}{\mathrm{d}t} = 0.$$

For a small spherical ball the space is essentially locally flat because even if $k = \pm 1$ the factor in the g_{rr} metric component

$$\frac{1}{1 - kr^2} \, dr^2 \simeq dr^2, \qquad r \ll 1.$$

So the proper volume of the spherical ball is approximately $V \simeq (4/3)\pi(rR)^3$ and its time rate of change is

$$\frac{dV}{dt} = \frac{d}{dt}\left(\frac{4}{3}\pi(rR(t))^3\right) = 4\pi r^3 R^2(t)\dot{R}(t) = \dot{V}.$$

So

$$\frac{\dot{V}}{V} = 3\frac{\dot{R}(t)}{R(t)}.$$

The time rate of rate of change of energy is

$$\frac{dE}{dt} = \frac{d}{dt}(\rho V) = \rho\dot{V} + \dot{\rho}V.$$

The pressure work term is

$$p\frac{dV}{dt} = p\dot{V}.$$

Combining terms gives

$$0 = \dot{\rho} + 3(\rho + p)\frac{\dot{R}}{R}. \tag{12.68}$$

SP 12.6 In §12.3 Schutz argues that the spatial components of the divergence of the stress–energy tensor must vanish identically because of isotropy. Show that this is indeed the case for a perfect fluid and that $T^{i\alpha}_{\ ;\alpha} = 0$ provides three trivial equations.

SP 12.7 The first relativistic cosmological model was by Einstein (1917), in which spacelike hypersurfaces of constant time had the geometry of a three-sphere. Here we will examine some key properties of this geometry (see Rindler, 2006, §8.2).
(a) Show that the volume of the three-sphere of radius R is $2\pi^2 R^3$.
(b) Geodesics in the three-sphere appear as locally straight lines in this space, but when represented in Euclidean space have the geometry of great circles on the surface of a normal sphere (i.e. a two-sphere) of Euclidean radius R. Consider a pair of two meridians or great circles passing through the north pole, separated in longitude by angle θ. Show that the perpendicular separation η between the pair measured at a geodesic distance r from the pole is given by $\eta = \theta R \sin(r/R)$. Thus distances have been shrunk relative to Euclidean geometry by a factor

$$\frac{\eta}{r\theta} = \frac{R}{r}\sin\left(\frac{r}{R}\right). \tag{12.69}$$

See fig. 12.1.

(c) A *geodesic sphere* is the surface equidistant from a given point as measured along geodesics issuing from that point. Argue that the surface area S of a geodesic sphere of radius r in a space with the geometry of a three-sphere of radius R is given by

$$S = 4\pi r^2 \left(\frac{\eta}{r\theta}\right)^2 = 4\pi R^2 \sin^2\left(\frac{r}{R}\right). \tag{12.70}$$

(d) Integrate the surface area S with respect to r to show that the volume enclosed by S is

$$V = 2\pi R^2 \left(r - \frac{R}{2}\sin\frac{2r}{R}\right). \tag{12.71}$$

(e) Show that the maximum surface area encloses only half the total volume of the three-sphere, while the surface that encloses the *entire* volume has *zero* area. This is analogous to the situation, at one dimension less, of geodesics circles drawn on a two-sphere. The maximum *perimeter* geodesic circle is the equatorial geodesic circle, which encloses half the *area*; while the circle that encloses the entire area has zero length.

Solution

(a) Recall from Exercise 6.33 that the three-sphere is the set of points equidistant from a given point in 4D Euclidean space. The volume element of this hypersurface in 4D Euclidean space with pseudo-Cartesian coordinates is easy, just

$$dV = dx\,dy\,dz,$$

though the limits of integration are a bit cumbersome. Fortunately the limits of integration are simple, if not immediately obvious, in our angular coordinates, χ, θ, ϕ also defined in Exercise 6.33. To find these limits we note that we want each of the Cartesian coordinates in eqn. (6.88) to extend continuously from $-R$ to R. For example, $0 \le \chi \le \pi$ is consistent with $w = R\cos\chi$, $-R \le w \le R$. The remaining angular coordinates are the same as those of a two-sphere in spherical coordinates: $0 \le \theta \le \pi$ and $0 \le \phi \le 2\pi$. And once we know the metric in our angular coordinates, see Exercise 6.33(b), it is easy to transform the volume element

$$\sqrt{|g|}\,d\chi\,d\theta\,d\phi = dx\,dy\,dz, \qquad\qquad \text{used eqn. (6.102)}$$

where

$$g = \det\left(g_{\mu\nu}(\chi,\theta,\phi)\right) = R^2 \cdot R^2 \sin^2\chi \cdot R^2 \sin^2\chi \sin^2\theta.$$

Finally the volume of the three-sphere then becomes

$$\int dV = \int_0^\pi \int_0^\pi \int_0^{2\pi} R^3 \sin^2\chi \sin\theta\,d\phi\,d\chi\,d\theta = 2\pi \int_0^\pi \int_0^\pi R^3 \sin^2\chi \sin\theta\,d\chi\,d\theta$$

$$= 4\pi \int_0^\pi R^3 \sin^2\chi\,d\chi = 4\pi R^3 \left[-\frac{\sin\chi\,\cos\chi}{2} + \frac{\chi}{2}\right]_0^\pi$$

$$= 2\pi^2 R^3. \tag{12.72}$$

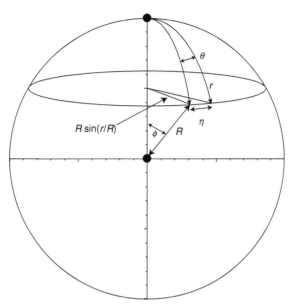

Figure 12.1 A two-sphere, in Euclidean space, of radius R with two geodesics, the pair of meridians passing through the north pole. They are separated by angle θ, which at distance r from the pole corresponds to a perpendicular distance $\eta = R\sin(r/R)\theta$. Figure adapted from (Rindler, 2006, Fig. 8.4).

Solution

(b) See fig. 12.1 herein from which it is clear that the perpendicular distance between the geodesics of length r is $\eta = R\sin(r/R)\theta$. Why? Because the polar angle $\phi = r/R$ radians. Thus $R\sin(r/R)$ is the radius of the disc that contains the line segment labeled η, which spans the angle θ.

In flat space the geodesics would be straight lines and their perpendicular separation would be just $r\theta$. Hence the ratio is

$$\frac{\eta}{r\theta} = \frac{R}{r}\sin\left(\frac{r}{R}\right),$$

as claimed in eqn. (12.69).

Solution

(c) Perpendicular distances between geodesics have been shrunk relative to Euclidean values by the ratio $\eta/(r\theta)$ given in (b). Now consider a second pair of geodesics close to the first pair but separated in a perpendicular direction. They too will be separated by η because the space is isotropic. So it is easy to see that an area element is shrunk relative to the Euclidean value by the square of this value $(\eta)^2/(r\theta)^2$. And the space is homogeneous so this shrinkage applies everywhere.

Hence we can obtain the area of the surface S enclosing a geodesic sphere of radius r by this factor times its corresponding Euclidean space value:

$$S = 4\pi r^2 \left(\frac{\eta}{r\theta}\right)^2 = 4\pi r^2 \frac{R^2}{r^2} \sin^2\left(\frac{r}{R}\right)$$

$$= 4\pi R^2 \sin^2\left(\frac{r}{R}\right), \tag{12.73}$$

in agreement with eqn. (12.70).

Solution

(d) To find the volume enclosed by S we simply integrate the expression in eqn. (12.73) over geodesic radius r:

$$V(r) = \int_0^r S(r')\mathrm{d}r' = \int_0^r 4\pi R^2 \sin^2\left(\frac{r'}{R}\right)\mathrm{d}r' \qquad \text{substituted eqn. (12.73)}$$

$$= 2\pi R^2 \left(r - R\cos(r/R)\sin(r/R)\right) \qquad \text{integration done in Maple}^{\text{TM}}$$

$$= 2\pi R^2 \left(r - \frac{R}{2}\sin\left(\frac{2r}{R}\right)\right), \qquad \text{simplified using trig identity}$$

$$\tag{12.74}$$

in agreement with eqn. (12.71).

Solution

(e) From eqn. (12.73) above it is immediately clear that $S(r)$ has a maximum when $r = R\pi/2$, for which

$$S(R\pi/2) = 4\pi R^2.$$

At $r = R\pi/2$, eqn. (12.74) gives for the enclosed volume,

$$V(R\pi/2) = \pi^2 R^3,$$

which is half the total volume of the three-sphere found in part (a) above, $V = 2\pi^2 R^3$, see eqn. (12.72).

To enclose the entire three-sphere volume it is clear from eqn. (12.74) above, or from fig. 12.1, that we must stretch the length of the geodesic to $r = R\pi$. But then S shrinks to zero!

SP 12.8 An experimentalist wishes to perform an experiment to test the role of the Ricci scalar on particle flux divergence; see Exercise 7.1(iii). Frustrated by the inability of a centrifuge, an electromagnetic field, or even a neutron star to produce a non-zero

Ricci scalar (see SP7.12, SP8.6, and SP10.12) she turns to cosmology. Show that for a Robertson–Walker universe without a cosmological constant, the Ricci scalar can be written as:

$$R^\alpha_{\ \alpha} = 8\pi(\rho - 3p).$$

SP 12.9 Apply the hypothetical law discussed in Exercise 7.1(iii),

$$\left(nU^\alpha\right)_{;\alpha} = q(R^\alpha_{\ \alpha})^2, \qquad\qquad \text{Schutz Eq. (7.3)} \qquad (12.75)$$

to a spherical ($k = +1$) Robertson–Walker universe. Derive an expression for the rate of increase/decay of the density of particles n. (Here $R^\alpha_{\ \alpha}$ is the Ricci scalar; not to be confused with the scale factor.)

Solution

Our strategy will be to integrate eqn. (12.75) over two spacelike 3D hypersurfaces Σ_1 and Σ_2 at two arbitrary instants of cosmic time t_0 and t that bound a volume of 4D spacetime V. First we write ($nU^\alpha \equiv N^\alpha$) because we will need the letter "n" for the unit normal one-form to the bounding surfaces n_α. Integrate eqn. (12.75) over the region between Σ_1 and Σ_2 using, as advised by Schutz in §6.2, the proper volume element $\sqrt{-g}\,d^4x$,

$$\int_V q(R^\alpha_{\ \alpha})^2 \sqrt{-g}\,d^4x = \int_V \left(N^\alpha_{\ ;\alpha}\right) \sqrt{-g}\,d^4x. \qquad (12.76)$$

Focus on the LHS of eqn. (12.76):

$$\int_V q(R^\alpha_{\ \alpha})^2 \sqrt{-g}\,d^4x = q \int_{t_0}^{t'} dt'(R^\alpha_{\ \alpha})^2(t') \int_{S^3} \sqrt{-\gamma(t')}\,d^3x \quad \text{used homogeneity}$$

$$= q \int_{t_0}^{t} dt'(R^\alpha_{\ \alpha})^2(t')2\pi^2 a^3(t'), \qquad \text{used eqn. (12.72)}$$

$$(12.77)$$

where $a(t')$ is the radius of the universe, that is the three-sphere S^3 at time t', and γ is the determinant of the metric induced on the three-sphere. Because the RW metric is diagonal, $\gamma_{ij} = g_{ij}$, cf. eqn. (7.78).

The RHS of eqn. (12.76) can be written

$$\int_V \left(N^\alpha_{\ ;\alpha}\right) \sqrt{-g}\,d^4x = \int_{\Sigma_2} n_\alpha N^\alpha \sqrt{-\gamma}\,d^3x - \int_{\Sigma_1} n_\alpha N^\alpha \sqrt{-\gamma}\,d^3x \quad \text{used eqn. (6.150)}$$

$$= 2\pi^2 \left(N^0(t)a^3(t) - N^0(t_0)a^3(t_0)\right), \qquad (12.78)$$

where N^0 is simply the number density n (but we leave it as N^0 to avoid confusion with n_α). Combining results eqns. (12.77) and (12.78) we have

$$q \int_{t_0}^{t} dt'(R^\alpha_{\ \alpha})^2(t')a^3(t') = \left(N^0(t)a^3(t) - N^0(t_0)a^3(t_0)\right). \qquad (12.79)$$

Differentiate this result with respect to t and rearrange to find

$$\dot{N}^0(t) = q(R^\alpha{}_\alpha)^2(t) - 3N^0 H(t), \tag{12.80}$$

where $H(t) = \dot{a}/a$ is the Hubble parameter. One could substitute for the Ricci scalar for the RW spacetime from eqn. (B.36). The value of $\dot{N}^0$ has a forcing term proportional to q. Even if $N^0 = 0$ at some initial time t_0, the q term will spontaneously generate new particles that in principle could be observed. Finally our persistent experimentalist has a way to test eqn. (12.75).

SP 12.10 Generalize the equation for the cosmological redshift

$$1 + z = \frac{R(t_0)}{R(t)} \qquad \text{Schutz Eq. (12.21)} \tag{12.81}$$

to the case where the source and receivor have four-velocities U_s and U_r respectively.

SP 12.11 The spectrum of cosmic microwave background (CMB) radiation energy very closely follows that of a black body, even though it has been out of equilibrium with matter for most of the age of the universe, more precisely since the time of last scattering (also called time of decoupling or recombination). Use the black-body spectrum and Wien's displacement law to argue that $R \propto 1/T$. Show that the solution found in Exercise 12.25(a), $T(t) = \beta/\sqrt{t}$ in the radiation-dominated era, is consistent with $R(t)$ during this era derived from the fluid equation, see Schutz Eq. (12.57).

Solution

Before recombination, radiation and matter were in equilibrium so the number of photons per unit volume per unit frequency $n(\nu, T)$ for a given temperature T is given by the black-body spectrum

$$n(\nu, T) = \frac{8\pi \nu^2}{e^{h\nu/k_B T} - 1}, \tag{12.82}$$

where h and k_B are the Planck and Boltzmann constants. To simplify, suppose the radiation went out of equilibrium in a sudden transition at cosmic time t_L, when the temperature was T_L and the scale factor was R_L. As the universe expands the frequency of a given photon is redshifted according to eqn. (12.26) so that at a latter time $t > t_L$

$$\nu(t) = \frac{R_L}{R(t)} \nu_L \implies d\nu = \frac{R_L}{R} d\nu_L. \tag{12.83}$$

Furthermore any cube $(dx^i)^3$ with proper volume at t_L of $dV_L = R_L^3 (dx^i)^3$ will increase in proper volume as the universe expands so that

$$dV(t) = R^3(t)(dx^i)^3 = \left(\frac{R}{R_L}\right)^3 dV_L. \tag{12.84}$$

Assuming photons are not created or distroyed, we can find the number of photons in a given volume at time $t > t_L$

$$n(v(t), T(t)) \, dV(t) \, dv(t) = n(v_L, T_L) \, dV_L \, dv_L \qquad \text{conservation of photons}$$

$$n(v(t), T(t)) = n(v_L, T_L) \left(\frac{R_L}{R(t)} \right)^2 \qquad \text{used eqns. (12.83, 12.84)}$$

$$= \frac{8\pi v_L^2}{e^{hv_L/k_B T_L} - 1} \left(\frac{R_L}{R(t)} \right)^2 \qquad \text{used eqn. (12.82)}$$

$$= \frac{8\pi v(t)}{e^{hv_L/k_B T_L} - 1}. \qquad \text{used eqn. (12.83)}$$

$$(12.85)$$

Notice that v_L still appears in the exponential term. Now we use Wien's displacement law, an implication of which is that the temperature associated with a black-body spectrum is proportional to the peak spectral density. Postulate a black-body spectrum for the CMB after recombination, and associate a temperature of the radiation $T(t) = T_L v(t)/v_L$. Substituting this into eqn. (12.85) we discover the internal consistency of this postulate:

$$n(v(t), T(t)) = \frac{8\pi v(t)}{e^{hv(t)/k_B T(t)} - 1}. \qquad \text{used Wien's displacement law} \qquad (12.86)$$

Thus the spectrum of the CMB, despite being out of equilibrium with matter for almost 14 billion years, corresponds almost perfectly to that one finds for radiation in equilibrium with the walls of a cavity at temperature T. The present temperature is known to be 2.72548 ± 0.00057 K (Fixsen, 2009).

This solution fills in some of the steps of the argument found in (Liddle, 2003, §10.1) and (Weinberg, 2008, §2.1).

SP 12.12   Do the galaxies move along geodesics of the RW metric? Consider just the galaxy motion due to the Hubble flow and use the standard cosmological model, a RW metric, and perfect fluid stress–energy tensor. Consider two cases: $p \neq 0$ and $p = 0$.

SP 12.13   In 1949 Kurt Gödel, most famous for his work on logic, introduced what is now called the Gödel universe (Gödel, 1949). The spacetime is stationary and homogeneous. The line element can be written

$$ds^2 = -a^2 \left(dt^2 - dx^2 + \frac{1}{2} e^{2x} dy^2 - dz^2 + 2e^x dt dy \right), \qquad (12.87)$$

where a is a constant. It does not admit a foliation of spacetime with spacelike hypersurfaces orthogonal to the time axis. This can be proven using an implication of Frobenius' theorem (Wald, 1984, Appendix B.3) that for a vector field ξ^μ to be orthogonal to any hypersurface, so-called hypersurface orthogonal, it must obey

$$\xi_{[\mu} \xi_{\nu;\alpha]} = 0. \qquad (12.88)$$

Let U^α be the four-velocity of the cosmic fluid (i.e. the four-velocity of the galaxies assuming zero random "peculiar" motion). Choose coordinate time to be the proper time of clocks at the galaxy centers, so that $\vec{U}$ is parallel to $\vec{e}_t$. Use eqn. (12.88) to show that $\vec{U}$ is not orthogonal to any hypersurfaces.

Solution

First expand eqn. (12.88), remembering that the square brackets mean taking the antisymmetric part:

$$\xi_{[\mu}\xi_{\nu;\alpha]} = (\xi_\mu(\xi_{\nu;\alpha} - \xi_{\alpha;\nu}) + \xi_\alpha(\xi_{\mu;\nu} - \xi_{\nu;\mu}) + \xi_\nu(\xi_{\alpha;\mu} - \xi_{\mu;\alpha}))/6$$
$$= (\xi_\mu(\xi_{\nu,\alpha} - \xi_{\alpha,\nu}) + \xi_\alpha(\xi_{\mu,\nu} - \xi_{\nu,\mu})$$
$$+ \xi_\nu(\xi_{\alpha,\mu} - \xi_{\mu,\alpha}))/6. \qquad \text{used } \Gamma^\alpha{}_{\mu\nu} = \Gamma^\alpha{}_{\nu\mu} \quad (12.89)$$

Choosing coordinate time to be the proper time of observers fixed to the galaxies we have $\vec{U} = (-g_{00})^{-1/2}\vec{e}_t = \vec{e}_t/a$. Using Gödel's metric eqn. (12.87) to lower the index we get

$$U_\alpha = g_{\alpha\beta}U^\beta = -(a, 0, ae^x, 0). \qquad (12.90)$$

It is a bit tedious to evaluate the $\binom{0}{3}$ tensor resulting from substituting U_α into eqn. (12.89). It's probably easiest to consider the contribution from each of the three terms in eqn. (12.89) separately, say $A_{\mu\nu\alpha}$ for the first term, etc., and then add them up as in

$$U_{[\mu}U_{\nu;\alpha]} \equiv A_{\mu\nu\alpha} + B_{\mu\nu\alpha} + C_{\mu\nu\alpha} \equiv D_{\mu\nu\alpha}. \qquad (12.91)$$

We find for the first term four non-zero components:

$$A_{\mu\nu\alpha} \equiv U_\mu(U_{\nu,\alpha} - U_{\alpha,\nu}), \quad A_{021} = -A_{012} = a^2e^x, \quad \cancel{A_{221}} = -\cancel{A_{212}} = a^2e^{2x}. \qquad (12.92)$$

Finding $B_{\mu\nu\alpha}$ and $C_{\mu\nu\alpha}$ is just a matter of playing with the indices of this first term:

$$B_{\mu\nu\alpha} \equiv U_\alpha(U_{\mu,\nu} - U_{\nu,\mu}) \quad B_{210} = -B_{120} = a^2e^x \quad \cancel{B_{212}} = -\underline{B_{122}} = a^2e^{2x},$$
$$C_{\mu\nu\alpha} \equiv U_\nu(U_{\alpha,\mu} - U_{\mu,\alpha}) \quad C_{102,} = -C_{201} = a^2e^x \quad \underline{C_{122,}} = -\cancel{C_{221}} = a^2e^{2x}. \qquad (12.93)$$

Many of the terms cancel (as indicated above with underline, and sloping "cancel" lines). The follow six non-zero terms remain in $U_{[\mu}U_{\nu;\alpha]} = D_{\mu\nu\alpha}$:

$$D_{012} = -a^2e^x \qquad D_{021} = a^2e^x \qquad D_{102} = a^2e^x$$
$$D_{120} = -a^2e^x \qquad D_{201} = -a^2e^x \qquad D_{210} = a^2e^x. \qquad (12.94)$$

This proves there are no (spacelike) hypersurfaces orthogonal to $\vec{U}$ and similarly there are no hypersurfaces orthogonal to $\vec{e}_t$ because $\vec{e}_t = a\vec{U}$.

Appendix A **Acronyms and definitions**

A.1 Acronyms

Table A.1 Acronyms and other abbreviations used by Schutz and this text

Acronym	Definition
BC	Boundary Condition (often for ODEs or PDEs)
CM	Center of Mass (reference frame)
CMB	Cosmic Microwave Background radiation
def	definition
Eq. $(n.m)$	equation number m of Chapter n of (Schutz, 2009)
eqn. $(n.m)$	equation number m of chapter n of *this*, not Schutz's, book
Exercise $n.m$	means exercise m from (Schutz, 2009, Chapter n)
EOS	Equation of State
Fig. $n.m$	figure number m of chapter n of (Schutz, 2009)
fig. $n.m$	figure number m of chapter n of this, not Schutz's book
FRW	Friedmann–Robertson–Walker (metric), see also RW
GR	general relativity
HOT	Higher Order Terms
LHS	left-hand side (of the equation)
LIF	local inertial frame (also Lorentz frame)
MCRF	momentarily co-moving reference frame
MCLIRF	momentarily co-moving local inertial reference frame
ODE	Ordinary Differential Equation
PDE	Partial Differential Equation
RHS	right-hand side (of the equation)
RW	Robertson–Walker (metric), see also FRW
SP$n.m$	Supplementary Problem number m of chapter n of this book
SR	special relativity
Schutz	the textbook (Schutz, 2009), see Reference section for citation
z	redshift parameter, see eqn. (12.25)

A.2   Mathematical and physical symbols

Table A.2   Symbols used by Schutz(2009) and this text

Symbole	Definition
$\in$	is an element of, e.g. $\alpha \in \{0, 1, 2, 3\}$ means α is in the set four numbers 0, 1, 2, and 3
$\forall$	for all, for each, e.g. $\alpha \geq 0 \ \forall \ \alpha \in \{0, 1, 2, 3\}$
3D	three dimensional
∂_α	coordinate basis vector $\vec{e}_\alpha$ used in advanced textbooks, see SP5.13
$A^{\alpha *}$	a "$*$" on the index indicates that we are considering one fixed value of the index
μ_0	permeability of free space, a physical constant
$\frac{D}{D\tau}$	intrinsic derivative along a curve parameterized by τ, see eqn. (6.136) in SP6.9
diag	diagonal matrix, e.g. $\mathrm{diag}(a, b, c) = \begin{pmatrix} a & 0 & 0 \\ 0 & b & 0 \\ 0 & 0 & c \end{pmatrix}$
$L(4)$	Lorentz group, see Exercise 3.33
$O(\ )$	of the order of, e.g. $O(x^2)$ means terms proportional to x^2, see Exercise 1.14
$O(1, 3)$	Lorentz group, see $L(4)$
$\mathbb{R}$	in mathematics, the reals
$\mathbb{R}^n$	in mathematics, n-dimensional Euclidean space
sgn	algebraic sign, $\mathrm{sgn}(x) = x/|x|$ for $x \in \mathbb{R}$
tr	trace, sum of diagonal elements of a matrix
U^T	transpose of matrix U, i.e. $U_{ij}^T = U_{ji}$
z	redshift parameter, see eqn. (12.25)

Appendix B **Useful results**

B.1 Linear algebra

The inverse of a 2×2 matrix can be found from (B.1). Note the final factor in parentheses is the determinant.

$$\begin{pmatrix} a & b \\ c & d \end{pmatrix} \begin{pmatrix} d & -b \\ -c & a \end{pmatrix} = \begin{pmatrix} 1 & 0 \\ 0 & 1 \end{pmatrix} (ad - bc). \tag{B.1}$$

B.2 Series approximations

The follow series are obtained via a Taylor series expansion about $x = 0$. The first one appears very often so we christen it "binomial series":

$$(1 + x)^a = 1 + ax + \frac{1}{2}a(a - 1)x^2 + \frac{1}{6}a(a - 1)(a - 2)x^3 \cdots \quad \text{binomial series} \tag{B.2}$$

$$\frac{1}{1 + x} = 1 - x + x^2 - x^3 + \cdots \tag{B.3}$$

Letting $x \rightarrow -x$ in (B.3) immediately gives

$$\frac{1}{1 - x} = 1 + x + x^2 + x^3 + \cdots \tag{B.4}$$

$$\sqrt{1 + x} = 1 + \frac{1}{2}x - \frac{1}{8}x^2 + \frac{1}{16}x^3 + \cdots \tag{B.5}$$

Letting $x \rightarrow -x$ in (B.5) immediately gives

$$\sqrt{1 - x} = 1 - \frac{1}{2}x - \frac{1}{8}x^2 - \frac{1}{16}x^3 + \cdots \tag{B.6}$$

$$\frac{1}{\sqrt{1 + x}} = 1 - \frac{1}{2}x + \frac{3}{8}x^2 - \frac{5}{16}x^3 + \frac{35}{128}x^4 - \frac{63}{256}x^5 \cdots \tag{B.7}$$

Letting $x \rightarrow -x$ in (B.7) immediately gives the series we will use repeatedly to approximate the Lorentz factor; just let $(x = v^2)$:

$$\frac{1}{\sqrt{1 - x}} = 1 + \frac{1}{2}x + \frac{3}{8}x^2 + \frac{5}{16}x^3 + \frac{35}{128}x^4 + \frac{63}{256}x^5 \cdots \tag{B.8}$$

$$\ln(1 + x) = x - \frac{x^2}{2} + \frac{x^3}{3} - \frac{x^4}{4} + \cdots \tag{B.9}$$

$$e^{ax} = 1 + ax + \frac{a^2}{2}x^2 + \frac{a^3}{6}x^3 + \frac{a^4}{24}x^4 + \cdots \tag{B.10}$$

B.3 Transformations between spherical polar and Cartesian coordinates

B.3.1 Upper indices: Cartesian to polar (lower indices polar to Cartesian)

$$r = \sqrt{x^2 + y^2 + z^2}$$
$$\theta = \arccos(z/r)$$
$$\phi = \arctan(y/x) \tag{B.11}$$

$$\frac{\partial r}{\partial x} = \sin\theta\,\cos\phi \qquad \frac{\partial \theta}{\partial x} = \frac{1}{r}\cos\theta\,\cos\phi \qquad \frac{\partial \phi}{\partial x} = -\frac{1}{r}\frac{\sin\phi}{\sin\theta}$$

$$\frac{\partial r}{\partial y} = \sin\theta\,\sin\phi \qquad \frac{\partial \theta}{\partial y} = \frac{1}{r}\cos\theta\,\sin\phi \qquad \frac{\partial \phi}{\partial y} = \frac{1}{r}\frac{\cos\phi}{\sin\theta}$$

$$\frac{\partial r}{\partial z} = \cos\theta \qquad \frac{\partial \theta}{\partial z} = -\frac{1}{r}\sin\theta \qquad \frac{\partial \phi}{\partial z} = 0 \tag{B.12}$$

B.3.2 Upper indices: polar to Cartesian (lower indices Cartesian to polar)

$$x = r\sin\theta\,\cos\phi$$
$$y = r\sin\theta\,\sin\phi$$
$$z = r\cos\theta \tag{B.13}$$

$$\frac{\partial x}{\partial r} = \sin\theta\cos\phi \qquad \frac{\partial x}{\partial \theta} = r\cos\theta\cos\phi \qquad \frac{\partial x}{\partial \phi} = -r\sin\theta\sin\phi$$

$$\frac{\partial y}{\partial r} = \sin\theta\sin\phi \qquad \frac{\partial y}{\partial \theta} = r\cos\theta\sin\phi \qquad \frac{\partial y}{\partial \phi} = r\sin\theta\cos\phi \tag{B.14}$$

$$\frac{\partial z}{\partial r} = \cos\theta \qquad \frac{\partial z}{\partial \theta} = -r\sin\theta \qquad \frac{\partial z}{\partial \phi} = 0$$

B.4 Selection of spacetimes

Below we summarize the important spacetimes studied herein by listing their line element, Christoffel symbols, and important tensors. Components that are not listed and are not related by symmetry to one of those listed are zero.

B.4.1 Rindler spacetime

The line element is

$$ds^2 = -a^2 d\lambda^2 + da^2. \tag{B.15}$$

This metric applies in flat (Minkowski) spacetime and was derived in Exercise 5.21 for a set of uniformly accelerating observers. Because it's a flat spacetime it follows that the Riemann tensor vanishes, and thus the Ricci tensor and scalar and Einstein tensor all vanish.

Christoffel symbols

$$\Gamma^\lambda{}_{\lambda a} = \frac{1}{a} \qquad\qquad \Gamma^a{}_{\lambda\lambda} = a. \tag{B.16}$$

B.4.2 Static spherically symmetric spacetimes

See Exercise 6.35. The line element is

$$ds^2 = -e^{2\Phi(r)} dt^2 + e^{2\Lambda(r)} dr^2 + r^2 d\theta^2 + r^2 \sin^2\theta \, d\phi^2. \tag{B.17}$$

Christoffel symbols

$$\Gamma^t{}_{tr} = \Phi' \qquad\qquad \Gamma^r{}_{tt} = e^{-2\Lambda} e^{2\Phi} \Phi' \qquad \Gamma^r{}_{rr} = \Lambda'$$

$$\Gamma^r{}_{\theta\theta} = -e^{-2\Lambda} r \qquad\quad \Gamma^r{}_{\phi\phi} = -e^{-2\Lambda} r \sin^2\theta \qquad \Gamma^\theta{}_{r\theta} = \frac{1}{r}$$

$$\Gamma^\theta{}_{\phi\phi} = -\sin(\theta)\cos(\theta) \qquad \Gamma^\phi{}_{r\phi} = \frac{1}{r} \qquad\qquad \Gamma^\phi{}_{\theta\phi} = \frac{\cos(\theta)}{\sin(\theta)}, \tag{B.18}$$

where $\Phi' \equiv d\Phi/dr$ and $\Lambda' \equiv d\Lambda/dr$.

Ricci tensor and Ricci scalar

See SP6.8:

$$R_{tt} = -e^{(2\phi-2\Lambda)} \left(\Lambda'\Phi' - \Phi'^2 - \Phi'' - \frac{2\Phi'}{r} \right) \qquad R_{rr} = -\left(-\Lambda'\Phi' + \Phi'^2 + \Phi'' - \frac{2\Lambda'}{r} \right)$$

$$R_{\theta\theta} = -\left(-1 + e^{-2\Lambda}[1 - r(\Lambda' - \Phi')] \right) \qquad R_{\phi\phi} = -\sin^2\theta \left(e^{-2\Lambda}[1 + r(\Phi' - \Lambda')] - 1 \right) \tag{B.19}$$

and

$$R = -2e^{-2\Lambda} \left(-\Lambda'\Phi' + \Phi'^2 + \Phi'' + \frac{2(\Phi' - \Lambda')}{r} + \frac{1 - e^{2\Lambda}}{r^2} \right). \tag{B.20}$$

Einstein tensor

$$G_{tt} = \frac{1}{r^2}e^{2\Phi}\frac{d}{dr}\left[r(1 - e^{-2\Lambda})\right], \qquad G_{rr} = \frac{1}{r^2}(1 - e^{2\Lambda}) + \frac{2}{r}\Phi',$$

$$G_{\theta\theta} = r^2 e^{-2\Lambda}\left(\Phi'' + (\Phi')^2 + \frac{\Phi'}{r} - \Phi'\Lambda' - \frac{\Lambda'}{r}\right), \quad G_{\phi\phi} = \sin^2\theta\, G_{\theta\theta}. \qquad \text{(B.21)}$$

as in Schutz Eq. (10.14)–(10.17).

B.4.3 Schwarzschild spacetime

The line element is

$$ds^2 = -\left(1 - \frac{2M}{r}\right)dt^2 + \left(1 - \frac{2M}{r}\right)^{-1}dr^2 + r^2 d\Omega^2, \quad \text{Schutz Eq. (10.36)} \quad \text{(B.22)}$$

with $d\Omega^2 = d\theta^2 + \sin^2\theta\, d\phi^2$ (same as metric (ii) of Exercise 7.7). This metric applies in the vacuum around a static spherically symmetric source. Because it's a *vacuum spacetime* it follows from the Einstein equations (see SP9.2) that $R_{\alpha\beta} = G_{\alpha\beta} = 0$ and $R = 0$.

Christoffel symbols

$$\Gamma^t{}_{tr} = \frac{M}{r^2}\left(1 - \frac{2M}{r}\right)^{-1} \quad \Gamma^r{}_{tt} = \frac{M}{r^2}\left(1 - \frac{2M}{r}\right) \quad \Gamma^r{}_{rr} = -\frac{M}{r^2}\left(1 - \frac{2M}{r}\right)^{-1}$$

$$\Gamma^r{}_{\theta\theta} = -r + 2M \qquad \Gamma^r{}_{\phi\phi} = (-r + 2M)\sin^2\theta \quad \Gamma^\theta{}_{r\theta} = \frac{1}{r}$$

$$\Gamma^\theta{}_{\phi\phi} = -\sin(\theta)\cos(\theta) \quad \Gamma^\phi{}_{r\phi} = \frac{1}{r} \qquad\qquad \Gamma^\phi{}_{\theta\phi} = \frac{\cos(\theta)}{\sin(\theta)}. \qquad \text{(B.23)}$$

B.4.4 Weak gravitational field

See Exercise 7.2. The line element is:

$$ds^2 = -(1 + 2\phi)dt^2 + (1 - 2\phi)\left(dx^2 + dy^2 + dz^2\right). \qquad \text{Schutz Eq. (7.8)} \qquad \text{(B.24)}$$

Christoffel symbols

$$\Gamma^i{}_{tt} = \phi_{,i} + O(\phi^2), \qquad\qquad \Gamma^i{}_{tj} = -\phi_{,t}\delta^i{}_j + O(\phi^2),$$

$$\Gamma^i{}_{jk} = \delta_{jk}\delta^{il}\phi_{,l} - \delta^i_j\phi_{,k} - \delta^i_k\phi_{,j} + O(\phi^2), \qquad\qquad \text{(B.25)}$$

where $i, j, k \in \{x, y, z\}$.

Ricci tensor and Ricci scalar

$$R_{tt} = 3\phi_{,tt} + \phi_{,xx} + \phi_{,yy} + \phi_{,zz} + O(\phi^2) \quad R_{ti} = 2\phi_{,ti} + O(\phi^2)$$
$$R_{ii} = -\phi_{,tt} + \phi_{,xx} + \phi_{,yy} + \phi_{,zz} + O(\phi^2) \quad R_{ij} = 0 + O(\phi^2) \qquad \text{when } i \neq j. \quad (B.26)$$

$$R = -6\phi_{,tt} + 2(\phi_{,xx} + \phi_{,yy} + \phi_{,zz}) + O(\phi^2). \tag{B.27}$$

Einstein tensor

$$G_{tt} = 2(\phi_{,xx} + \phi_{,yy} + \phi_{,zz}) + O(\phi^2) \quad G_{ti} = 2\phi_{,it} + O(\phi^2)$$
$$G_{ii} = 2\phi_{,tt} + O(\phi^2) \qquad\qquad\qquad G_{ij} = 0 + O(\phi^2) \qquad \text{when } i \neq j. \quad (B.28)$$

B.4.5 Post-Newtonian spherical rotating star

The line element was derived in Exercise 8.19, see eqn. (8.59):

$$ds^2 = -\left(1 - \frac{2M}{r}\right) dt^2 - 4J \frac{\sin^2\theta}{r} \, dt \, d\phi + \left(1 + \frac{2M}{r}\right) (dr^2 + r^2 \, d\theta^2 + r^2 \sin^2\theta \, d\phi^2). \tag{B.29}$$

This metric applies in the vacuum around a spherical source that rotates. Because it's a *vacuum spacetime* it follows from the Einstein equations (see SP9.2) that $R_{\alpha\beta} = G_{\alpha\beta} = 0$ and $R = 0$.

Christoffel symbols

$$\Gamma^t{}_{tr} = -\frac{2J^2 \cos^2\theta + 2M^2 r^2 + r^3 M - 2J^2}{r(4J^2 \cos^2\theta + 4M^2 r^2 - r^4 - 4J^2)} \quad \Gamma^t{}_{t\theta} = -\frac{4J^2 \sin\theta \cos\theta}{4J^2 \cos^2\theta + 4M^2 r^2 - r^4 - 4J^2}$$

$$\Gamma^t{}_{r\phi} = \frac{r J \sin^2\theta (3r + 4M)}{4J^2 \cos^2\theta + 4M^2 r^2 - r^4 - 4J^2}$$

$$\Gamma^r{}_{tt} = \frac{M}{r(r + 2M)} \qquad\qquad \Gamma^r{}_{t\phi} = -\frac{J \sin^2\theta}{r(r + 2M)}$$

$$\Gamma^r{}_{rr} = -\frac{M}{r(r + 2M)} \qquad\qquad \Gamma^r{}_{\theta\theta} = -\frac{r(M + r)}{r + 2M}$$

$$\Gamma^r{}_{\phi\phi} = -\frac{r(M + r)\sin^2\theta}{r + 2M} \qquad \Gamma^\theta{}_{t\phi} = \frac{2J \sin\theta \cos\theta}{r^2(r + 2M)}$$

$$\Gamma^\theta{}_{r\theta} = \frac{M + r}{r(r + 2M)} \qquad\qquad \Gamma^\theta{}_{\phi\phi} = -\sin\theta \, \cos\theta$$

$$\Gamma^\phi{}_{tr} = \frac{J}{4(J^2 \sin^2\theta - M^2 r^2) + r^4} \qquad \Gamma^\phi{}_{t\theta} = \frac{2J \cos\theta(2M - r)}{\sin\theta[4(J^2 \sin^2\theta - M^2 r^2) + r^4]}$$

$$\Gamma^\phi{}_{r\phi} = -\frac{2J^2 \sin^2\theta + 2M^2 r^2 + r^3 M - r^4}{r[4(J^2 \sin^2\theta - M^2 r^2) + r^4]} \qquad \Gamma^\phi{}_{\theta\phi} = \cot\theta \tag{B.30}$$

B.4.6   Kerr spacetime

The line element is:

$$ds^2 = -\frac{\Delta - a^2 \sin^2\theta}{\rho^2} dt^2 - 2a\frac{2Mr \sin^2\theta}{\rho^2} dt\, d\phi$$

$$+ \frac{(r^2 + a^2)^2 - a^2\Delta \sin^2\theta}{\rho^2} \sin^2\theta\, d\phi^2 + \frac{\rho^2}{\Delta}dr^2 + \rho^2 d\theta^2, \qquad \text{(B.31)}$$

where M and a are constants and $\Delta \equiv r^2 - 2Mr + a^2$, $\rho^2 \equiv r^2 + a^2\cos^2\theta$. (This is metric (iii) of Exercise 7.7). This metric applies in the vacuum around a source that rotates. Because it's a *vacuum spacetime* it follows from the Einstein equations (see SP9.2) that $R_{\alpha\beta} = G_{\alpha\beta} = 0$ and $R = 0$.

Christoffel Symbols

$$\Gamma^t_{tr} = \frac{M(r^2 - a^2\cos^2\theta)(r^2 + a^2)}{\rho^4\Delta} \qquad\qquad = \frac{-pb^2}{\Delta}$$

$$\Gamma^t_{t\theta} = -\frac{2a^2\sin\theta\cos\theta\, Mr}{\rho^4} \qquad\qquad = \frac{qa^2\sin(2\theta)}{\Sigma}$$

$$\Gamma^t_{r\phi} = \frac{aM\sin^2\theta[a^4\cos^2\theta - a^2r^2(1+\cos^2\theta) - 3r^4]}{\rho^4\Delta} \qquad = \frac{a\sin^2\theta}{\Delta}(pb^2 + 2qr)$$

$$\Gamma^t_{\theta\phi} = \frac{2\sin^3\theta\,\cos\theta\, Ma^3 r}{\rho^4} \qquad\qquad = -\frac{qa^3\sin^2\theta\,\sin(2\theta)}{\Sigma}$$

$$\Gamma^r_{tt} = \frac{M\Delta(r^2 - a^2\cos^2\theta)}{\rho^6} \qquad\qquad = -\frac{\Delta p}{\Sigma}$$

$$\Gamma^r_{t\phi} = -a\sin^2\theta\,\Gamma^r_{tt} \qquad\qquad = \frac{a\Delta p\sin^2\theta}{\Sigma}$$

$$\Gamma^r_{rr} = \frac{M(a^2\cos^2\theta - r^2) + a^2 r\sin^2\theta}{\rho^2\Delta} \qquad\qquad = \frac{r}{\Sigma} + \frac{M-r}{\Delta}$$

$$\Gamma^r_{r\theta} = -\frac{a^2\sin\theta\,\cos\theta}{\rho^2} \qquad\qquad = -\frac{a^2\sin(2\theta)}{2\Sigma}$$

$$\Gamma^r_{\theta\theta} = -\frac{r\Delta}{\rho^2} \qquad\qquad = -\frac{r\Delta}{\Sigma}$$

$$\Gamma^r_{\phi\phi} = -\frac{\Delta\sin^2\theta[r\rho^4 + M(a^2\cos^2\theta - r^2)a^2\sin^2\theta]}{\rho^6} \qquad = -\frac{\Delta\sin^2\theta}{\Sigma}(r + pa^2\sin^2\theta)$$

$$\Gamma^\theta_{tt} = -\frac{2Ma^2 r\cos\theta\,\sin\theta}{\rho^6} \qquad\qquad = \frac{qa^2\sin(2\theta)}{\Sigma^2}$$

$$\Gamma^\theta_{t\phi} = -\frac{a^2 + r^2}{a}\Gamma^\theta_{tt} \qquad\qquad = -\frac{qab^2\sin(2\theta)}{\Sigma^2}$$

$$\Gamma^{\theta}_{\,rr} = \frac{a^2 \sin\theta \, \cos\theta}{\rho^2 \Delta} \qquad\qquad = \frac{a^2 \sin(2\theta)}{2\Sigma\Delta}$$

$$\Gamma^{\theta}_{\,r\theta} = \frac{r}{\rho^2} \qquad\qquad = \frac{r}{\Sigma}$$

$$\Gamma^{\theta}_{\,\theta\theta} = \Gamma^{r}_{\,r\theta} \qquad\qquad = \Gamma^{r}_{\,r\theta}$$

$$\Gamma^{\theta}_{\,\phi\phi} = -\frac{\sin\theta \, \cos\theta}{\rho^4}[\rho^4(r^2 + a^2) \qquad\qquad = -\frac{\sin(2\theta)}{2\Sigma}\left[b^2 - 2a^2 \sin^2\theta \, q\right.$$

$$+ 2Mra^2 \sin^2\theta(2\rho^2 + a^2 \sin^2\theta)] \qquad\qquad \left. \times \left(2 + \frac{a^2 \sin^2\theta}{\Sigma}\right)\right]$$

$$\Gamma^{\phi}_{\,tr} = \frac{aM(r^2 - a^2 \cos^2\theta)}{\rho^4 \Delta} \qquad\qquad = -\frac{ap}{\Delta}$$

$$\Gamma^{\phi}_{\,t\theta} = -\frac{2aMr}{\rho^4}\cot\theta \qquad\qquad = \frac{2qa\cot\theta}{\Sigma}$$

$$\Gamma^{\phi}_{\,r\phi} = \frac{r\rho^4 - 2Mr^2\rho^2 + a^2M\sin^2\theta(a^2\cos^2\theta - r^2)}{\rho^4\Delta} \qquad\qquad = \frac{r}{\Delta}(1 + 2q) + a^2 p\frac{\sin^2\theta}{\Delta}$$

$$\Gamma^{\phi}_{\,\theta\phi} = \frac{\cot\theta}{\rho^4}(\rho^4 + 2Mra^2 \sin^2\theta) \qquad\qquad = \frac{\cot\theta}{\Delta}\left[(1+2q)(b^2-2qa^2\sin^2\theta)\right.$$

$$\left. -\frac{2qa^2b^2\sin^2\theta}{\Sigma}\right]$$

$$(B.32)$$

The second equality for each $\Gamma^{\alpha}_{\,\mu\nu}$ in eqn.(B.32) uses the notation of Frolov and Novikov (1998, Appendix D), with $\Sigma = \rho^2, q = -Mr/\Sigma, p = M(a^2 \cos^2\theta - r^2)/\Sigma^2, b^2 = r^2 + a^2$ when the total charge $Q = 0$.

B.4.7 Robertson–Walker spacetime

The line element is:

$$ds^2 = -dt^2 + R^2(t)\left(\frac{1}{1 - kr^2}dr^2 + r^2(d\theta^2 + \sin^2\theta \, d\phi^2)\right) \qquad \text{Schutz Eq. (12.13)}$$

$$(B.33)$$

with three possible values of k:

$$k = 1 \qquad\qquad \text{"closed" or "spherical" universe}$$
$$k = 0 \qquad\qquad \text{spatially "flat" universe}$$
$$k = -1 \qquad\qquad \text{"open" or "hyperbolic" universe}$$

(This is metric (iv) of Exercise 7.7.)

Christoffel symbols

$$\Gamma^t_{\ rr} = \frac{R\dot{R}}{1 - kr^2} \qquad\qquad \Gamma^t_{\ \theta\theta} = R\dot{R}r^2 \qquad \Gamma^t_{\ \phi\phi} = R\dot{R}r^2\sin^2\theta$$

$$\Gamma^r_{\ tr} = \frac{\dot{R}}{R} \qquad\qquad \Gamma^r_{\ rr} = \frac{kr}{1 - kr^2} \qquad \Gamma^r_{\ \theta\theta} = -r(1 - kr^2)$$

$$\Gamma^r_{\ \phi\phi} = -r(1 - kr^2)\sin^2\theta$$

$$\Gamma^\theta_{\ t\theta} = \frac{\dot{R}}{R} \qquad\qquad \Gamma^\theta_{\ r\theta} = \frac{1}{r} \qquad \Gamma^\theta_{\ \phi\phi} = -\sin\theta\ \cos\theta$$

$$\Gamma^\phi_{\ t\phi} = \frac{\dot{R}}{R} \qquad\qquad \Gamma^\phi_{\ r\phi} = \frac{1}{r} \qquad \Gamma^\phi_{\ \theta\phi} = \cot\theta \qquad (B.34)$$

Ricci tensor and Ricci scalar

Ricci tensor:

$$R_{tt} = -\frac{3\ddot{R}}{R} \qquad\qquad\qquad R_{rr} = \frac{R\ddot{R} + 2\dot{R}^2 + 2k}{1 - kr^2}$$

$$R_{\theta\theta} = r^2\left(R\ddot{R} + 2\dot{R}^2 + 2k\right) \qquad\qquad R_{\phi\phi} = R_{\theta\theta}\sin^2\theta. \qquad (B.35)$$

Ricci scalar:

$$R^\alpha_{\ \alpha} = 6\left(\frac{\ddot{R}}{R} + \frac{\dot{R}^2}{R^2} + \frac{k}{R^2}\right). \qquad (B.36)$$

Einstein tensor:

$$G_{tt} = 3\left(\frac{\dot{R}^2}{R^2} + \frac{k}{R^2}\right) \qquad\qquad G_{rr} = \frac{-2R\ddot{R} - \dot{R}^2 - k}{1 - kr^2}$$

$$G_{\theta\theta} = -r^2\left(2R\ddot{R} + \dot{R}^2 + k\right) \qquad\qquad G_{\phi\phi} = G_{\theta\theta}\sin^2\theta. \qquad (B.37)$$

References

Aldrovandi, R., Cuzinatto, R. R., and Medeiros, L. G. 2007. Primeval symmetries. *Gen. Relat. Gravit.*, **39**(11), 1813–1832.

Batchelor, George Keith. 1969. Computation of the energy spectrum in homogeneous two-dimensional turbulence. *Phys. Fluids Suppl. II*, **12**, 233–239.

Baumgarte, T. W., and Shapiro, S. L. 2010. *Numerical Relativity: Solving Einstein's Equations on the Computer*. Cambridge University Press.

Berzi, V. and Gorini, V. 1969. Reciprocity principle and Lorentz transformations. *J. Math. Phys.*, **10**(8), 1518–1524.

Boas, Mary L. 1983. *Mathematical Methods in the Physical Sciences*. New York: John Wiley and Sons.

Carroll, Sean. 2004. *Spacetime and Geometry: An Introduction to General Relativity*. San Francisco: Addison Wesley.

Davis, Harry F. and Snider, Arthur D. 1979. *Introduction to Vector Analysis*. 4th edition. Allyn and Bacon.

Earman, John and Glymour, Clark. 1980. The gravitational redshift as a test of general relativity: history and analysis. *Stud. Hist. Philos. Sci.*, **11**(3), 175–214.

Einstein, A. 1905. On the electrodynamics of moving bodies. *Annalen der Physik*, **17**, 891–921. Translation by W. Perrett and G. B. Jeffery, *The Principle of Relativity* (New York: Dover, 1952).

Einstein, A. 1917. Cosmological considerations on the General Theory of Relativity. *Sitzungsberichte der Preussischen Akad. d. Wissenschaften*, **8**, 142–152. Translation by W. Perrett and G. B. Jeffery, *The Principle of Relativity* (New York: Dover, 1952).

Eisberg, Robert and Resnick, Robert. 1985. *Quantum Physics of Atoms, Molecules, Solids, Nuclei, and Particles*. 2nd edition. New York: John Wiley and Sons.

Faber, Richard L. 1983. *Differential Geometry and Relativity Theory: An Introduction*. New York: Marcel Dekker.

Felder, Gary and Felder, Kenny. 2014. *Math Methods for Engineering and Physics*. Wiley.

Fixsen, D. J. 2009. The temperature of the Cosmic Microwave Background. *Astrophys. J.*, **707**(2), 916.

Friedlander, F. G., and Joshi, M. S. 1999. *Introduction to the Theory of Distributions*. Cambridge University Press.

Frolov, Valeri P., and Novikov, Igor D. 1998. *Black Hole Physics: Basic Concepts and New Developments*. Kluwer Academic Publishers, Dordrecht.

Gödel, Kurt. 1949. An example of a new type of cosmological solution of Einstein's field equations of gravitation. *Rev. Mod. Phys.*, **21**(Jul), 447–450.

Griffiths, R. W. 2000. The dynamics of lava flows. *Annu. Rev. Fluid Mech.*, **32**(1), 477–518.

Hassani, Sadri. 1999. *Mathematical Physics: A Modern Introduction to Its Foundations.* New York: Springer.

Hassani, Sadri. 2008. A heuristic derivation of Minkowski distance and Lorentz transformation. *Eur. J. Phys.*, **29**(1), 103–111.

Hawking, Stephen W., and Ellis, G.F.R. 1973. *The Large Scale Structure of Space-Time.* Cambridge University Press.

Hobson, M. P., Efstathiou, G. P., and Lasenby, A. N. 2006. *General Relativity: An Introduction for Physicists.* Cambridge University Press.

Kibble, Tom W. B. and Berkshire, Frank H. 2004. *Classical Mechanics.* London: Imperial College Press.

Landau, L. D. and Lifshitz, E. M. 1966. *Fluid Mechanics.* 3rd impression of English translation edition. Course of Theoretical Physics, vol. 6. Oxford: Pergamon Press.

Lawden, D. F. 2002. *Introduction to Tensor Calculus, Relativity and Cosmology.* 3rd edition. Mineola, NY: Dover.

Liddle, Andrew. 2003. *An Introduction to Modern Cosmology.* Chichester and Hoboken, NJ: Wiley & Company.

Mana, PierGianLuca Porta and Zanna, Laure. 2014. Toward a stochastic parameterization of ocean mesoscale eddies. *Ocean Model.*, **79**(0), 1 – 20.

Maudlin, Tim. 2012. *Philosophy of Physics: Space and Time.* Princeton University Press.

Misner, C. W., Thorne, K. S., and Wheeler, J. A. 1973. *Gravitation.* San Francisco: W.H. Freeman.

Möller, Christian. 1952. *The Theory of Relativity.* Oxford: Clarendon Press.

Mukhanov, V. 2005. *Physical Foundations of Cosmology.* Cambridge University Press.

Poisson, Eric. 2004. *A Relativist's Toolkit: The Mathematics of Black-hole Mechanics.* Cambridge University Press.

Price, H. E. 1974. Gravitational red-shift formula. *Am. J. Phys.*, **42**(4), 336–337.

Riley, K. F., Hobson, M. P., and Bence, S. J. 2006. *Mathematical Methods for Physics and Engineering.* 3rd edition. Cambridge University Press.

Rindler, Wolfgang. 2006. *Relativity: Special, General and Cosmological.* 2nd edition. Oxford University Press.

Schutz, Bernard. 1980. *Geometrical Methods of Mathematical Physics.* Cambridge University Press.

Schutz, Bernard. 1985. *A First Course in General Relativity.* Cambridge University Press.

Schutz, Bernard. 2009. *A First Course in General Relativity.* 2nd edition. Cambridge University Press.

Sparke, L. S. and Gallagher III, J. S. 2007. *Galaxies in the Universe.* 2nd edition. Cambridge University Press.

Vallis, G. K. 2006. *Atmospheric and Oceanic Fluid Dynamics: Fundamentals and Large-scale Circulation.* Cambridge University Press.

Wald, Robert M. 1984. *General Relativity.* The University of Chicago Press.

Weinberg, Steven. 1972. *Gravitation and Cosmology: Principles and Applications of the General Theory of Relativity.* New York: John Wiley & Sons.

Weinberg, Steven. 2008. *Cosmology.* Oxford University Press.

Weinberg, Steven. 2013. *Lectures on Quantum Mechanics.* Cambridge University Press.

Index

Printed in the United States
by Baker & Taylor Publisher Services